HVAC

Control

Systems:

Modelling, analysis and design

Other titles from E & FN Spon

Facilities Management: Theory and practice
edited by Keith Alexander

Ventilation of Buildings
Hazim B. Awbi

Engineering Noise Control
David A. Bies and Colin H. Hansen

Air Conditioning: A practical introduction
David V. Chadderton

Building Services Engineering
David V. Chadderton

Building Services Engineering Spreadsheets
David V. Chadderton

Naturally Ventilated Buildings
edited by Derek Clements-Croome

Active Control of Noise and Vibration
Colin H. Hansen and Scott D. Snyder

Illustrated Encyclopedia of Building Services
David Kut

Spon's Mechanical and Electrical Services Price Book
Davis Langdon & Everest

Building Energy Management Systems
Geoff Levermore

Energy Management and Operating Costs in Buildings
Keith Moss

Heat and Mass Transfer in Building Services Design
Keith Moss

Heating and Water Services Design in Buildings
Keith Moss

For more information about these and other titles please contact:
The Marketing Department, E & FN Spon, 11 New Fetter Lane, London,
EC4P 4EE

HVAC

Control

Systems:

Modelling, analysis and design

C. P. Underwood

 Spon Press
Taylor & Francis Group

LONDON AND NEW YORK

First published 1999
by Spon Press
11 New Fetter Lane, London EC4P 4EE

Simultaneously published in the USA and Canada
by Routledge
29 West 35th Street, New York, NY 10001

Transferred to Digital Printing 2004

Spon Press is an imprint of the Taylor & Francis Group

Typeset in 10/12pt Baskerville by Best-set Typesetter Ltd, Hong Kong
Printed and bound in Great Britain by Biddles Ltd, King's Lynn, Norfolk

British Library Cataloguing in Publication Data
A catalogue record for this book is available from the British Library

Library of Congress Cataloging-in-Publication Data
A catalog record for this book is available on request

ISBN 0 419 20980 8

Dedicated to
My wife, Allison, and our children
Beatrice and Andrew

Contents

Preface

This book concerns itself with the *regulatory* control of heating, ventilating and air conditioning (HVAC) processes in buildings. It attempts to bridge the gap between the handful or so of excellent practical books on the subject, and the numerous and equally excellent existing texts which deal with process control generically and frequently in a highly analytical manner. The net result is a book which, it is hoped, treats the subject of HVAC control analytically, and yet is useful to a wide range of students, researchers and specialist practitioners and engineers concerned with HVAC control system design, installation and operation in buildings.

HVAC control problems are not trivial. Non-stationary plant operating conditions resulting from low-frequency (seasonal) and high-frequency (diurnal) climate changes coupled with complex patterns of user demand combine with the intrinsically non-linear characteristics endemic in HVAC plant to give one of the more complex control applications known. It is only because the consequences of failure of HVAC plant through bad control are rarely catastrophic that we have been able to treat many problems almost as an art in the past. This has sometimes led to surprisingly good results, but frequently to a failure to satisfy all the essential tenets of good HVAC control – *to achieve good comfort at minimum energy use, operating cost and initial cost.*

The gradual introduction of *discrete-time control* through the use of building energy management systems and stand-alone digital controllers has opened up immense opportunity in new methods, frequently referred to as *intelligent* control. This opportunity now makes the essential tenets of good HVAC control highly realisable in all applications – not just those lucky enough to have a patient energy manager at their disposal. The primary aim of the book therefore is a better *understanding* of HVAC control problems from which improved design, installation and operation are bound to follow.

Chapter 1 deals with practical HVAC control systems with reference to established practice, and particular problems arising from these are identified and discussed. Special consideration is given to distributed control and building energy management systems in this chapter. Chapter 2 looks at control hardware with an emphasis on sensors and control elements. Chapter 3 treats the dynamic modelling of building elements and plant components with particular reference to later consideration of linear and linearised plant modelling. Chapter 4 looks at stability tests applied to plant modelled in continuous time and takes a special look at methods of reducing the frequently high-order models arising in HVAC control problems. Acknowledging that modern electronic HVAC control is mostly digital,

Chapter 5 deals with discrete-time principles and revisits the issue of stability tests for the discrete-time case, as well as looking at some of the simple options available for designing fixed-parameter discrete-time controllers. Chapter 6 takes a look at multiple-input multiple-output (MIMO) control with special reference to the simultaneous control of temperature and humidity in close control air conditioning. Chapter 7 considers methods of fitting models to experimental data (system identification) and uses techniques based on these methods to 'tune' fixed-parameter PID controllers. Chapter 8 is devoted to relatively recent developments in adaptive control and considers direct/indirect on-line methods, model-reference adaptive control and gain scheduling. Finally, Chapter 9 looks at some of the recent new technologies and methods: robust control design, fuzzy logic control and artificial neural networks.

The book is underpinned throughout with worked examples from which, it is hoped, solutions to realistic practical problems can be understood and replicated by the reader. Because few practical problems are truly amenable to manual calculation, considerable use is made of the computational environment, MATLAB, and a number of its specialist toolboxes, in the book. MATLAB/Simulink is especially frequently used throughout the book and, whilst many readers will be able to replicate results to many of the worked examples using other simulation programs, the student edition of MATLAB/Simulink is particularly recommended for the serious reader.

Whilst it is hoped that there is something for everyone who has an interest in HVAC control in this book, serious readers will require a knowledge of some of the key mathematical methods used in control. In particular, a knowledge of classical calculus, Laplace transforms and basic matrix algebra are essential for the serious reader.

C. P. Underwood
February 1998

Acknowledgements

The author is grateful to the following for permission to reproduce material.

Satchwell Control Systems Ltd – Figure 1.26.

Cylon Controls (UK) Ltd – Figures 1.27, 1.28, 1.29.

Landis & Staefa Ltd – Figures 1.30, 2.4.

CMR Controls Ltd – Figures 2.5, 2.6.

Eurotherm Controls – Figure 2.22.

MAP – Motorised Air Products Ltd – Figure 2.14.

Caradon Trend Ltd – Figures 2.9, 2.18.

Figures 2.2 and 2.3 reproduced by permission of John Wiley & Sons Inc.

Table 7.1 reproduced by permission of the American Society of Heating, Refrigerating and Air Conditioning Engineers Inc.

Figure 2.8 reproduced with permission of the Australian Journal of Refrigeration, Air Conditioning and Heating.

Figure 2.7 reproduced with permission of the International Council for Building Research Studies and Documentation.

Figure 2.16 reproduced with permission of the Department of Environment, Transport and the Regions.

Figure 2.15 and Table 2.1 reproduced with permission of the Chartered Institution of Building Services Engineers.

All graphs (apart from Figures E4.3.3, E5.4.1 and 9.6) were created with Origin 4.1 from Microcal Software. The author is indebted to The Mathworks Inc., whose products Matlab, Simulink and a number of the associated toolboxes made preparation of parts of the book considerably easier than would otherwise have been the case.

1 HVAC control systems

1.1 Basic concepts

It is not frequently acknowledged that heating, ventilating and air conditioning (HVAC) systems present one of the most challenging situations to deal with from the point of view of control. Swings in day-to-day, week-to-week and season-to-season energy demand together with the infinitely complex combination of user needs at the human interface contribute to a highly non-stationary 'environment' within which control takes place. It is little wonder then that much of HVAC control is about compromise; a compromise that usually amounts *to reasonable comfort at minimum energy use and cost.*

HVAC control in common with all process system control requires the governance of two distinct actions; those of *switching* or 'enabling' and *regulating* or 'adjustment'. Switching in the majority of applications amounts to ensuring that the plant is available at certain times of the day (generally, those times of the day in which a building is in use). This will be essentially clock-based, occupancy-sensor-based or indeed based on some other logical two-state condition such as an alarm state. Regulation, which is what this book in essence is about, amounts to ensuring that the plant capacity is matched to the demands placed upon the system.

Let us consider some basic concepts of HVAC control. Figures 1.1, 1.2 and 1.3 show three quite contrasting ways of controlling the output of a simple heat exchanger. The actions of these three systems are quite different.

Under *analogue* control, the controller positions the valve anywhere between fully open and fully closed (Figure 1.4) using a smooth continuous signal from the controller. Control is defined in terms of a reference condition or *set point* (r) and a *proportional band* (p) which represents the range of the controlled condition for which the output of the controller is proportional. Figure 1.4 for instance shows that the valve will settle at the set point at 50% of its range and, during fluctuating conditions, the temperature will drift either side of the set point until control action restores its value at the latter. This was the predominant method of HVAC system control until the early 1980s and still exists today in the form of mechanical direct-acting control devices. For example, the 'thermostatic' radiator valve (TRV) is one case of this type of control.

Under *thermostatic* control, control is a two-state process. The thermostat is a switching device and positions the valve periodically in either a fully closed position or a fully open position; there are no in-betweens. Control action is defined in terms of a set point and *dead band* (d) which represents a band of inactivity the limits of which form switching events (Figure 1.5).

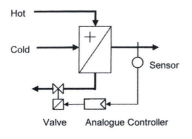

Figure 1.1 Analogue control.

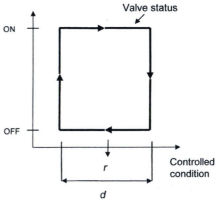

Figure 1.2 Thermostatic control.

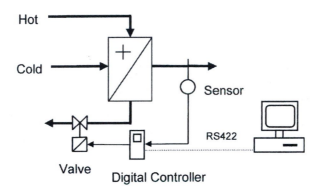

Figure 1.3 Digital control.

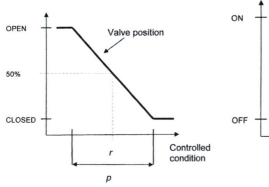

Figure 1.4 Proportional control action.   Figure 1.5 Thermostatic control action.

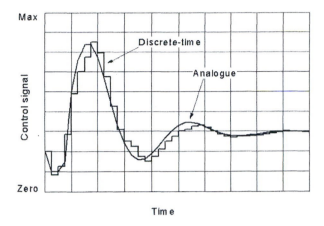

Figure 1.6 Typical analogue and digital control signals.

This method of control is widespread; there are numerous examples in the home such as the domestic iron and the portable electric fan heater. It has the advantage of achieving reasonably effective control with great simplicity but the disadvantages of a fluctuating controlled condition (within the limits of the dead band) and the frequent switching of equipment which in some cases can cause excessive wear on components. For these reasons it is not used commonly in the UK for conventional HVAC control but is used widely for domestic heating, unitary air conditioners and small-scale atmospheric boilers.

Under *digital* control, control action takes place at discrete points in time, each point separated by a time interval, T. The valve position will assume a position anywhere between fully open and fully closed according to a calculation carried out by the controller at each discrete point in time (Figure 1.6). In between these times, control action freezes. Digital control offers enormous flexibility particularly when the controller forms part of a network linked to a supervisory computer. The controller can be designed as a 'stand-alone' for the specific task, merely requiring setting up and adjustment for precise plant conditions – *application-specific controller* (ASC) – or freely programmable for whatever task is intended – *universal controller* (UC). Alternatively, parametric control decisions and actions can be performed by the central supervising computer with local control decision-making and signal management carried out at the plant 'level' by programmable *intelligent outstations.*

Whether networked or stand-alone application-specific, this type of control is now the predominant method for conventional non-domestic HVAC systems and plant.

We will now move on to take a look at some of the established applications of HVAC control.

1.2 HVAC control strategies

Constant-volume air handling

Figure 1.7 shows the general arrangement for the four-channel control of an air handling system serving one zone in which the four channels control, in sequence, heating → heat recovery → fresh air ('free' cooling) → chilled water-based cooling.

Note that the heat recovery is optional and would only tend to be used in this application when the minimum fresh air in total air handled is high due, for instance, to high occupancy density in the zones served by the air handling system. Note also that the controlled condition becomes supply air temperature in situations where the air handling system's function is to provide pre-conditioned air to rooms or zones which themselves have local sequenced heating and cooling control.

Operation The controlled condition is the space temperature, θ_r. It is usual to measure this in the return air duct, provided that there are no intervening heat gains, such as light fittings, in the return air path. The return air will be well-mixed and the temperature sensor at the higher in-duct air velocities will tend to be more responsive than in the comparatively still air conditions of the space. For space or room temperature control, the manipulated variables will be:

- the flow rate of heating water by means of diverting control valve, V_H;
- the proportion of direct and bypass air at the heat recovery heat exchanger, HRX, by means of face and bypass dampers, D_{HR};

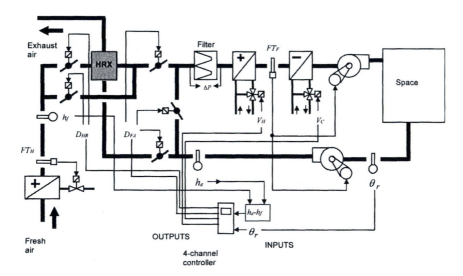

Figure 1.7 Control for constant-volume air handling.

- the proportion of fresh and recirculating air at the mixing dampers, D_{FA}; and
- the flow rate of chilled water by means of diverting control valve, V_C.

During very warm summer weather, the fresh air damper will be fully open and the heat recovery face and bypass dampers positioned for full bypass in order to achieve 'free cooling'. The limit to free cooling is reached when the fresh air intake enthalpy exceeds the enthalpy of the return air at a point downstream of the return air fan, a condition that takes place during the peak summer season in the UK though will often be short-lived. At this condition, it is necessary to signal to the fresh air damper a return to its minimum position and for the heat recovery face and bypass dampers to reposition so as to allow full heat recovery (which will now, in effect, provide pre-cooling of the fresh air from the cool building exhaust air). Enthalpy sensors h_e and h_f provide this and most control system manufacturers will permit additional plug-in modules for their multi-channel controllers to achieve these summer reset conditions. In the arrangement shown, when $h_e - h_f$ produces a negative signal, the reset conditions described are enacted.

To provide frost protection, there are two features. Firstly, the heat recovery heat exchanger (if present) is protected from icing during very cold weather by an upstream frost coil on the fresh air side. Because of the highly temporal nature of calls on the frost coil, a simple thermostatic control through a two-position two-port control valve is usually sufficient. Frost thermostat FT_H provides for this. Secondly, should any primary heating plant such as central boilers or pumps fail, all coils in the arrangement are vulnerable to freezing of their respective water contents during very cold weather, especially the frost coil itself. To provide last-line protection against this, frost thermostat FT_F acts to trip both fans (a further precaution might be to provide a two-position isolating damper at the fresh air intake which coincidentally closes in these conditions thereby eliminating any natural draught through the plant when idle). Typically, FT_H would have a set point of 5°C, and FT_F, 3°C.

Sequencing Figure 1.8 shows the sequencing of this arrangement. Essentially, sequencing ensures that all control actions are mutually exclusive. From a very low controlled condition value, heating modulates as heat recovery is sustained at maximum, fresh air is a minimum and cooling is off. With the controlled condition value rising, the heating eventually reaches its off position. Heat recovery is then allowed to modulate with minimum fresh air and cooling off, and free cooling is allowed to proceed when the heat recovery reaches its off position. Only when maximum free cooling is effected is the chilled water cooling allowed to modulate at the high controlled condition value. By this sequence of events, minimum energy use is assured.

Dead zones may be programmed into the settings of each control channel, since each channel will be allocated a band of the control variable value

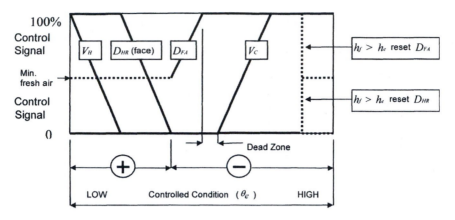

Figure 1.8 Sequencing for constant-volume air handling.

within which it is active. A dead zone for instance between a heating and cooling action can help to ensure not only the mutual exclusivity of heating and cooling, but also that a gap exists between winter heating set point and summer cooling set point thereby minimising energy use throughout the seasons.

Variable-volume air handling

Figure 1.9 shows the general arrangement for variable-air-volume air handling (VAV) in which a number of zones or rooms are served (shown this time without the heat recovery option). Many of the principles of the previous case are applicable here as far as thermal control is concerned but, in essence, the supply air temperature becomes the controlled condition.

Operation There are a number of essential features and some optional features shown in this arrangement. First the essentials.

1   The supply fan flow rate is adjusted in order to keep the in-duct static pressure (measured by static pressure sensor, P) constant. The flow rate is adjusted via variable-speed fan drive, VSD_S. Though there are several methods of fan capacity modulation, the variable-speed drive (typically using a frequency inverter (section 2.2)) is becoming almost universal practice due, potentially, to high part-load power savings. The location of P is an issue (Shepherd *et al.*, 1993). Located in the branch duct to that zone with the highest heat gain profile will mean that some branches will receive more air than they need and, though these branches can adjust to their required conditions by means of their VAV terminal devices, the full power saving of the fan will not be realised. Conversely, if P is located in a branch with a low heat gain profile, many branches will be air-starved. A commonly practised compromise is to locate P two-thirds of the way down the index path from the supply fan.

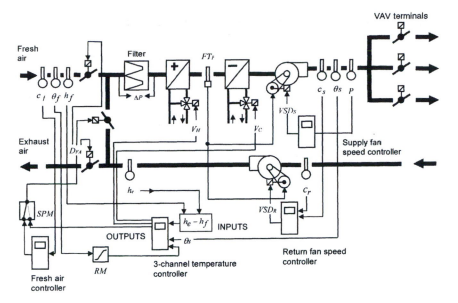

Figure 1.9 Control for variable-air-volume air handling.

Networked digital control can also help substantially here, for in such systems all VAV terminal device signals are known at a particular time and the fan speed can be selected to satisfy a majority of these.

2　The return fan is required to track the supply fan so that some sort of balance exists between total air delivered and returned. There are at least three options for achieving this. The first and simplest is to position the return fan using the same signal as that used by the supply fan, i.e. both fans modulate with the same speed profile. This method, though simple, does not work well in conventional VAV because the supply fan seeks to maintain a constant in-duct static pressure whilst terminal devices close down, with the result that a return fan which precisely tracks the accompanying supply fan will tend to handle much more air at part-load than the supply fan. One remedy is to 'calibrate' the return fan system against the supply fan system on site but this is complicated. The second method is depicted in Figure 1.9 and achieves a more satisfactory tracking performance. Velocity sensors (c_s, c_r) are positioned in supply and return ducts close to the two fans. A separate return fan controller tracks the supply duct velocity (and hence volume flow rate) and seeks to match this velocity with the velocity in the return air duct by adjusting the return fan speed. A third method is to position the return fan in order to maintain a positive static pressure within the building with reference to ambient conditions using a remote differential pressure sensor (Smith, 1990). The difficulty with this method is the choice of location of the sensor in multi-zone applications and it is not commonly adopted in UK practice for conventional air conditioning applications, though it does have some merit in single-zone situations where internal pressure conditions are critical (e.g. laboratories).

There are two optional features shown in Figure 1.9.

3   One of the major difficulties with VAV systems lies in ensuring that minimum fresh air rates are maintained at all times (Janu *et al.*, 1995). The distribution of minimum fresh air among zones is indeed a problem which cannot be resolved by control alone; the cautious practitioner may well opt for a fixed fresh air strategy with heat recovery in applications where indoor air quality is a paramount concern. Where control can help however is ensuring that minimum *total* fresh air is always maintained. The velocity sensor, c_f, mounted in the fresh air intake duct, and fresh air controller, generate a minimum fresh air damper position to achieve the target fresh air rate. This damper signal is compared with the damper signal coming from the appropriate channel of the temperature sequence controller by signal priority module (*SPM*) which feeds forward the larger of the two received signals. This guarantees minimum total fresh air (provided of course that the supply fan turndown has been correctly selected).

4   Another potential problem can occur in winter conditions. Conventionally, VAV seeks to maintain a constant supply temperature (for example 12°C) throughout the year. This can lead to draughts during winter even at turndown conditions which can be alleviated by resetting the set point of the supply temperature to a higher value as winter conditions develop; a process sometimes called *set point scheduling* or reset (see e.g. Mutammara & Hittle, 1990). The reset module (*RM*) schedules the required supply temperature set point from the prevailing fresh air temperature, θ_f. A typical scheduling range might be 12°C (summer peak) $\rightarrow$ 17°C (winter peak). Care has to be taken in the selection of the latter value; too high a value will tend to restrict VAV terminal device travel towards minimum position thereby restricting full fan power economy. Figure 1.10 gives a typical set point schedule for this.

Sequencing   Temperature control for the arrangement of Figure 1.9 is very similar to that of the constant-volume case and is shown in Figure 1.11.

Supply fan control is governed by the static pressure sensor, *P*, which in turn is influenced by collective VAV terminal device activity. The VAV terminals will never shut off completely but will always pass sufficient air at turndown to ensure that minimum fresh air and reasonable air distribution in the space are maintained. Most practical systems will not turn down below 25%. Because a conventional VAV supply fan will seek to maintain a constant in-duct static pressure, fan power will be proportional to volume flow rate which, in turn, will be proportional to the cube of the speed. It follows that at 25% volume turndown, the supply fan will need to run at 63% of its design-rated speed at which point 25% of fan power will be drawn. Thus practical speed range is limited in these systems. The use

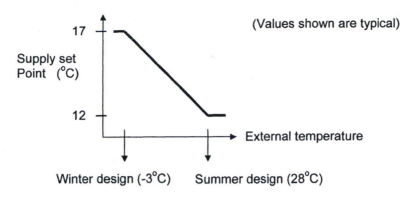

Figure 1.10 Supply temperature set point schedule.

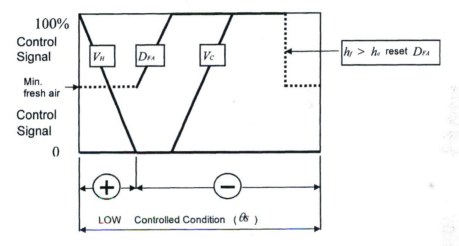

Figure 1.11 Temperature control sequencing for variable-volume air handling.

of *terminal-regulated air volume* (TRAV – Hartman, 1993) obtains more fan power saving than in conventional VAV by allowing duct pressures to fall while VAV terminal dampers remain open, through the use of terminal box feedback. See also Li *et al.* (1996), Tung & Deng (1997) and Englander & Norford (1992).

Room heating

Two strategies for the control of space heating in a room are considered in the following.

Compensated heating

This is a very common method of heating control and is, in effect, a form of *feedforward compensation* – a method discussed in detail in section 4.3.

Figure 1.12 shows the arrangement for base compensation with external temperature and supplementary compensation with solar intensity and wind speed.

Operation Control is with respect to the heating flow water temperature, θ_{wf}. The control valve in this case is connected to mix with respect to the heating system (and divert with respect to the heating source) with a pump in the mixed flow. The controller seeks to maintain the flow water temperature at a set point which is reset from the external temperature, θ_o, via the reset module, *RM*, as illustrated in Figure 1.13. Optionally, set point scheduling can be further compensated with solar intensity (I_s) and/or wind speed (c_w).

The balance point temperature is that value of prevailing external temperature at which a combination of heat gains to the building and its thermal capacity are such that the building is in a state of thermal equilibrium. The balance point value will vary from building to building but a reasonable typical value is 15°C. This method of control has the

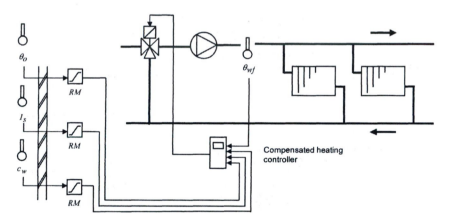

Figure 1.12 Compensated heating.

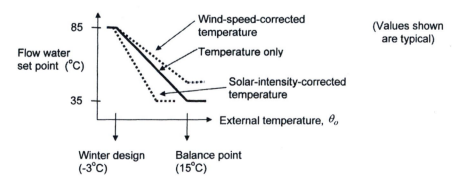

Figure 1.13 Typical compensated heating schedules.

advantage of being inexpensive in that, in theory, it can satisfy control for an entire building with temperature compensation, or an entire orientation of a building with temperature and solar compensation plus optional wind-speed correction. In practice, however, the method is seriously flawed since, even with full weather parameter compensation, internal casual heat gains are ignored leading to the possibility of over-heating (and consequential energy wastage).

Some measure of improvement can be obtained by including local thermostatic (radiator) valves (TRVs) at individual emitters though this presents other complications due to the varying sensitivity of valve sensing element to prevailing flow water temperature (Fisk (1981) gives a detailed discussion of this problem and possible remedies).

Feedback heating control

The conventional method which eliminates the over-heating tendency of the previous system requires a control loop for each room or group of similar rooms. Whilst more costly, feedback control provides the only solution for potentially good control and energy efficiency. The (three-port) control valve in this case is connected to divert with respect to the heating system using a single pump, as shown in Figure 1.14. An alternative, which is beginning to become popular with some practitioners, is to use two-port modulating valves and a variable-capacity pump.

Sequenced room heating and cooling

The need to sequence heating and cooling in the thermal control of a space arises in many situations, most commonly:

- four-pipe fan coil unit air conditioning,
- VAV with zone reheat,

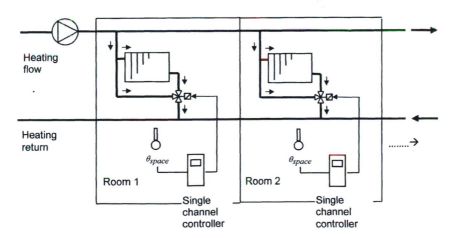

Figure 1.14 Feedback heating control.

- VAV with local perimeter heating,
- chilled ceiling/beam with local perimeter heating, and
- constant-volume air conditioning with terminal heating and cooling.

Sequenced feedback room control

Figure 1.15 depicts the arrangement for the constant-volume case, which is typical. A dead zone in the control sequence for this arrangement will ensure a gap between seasonal control set points. This is desirable to minimise heating and cooling energy use (e.g. 20°C winter, 24°C summer) – see for example Boyens & Mitchell (1991) for a treatment of minimised cost of operation with reference warm air heating plant. Figure 1.16 shows the typical control sequence.

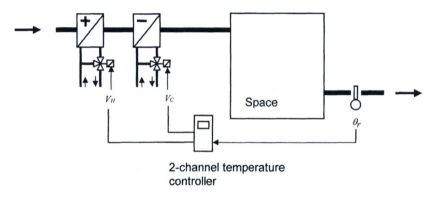

Figure 1.15 Sequenced room heating/cooling.

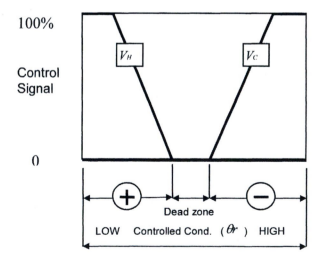

Figure 1.16 Control sequence for room heating and cooling.

Cascade room temperature control

In the previous case, there are advantages in control quality and response if the supply temperature is 'cascaded' from the room temperature by resetting the set point of the former from the latter. Supply temperature therefore becomes the controlled condition but room temperature is fed forward and fully participates in the control. With this strategy, disturbances that result from plant interaction can be eliminated before they disturb the room condition, leading to a smoother response in the latter. This method is therefore often used in close control applications. Figure 1.17 shows the arrangement and Figure 1.18 gives a typical schedule for the reset module, *RM.*

Humidity and air quality

Year-round humidity control is generally only considered for close control applications – see for example Atkinson & Martino (1989), Mamula *et al.* (1989) and Clemens (1996). Many humidity and air quality control schemes will require the use of *signal prioritisation.* We will consider each in turn.

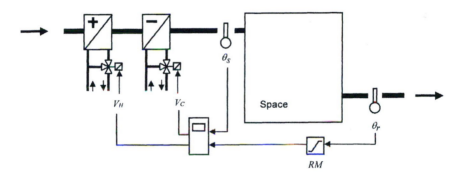

Figure 1.17 Cascade room control.

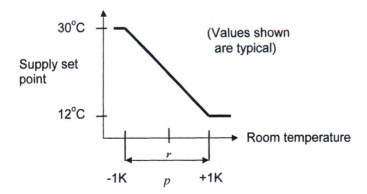

Figure 1.18 Typical reset schedule for cascade room control.

Humidity control

Steam humidification has become the principal choice today because of the potential for *Legionella* propagation in spray humidifier systems. In fact steam humidification offers one major advantage over spray humidification in any case; it can be readily modulated.

Dehumidification is less straightforward because, in comfort air conditioning at least, it is effected by condensing action at the cooling coil; but the cooling coil also has a role to play in temperature control. This dual role for the cooling coil leads to temperature and humidity control variables becoming 'coupled' – something which will be given special consideration in Chapter 6. Meanwhile there are two practical methods for tackling combined temperature and humidity control in full air conditioning systems.

1   Adopt a *constant dew point–reheat* strategy. The design coil leaving air condition is fixed at an apparatus dew-point condition such that its temperature and moisture content are lower than is required to satisfy zone cooling and dehumidifying needs for all but the instant at which design summer conditions prevail. This leads to a reheat and humidification requirement at the room at all times, effectively decoupling the temperature and humidity roles of the cooling coil. The method leads to close control but is energy-inefficient. For a further discussion on this, see section 6.3.
2   *Prioritisation* of the cooling coil. The cooling coil is controlled by the larger of two input signals from temperature control and humidity control loops respectively. This is a more widely practised method for applications where energy efficiency is important. However, the temperature and humidity control loops tend to compete for the cooling coil, which means that close control over both conditions is not possible. Figure 1.19 shows the general arrangement for the prioritisation.

Operation In winter a need for humidification and heating will exist and, accordingly, heating valve V_H and steam humidifier valve V_S will be sanctioned by the respective channels of the two controllers. In summer, the heating and steam humidifier valves will be closed and a need for cooling and dehumidification will exist. The signal priority module (*SPM*) receives dehumidifying and temperature control signals from the respective channels of the two controllers and selects the *larger* of the two signals for onward transmission to the cooling valve, V_C.

Sequencing Temperature sequencing is as before (Figure 1.16) and the humidification/dehumidification sequence will typically follow the pattern given in Figure 1.20.

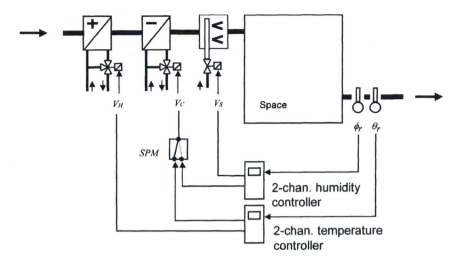

Figure 1.19 Sequenced room heating/cooling/humidity control.

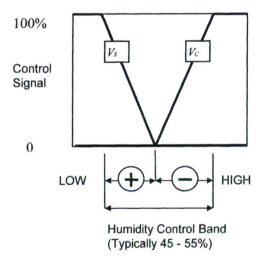

Figure 1.20 Room humidity control sequence.

Indoor air quality control

Control over indoor air quality (IAQ) generates some interest today since, *inter alia*, the possible link between 'sick building syndrome' and air quality was established in the 1980s. In HVAC plant, IAQ is effected by diluting sources of contaminant and 'used' air with fresh air. In situations where mixing dampers are in use it is necessary to prioritise the fresh air damper control with respect to free cooling demand and IAQ demand. The latter is established using either a selective gas sensor (e.g. carbon dioxide – see Warren & Harper, 1991) or a non-selective IAQ sensor (more about these in section 2.1). Figure 1.21 shows a possible arrangement.

Operation The temperature control sequence follows the now familiar heating → damper (free cooling) → cooling pattern. For air quality control, an air quality signal is generated by sensor q_r and the IAQ controller determines a damper signal based on this. The signal priority module (*SPM*) compares this signal with the coincident damper signal generated for free cooling and feeds forward the *larger* of the two signals for damper positioning.

Primary plant

Control of single units of primary plant, such as boilers, chillers and heat pumps, tends to be quite simple, based on a single- or dual-stage thermostat (e.g. for high → low → off control of a boiler). With such control, the only area of real concern lies in switching frequency; a rapid switching rate will tend to lead to wear on components with consequential maintenance problems. A judiciously sized buffer store can relieve this problem. For example, Wong & James (1990) report on the design of buffer stores for chilled water plant. See also Kintner-Meyer & Emery (1995) and Braun *et al.* (1989).

From the point of view of control, the real interest lies in multiple arrangements of primary plant where sequencing becomes an issue. Whilst

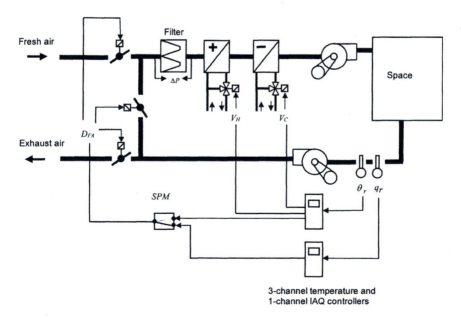

3-channel temperature and
1-channel IAQ controllers

Figure 1.21 Sequenced room temperature and air quality control.

there are various strategies for sequencing this type of plant, the common premise is to ensure that the minimum number of active units are on-line at any one time. This ensures the minimum use of energy through the operation of the plant as high up its efficiency profile as the prevailing demand allows.

There are two alternatives: *stepwise-controlled* plant and *modulating* plant.

Stepwise control

Examples include atmospheric heating boilers with on:off or high:low:off firing combinations such as modular boiler packages and small- to medium-scale refrigeration and heat pump systems with cylinder-unloading control.

Under stepwise control, a permanent offset in flow water temperature arises because at any one time there will be a mismatch between what the heating system wants and what the plant can supply. The size of the offset will depend on the number of step stages or the step resolution. For example, three four-cylinder refrigeration machines with individual cylinder unloading offer 12 steps of capacity. Assuming equal capacity steps, the resulting control of chilled water for a circuit with control band equal to its design temperature difference of 5K will operate with a minimum offset with respect to set point of $5/12 = 0.42K$. This is quite tolerable, but three single-stage machines then slip to a minimum offset of 1.67K, which may not be acceptable in some cases. For this reason, the control sensor is often placed in the water return connection to the plant where the signal tends to give a better reflection of the demand from the system.

Figure 1.22 gives a typical arrangement for a stepwise multiple boiler plant with high:low:off burners (note that many of the principles apply equally to stepwise refrigeration plant).

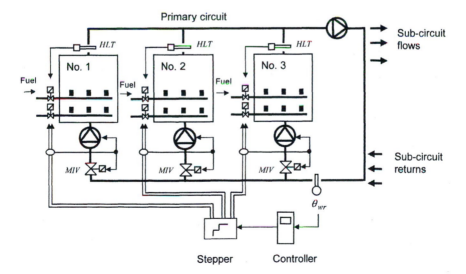

Figure 1.22 Stepwise multiple boiler control.

Operation The controller generates a control signal based on the signal from the return water temperature sensor, θ_{wr}. The size of this signal determines how many stages of boiler burner activity are stepped in. Generally, control in these arrangements will take place within a proportional band equal to the design circuit temperature difference with set point at the mid-position. A simple proportional controller is used since, with step offsetting, it is argued that it would be counter-productive to use anything more sophisticated (CIBSE, 1985). For example, a hot water heating system with a design flow water temperature of 80°C and return water temperature of 70°C will have a (return water) temperature set point of 75°C and a control proportional band of ±5K. Hence at 70°C, all boiler stages will be at high fire and at 80°C all stages will be off. This simple scheduling approach is common but 'near-optimal' performance can be obtained using *heuristic* schedules (Zaheer-uddin *et al.*, 1990).

As each boiler is brought off-line, the motorised isolating valve (*MIV*) and boiler shunt pump are positioned 'off'. This prevents energy wastage by circulating hot return water through the idle boiler waterways but generates disturbances in overall circuit flow rate for which there are various remedies. One method is to oversize the primary circuit by 15–20% and include a large-capacity *balancing header* to which the various sub-circuits are connected.

Each boiler has a high limit trip thermostat (*HLT*) which is usually a feature of the integral controls provided by the boiler manufacturer. This will be set typically at 90–95°C for hot water heating operating at a nominal flow temperature of 80°C. Another common option is to locate a corrosion-protection thermostat in the return water connection to each boiler with a local bypass pipe and diverting valve around the boiler. This allows the shunt pump to short-circuit water around the boiler waterways immediately prior to bringing the boiler on-line, thereby rapidly heating the boiler waterways and preventing cool system return water from forcing flue gas dewing at the rear of the combustion chamber which can lead to corrosion. The boiler is then offered on-line as soon as the corrosion thermostat is satisfied.

Sequencing Figure 1.23 gives an example of the sequencing pattern for the arrangement of Figure 1.22. In the firing sequence shown, boiler 1 is the *lead* boiler (i.e. the first called), boiler 2 is the *first lag* and boiler 3 the *second lag*. An important feature of good sequence control for multiple primary plant is the ability of the control system automatically to rotate the lead boiler periodically to ensure equal use of the available plant.

Modulating primary plant

Modulating control of primary plant has, until recently, mainly been restricted to large-capacity plant; mostly in the multi-MW capacity range – see for example Lewis (1990). More recently, advances in variable-speed drives

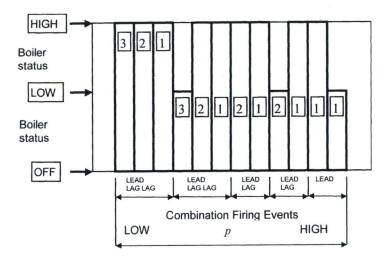

Figure 1.23 Typical stepwise boiler sequence.

and burner technology have resulted in some manufacturers developing modulating control for lower-capacity plant. Under modulating control, it is possible to match plant capacity with system demand at all times, leading to more precise control. The plant remains on-line continuously except for periods of very light load and this prevents the stresses associated with frequent switching from 'cold' inherent in on:off controlled equipment. Examples of modulating primary plant include variable-speed drives in rotary compressor-driven refrigeration and heat pump plant, inlet guide vanes in centrifugal refrigeration plant, and variable air/fuel controls in pressurised fuel burners associated with boiler plant.

Figure 1.24 gives a typical multiple control arrangement for water-cooled modulating refrigeration plant.

Operation For water-cooled chillers, the usual objective is to maintain constant flow rate and constant temperature condenser water conditions at each condenser in order to maintain stable condensing gas pressure. In Figure 1.24 this is achieved with a three-port valve connected to mix warm return water with cool flow water from the source (cooling tower or naturally occurring source). The set point of the flow temperature, θ_{wc}, will typically be a little above summer design wet bulb temperature (e.g. 30–35°C for UK conditions) for cooling tower applications. For air-cooled machines, each condenser will be equipped with one or more variable- or stepped-speed propeller fans which adjust according to a signal, typically, via a pressure sensor in the condenser.

On the evaporator side, note that the arrangement shown is a *series* connection. There are certain advantages with this, in particular in maintaining stable overall circuit flow conditions irrespective of the number of

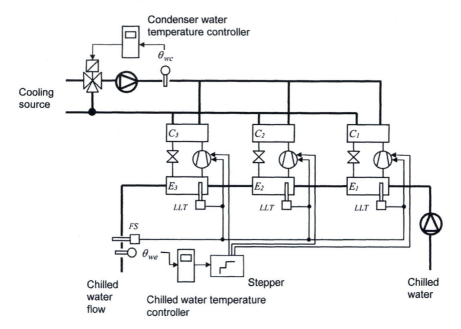

Figure 1.24 Typical multiple modulating chiller control.

machines on-line and there are also performance advantages due to higher evaporating pressures in the first-pass machine.

There are two essential safety control features, besides the usual internal safety controls (not shown). Each machine has a low limit thermostat (*LLT*) to protect against icing in the evaporator, and as a further safeguard against this, a single flow switch (*FS*) is usually essential to protect all machines in the event of a chilled water pump failure.

As to capacity control, since with modulating plant a precise balance between refrigeration delivered and that demanded by the system can be assured, the temperature sensor, θ_{we}, is located in the flow. The stepper in this case is responsible for switching each machine on- and off-line as well as to relay the appropriate positioning control signal from the controller.

Sequencing Figure 1.25 shows a possible sequence in which, again, the essential philosophy is to minimise the number of on-line machines at any one time. Few modulating machines are capable of modulating to zero capacity without significant loss in efficiency. Typical turndown however is low for most types of compressor and for centrifugal compressors under inlet guide vane control a turndown of 10% can be expected.

At very light load, the lead machine modulates until, with increasing controlled condition value (θ_{we}), it reaches its rated capacity (one-third of the overall rated plant capacity if all machines are equally sized). At this point, the first lag machine is called and both machines can satisfy the

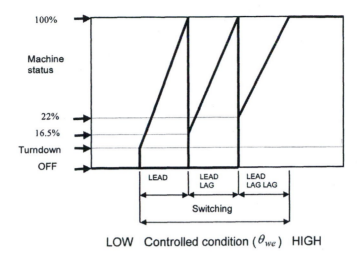

Figure 1.25 Sequencing of modulating chillers.

prevailing overall load at 16.5% of their rated capacities. A further increase in controlled condition value causes both machines eventually to reach rated capacity, equivalent to 66% of the overall plant load, and at this point the second lag machine is called. All three machines now have an equal share of the load; equivalent to 22% of each machine's rated capacity. All three now modulate to capacity as the controlled condition value increases. In this way, only the lead machine is ever likely to operate periodically at, or close to, full turndown and at any position along the load continuum a minimum number of machines will be on-line.

So ends our consideration of control strategy. We have looked at a variety of fairly well established control schemes for HVAC applications with a major emphasis on capacity control. It is now time to move on and take a look at control system configuration.

1.3 Distributed control

The earliest 'distributed' HVAC control systems arose in the early 1970s from the *supervision* of a large number of conventional analogue controllers by a single central 'supervisory' computer. The control in these systems was carried out at the plant 'level' by these conventional controllers but the central computer enabled monitoring, remote switching and control parameter adjustment (e.g. set points and controller settings) to be carried out in one location.

Developments in microprocessors through the 1980s resulted in the evolution of the stand-alone microprocessor-based *intelligent outstation*. The outstation contained all of the necessary communications ports, signal management and residential software to cater not only for all HVAC control and monitoring requirements (e.g. Norford & Rabi, 1987; Hartman,

1990; Akbari *et al.*, 1987), but also for other building management functions such as fire and security for an entire zone or floor of a building (e.g. Lute & van Paassen, 1990; Honda *et al.*, 1993). All outstations would then be networked forming a dedicated *local area network* (LAN). Thus the *building energy management system* (BEMS) evolved (Birtles, 1985).

The motives for this were flexibility (mainly in terms of plant maintenance management in early systems) and, later, improvements in control through the infinite flexibility of software-based control and what became known among HVAC practitioners as *direct digital control* (DDC). The flexibility advantage then began to see DDC replace the previously popular choice of pneumatic control on large projects (Clark *et al.*, 1991; Peterson & Sosoka, 1990). There were also advantages in energy economics, mainly through the remote switching of plant when not needed and the ability to supervise control set points. In a before-and-after study for instance, Birtles *et al.* (1985) concluded that payback periods of 3.6–4.2 years could be expected for a BEMS installation at a time when these systems were considerably more costly to install than today (see also Lowry, 1996).

The fall in cost and dramatic improvement in capability of microprocessors since the late 1980s has meant that all control and monitoring functions can now be handled economically at the immediate plant level. Thus we now have smaller programmable *universal controllers* (UCs) and purpose-designed *application-specific controllers* (ASCs) bringing the powerful BEMS advantages to smaller buildings. For discussions on the use of programmable ASCs and UCs, see for example Sosoka & Peterson (1988), Payne (1988) and Cole & Holness (1989).

Thus a distributed hierarchic control system of between three and five levels has emerged:

- Level 1 – supervision → computer
- Level 2 – communication → LAN and server or communication controllers
- Level 3 – wide area control → outstations for larger distributed installations
- Level 4 – local control → universal or application-specific controllers
- Level 5 – plant → sensors, actuators and drives

Work remains to be done on system compatibility and communication protocols within networked HVAC control – see Bushby (1988, 1990a,b). However, much work has been done on the testing and performance evaluation of BEMS using *emulators* in which a computer model of, for instance, the building is integrated with physical control equipment (and in some cases plant). For progress on the development of emulators for BEMS applications, see for example Kelly & May (1990), Wang *et al.* (1994) and Kärki & Lappalainen (1994), and progress on BEMS performance evaluation can be found in Kelly *et al.* (1994), Peitsman *et al.* (1994) and Visier *et al.* (1994).

Case studies

In the following we will take a look at three commercially available examples for small-, medium- and large-scale distributed control.

Satchwell MMC

Figure 1.26 shows the *Satchwell Micro-Management Controller* (MMC). The MMC is a universal controller which can be programmed for a wide variety of HVAC applications. Each MMC can optionally be networked via RS422/485 serial link to a supervisory computer for remote system management, thus forming a three-level hierarchy. The Satchwell MMC2452/2453 gives three channels of 0–10V(dc) output for, typically, sequenced air handling control, whilst the MMC2451 has a single pulsed output. Up to 32 MMCs can be networked, hence this application is suitable for relatively small-scale applications.

Cylon Unitron

As a highly modular system, the *Cylon Unitron* application offers scope for both small and very large distributed control and has been developed specifically for HVAC applications. Figure 1.27 shows the 'system architecture'. The network is built up as a set of LANs – each LAN covering a building on a large site, or a zone of a large building. Each LAN has a Unitron UCC4 (Figure 1.28) communication controller which coordinates network traffic and provides an essential communication port for supervisory computers and other interface devices. From each UCC4, a family

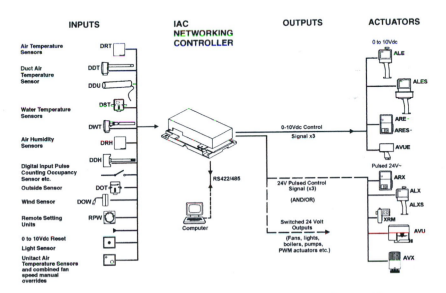

Figure 1.26 The Satchwell MMC distributed control system. (Courtesy: Satchwell Control Systems Ltd.)

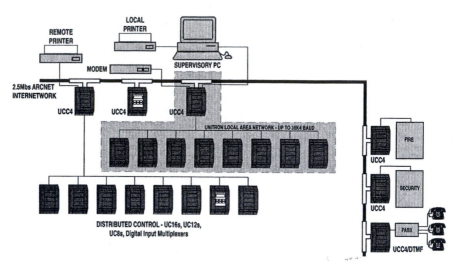

Figure 1.27 The Unitron system architecture. (Courtesy: Cylon Controls (UK) Ltd.)

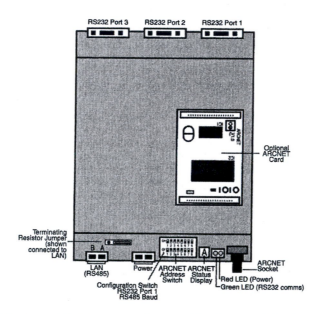

Figure 1.28 The Unitron UCC4 communication controller. (Courtesy: Cylon Controls (UK) Ltd.)

of 32 universal UC12 controllers (each with a capacity of 12 input/output (I/O) *signal points*) or eight UC16 controllers (each with a capacity of 16 I/O points) can be connected via RS485 serial link (Figure 1.29). Each LAN becomes, in effect, a 'distributed' outstation and so this application is very flexible indeed.

The UCC4s are networked to a supervisor via a high-speed network which is based on a dedicated medium – the long-established LAN standard

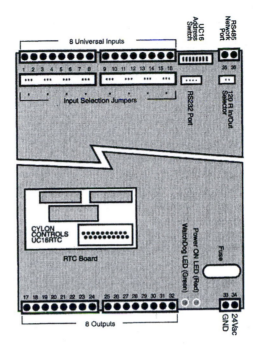

Figure 1.29 The Unitron UC16 universal controller. (Courtesy: Cylon Controls (UK) Ltd.)

ARCNET. Up to eight UCC4 LANs may be linked to a single ARCNET bus but a larger 'star' configuration based around eight-port active hubs can enable up to 255 UCC4 networks to be accommodated. Thus up to 8160 of the smaller UC12 controllers or 2040 of the larger UC16s could be accommodated in such a network providing for the largest of applications in a four-level hierarchy.

Typically, the 12 I/O signal points of the UC12 have four universal inputs and eight outputs split between analogue 0–10 V(dc) and Triac-driven two-motor actuator types. Alternatively, a configuration with eight universal inputs and four analogue/Triac outputs is possible. The larger UC16 has eight universal inputs and eight analogue outputs (Figure 1.29). Although it is a universal controller, the UC12 has been developed with air conditioning terminal equipment control in mind (VAV terminals and fan coil units).

Landis & Gyr BEMS

Figure 1.30 shows the main features of the five-level Landis & Gyr building energy management system (plant level not shown). This forms one of the traditional types of BEMS available today and shares most of the features of the Unitron system with the main exception that intelligent peer-to-peer

Client Server Architecture

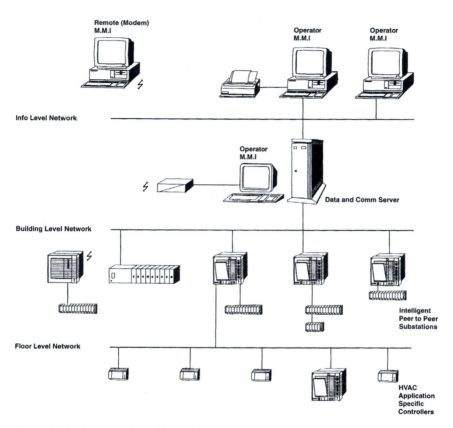

Figure 1.30 The Landis & Gyr BEMS. (Courtesy: Landis & Staefa Ltd.)

substations supported where necessary with ASCs replace the UCC4/UC16/UC12 combination of the Unitron case. All network communications management is handled by a server at the 'information level' in the hierarchy. This type of distributed system is very easy to add to and there are no practical limits to the physical scale of the application.

1.4 System configuration

At the strategic level, control systems are configured in four stages.

1 *Stage 1.* Draw-up a *schematic* representation of the system or subsystem to be controlled.
2 *Stage 2.* Add all *measuring and adjusting* control devices and identify the signals associated with each device:

   a Sensors – for the measurement of temperature, pressure, flow, air quality, etc.

   b Switching devices – thermostats, differential pressure switches and status relays for monitoring plant condition and alarms.

   c Actuators – for driving final control elements such as valves, dampers, etc.

   d Contactors/relays – for switching motor loads.

3 *Stage 3*. Prepare a *points schedule* which lists the total number of control and monitoring I/O points for the system and which can be used to determine the number or required usage of the local outstation or UC. The following should be identified:

   a Analogue input (AI) – usually $0-10V(dc)$ or $4-20mA(dc)$ or a bridge resistance (e.g. $100-200\Omega$); these form input signals from sensors.

   b Analogue output (AO) – usually $0-10V(dc)$ or $0-5V(dc)$ which form positioning signals to actuators.

   c Binary state input (BI) – binary state 0 or 1 from status relays, thermostats, switches.

   d Binary state output (BO) – binary state 0 or 1 for switching contactors (normally via relays).

Note that in HVAC control, binary state signals are usually at two voltage levels such as $0V(dc)$ to represent an *off* state, or $5V(dc)$ to represent an *on* state. The voltage value itself is immaterial and a fairly wide tolerance is anticipated to allow for line losses.

4 *Stage 4*. Specify the logical and functional requirements of the control system which will, in turn, be used for specifying any specialist software requirements and to determine the need for any ASCs as well as for the programming of UCs and outstations. There are four logical/functional outcomes to consider:

   a Control – switching.

   b Control – positioning.

   c Status – monitoring.

   d Status/action – alarm.

The various stages are illustrated in Example 1.1. For a more substantial treatment, see Newman (1994), upon which the following is based. Other work on the planning, configuration and software specification of BEMS can be found in Bynum (1991) and Mumma & Tsai (1991).

Example 1.1

The air handling arrangement of Figure 1.7 is used as an illustrative example of control system configuration.

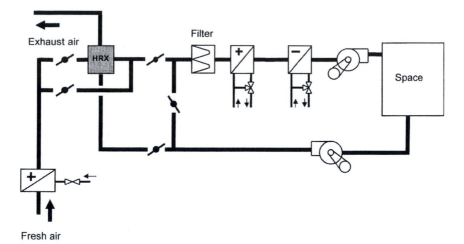

Figure E1.1.1 Basic system schematic.

Stage 1

Figure E1.1.1 shows the basic system schematic – the essential starting point for control system planning. At this point, air flow rates and water flow rates will be known, as will all plant duties (fans, coils and the heat recovery device, *HRX*).

Stage 2

In Figure E1.1.2, all control measuring and positioning devices have been added together with their associated I/O signal types. The numbering system for the resulting 'points' is of a form that could be used as practical system addresses. The preceding letters represent the signal type, the first digit signifies a plant reference (1 has been used in this example) and the remaining two digits represent the point reference. Note that the point reference number is incremented in 10s which allows additional points to be added later if needed.

Stage 3

We are now in a position to draw up the points schedule. This is given in Table E1.1.1. The points schedule is now used to select the number and disposition of outstations or UCs. We have a need for nine inputs and 10 outputs.

For example, suppose we chose to adopt the Unitron system based on UC16 controllers. The UC16 has eight universal inputs and eight analogue outputs. For the plant considered, we would need two UC16s leaving us with seven spare input points and six spare outputs (the switched voltage binary outputs can be done from the analogue board).

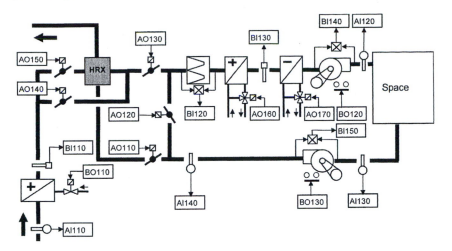

Figure EI.I.2 Control devices and I/O signals.

Table EI.I.I Points schedule

Point	Description	Outstation: I			
		AO	BO	AI	BI
AO110	Control signal — exhaust air damper	X			
AO120	Control signal — recirculating air damper	X			
AO130	Control signal — fresh air damper	X			
AO140	Control signal — heat recovery device bypass damper	X			
AO150	Control signal — heat recovery device face damper	X			
AO160	Control signal — heating coil valve	X			
AO170	Control signal — cooling coil valve	X			
BO110	Control state — frost coil valve		X		
BO120	Control state — supply fan motor		X		
BO130	Control state — return fan motor		X		
AI110	Fresh air temperature/humidity/enthalpy signal			X	
AI120	Supply air temperature signal			X	
AI130	Space air temperature signal			X	
AI140	Return air temperature/humidity/enthalpy signal			X	
BI110	Frost coil thermostat open/closed				X
BI120	Filter differential pressure switch open/closed				X
BI130	Plant frost protection thermostat open/closed				X
BI140	Supply fan differential pressure switch open/closed				X
BI150	Return fan differential pressure switch open/closed				X
	Total I/O	7	3	4	5

Stage 4

The following control logic and functionality can now be identified.

a   Control – switching

- IF (preset run times prevail) THEN (switch fans to run (BO120 and BO130)) OTHERWISE (off).

- IF (frost coil thermostat opens due to low limit (BI110)) THEN (open heating valve (BO110)).

b Control – positioning

- Maintain space temperature (AI130) by controlling the heating coil (valve; AO160), heat recovery device (dampers; AO140, AO150), mixing dampers (AO110, AO120, AO130) and cooling coil (valve; AO170) in sequence.
- IF (supply air temperature (AI120) falls outside preset maximum and minimum limits) THEN (this signal shall take priority over space temperature (AI130)).
- IF (fresh air enthalpy signal (AI110) is greater than return air enthalpy signal (AI140)) THEN (position heat recovery device dampers to closed (AO140) and open (AO150)) AND (fresh/return air dampers to minimum (AO130, AO110) and recirculating damper to open (AO120)).

c Status – monitoring

- Monitor and store key plant performance parameters likely to be of interest for long-term energy management. For example, fresh air temperature (AI110), supply air temperature (AI120), space air temperature (AI130) and return air humidity (AI140).

d Status – alarm

- IF (filter differential pressure switch (BI120) closes) THEN (report alarm condition 'FILTER EXHAUSTED').
- IF (plant frost protection thermostat (BI130) opens) THEN (shut down both fans (BO120, BO130)) AND (close heat recovery device dampers (AO140, AO150)) AND (report alarm condition 'PLANT SHUT DOWN ON FROST PROTECTION').
- IF (supply fan differential switch (BI140) open) AND (fan is signalled 'on' (BO120)) THEN (report alarm condition 'SUPPLY FAN FAILURE').
- IF (return fan differential pressure switch (BI150) open) AND (fan is signalled 'on' (BO130)) THEN (report alarm condition 'RETURN FAN FAILURE').

In this chapter we have taken a fairly broad look at many of the practical HVAC control systems; how they are configured and what, in principle, they seek to achieve. These are actually the routine considerations of HVAC control in practice. Having selected a system and identified what we want it to do the fundamental question remains: Will it work? In the next chapter we will move one step closer to equipping ourselves to answer this question by taking a look at the characteristics and selection procedures for the various components that make up a control system.

References

Akbari, H., Warren, M., Harris, J. (1987) Monitoring and control capabilities of energy management systems in large commercial buildings. *ASHRAE Transactions*, **91** (1), 961–973.

Atkinson, G.V., Martino, P.E. (1989) Control of semiconductor manufacturing cleanrooms. *ASHRAE Transactions*, **95** (1), 477–482.

Birtles, A.B. (1985) *Selection of Building Management Systems.* BRE Information Paper IP 6/85, Building Research Establishment, Garston.

Birtles, A.B., John, R.W., Smith, J.T. (1985) *Performance of a PSA Trial Energy Management System.* BRE Information Paper IP 2/85, Building Research Establishment, Garston.

Boyens, A., Mitchell, J.W. (1991) Experimental validation of a methodology for determining heating system control strategies. *ASHRAE Transactions*, **97** (2), 24–35.

Braun, J.E., Klein, S.A., Beckman, W.A., Mitchell, J.W. (1989) Methodologies for optimal control of chilled water systems without storage. *ASHRAE Transactions*, **95** (1), 652–662.

Bushby, S.T. (1988) Application layer communication protocols for building energy management and control systems. *ASHRAE Transactions*, **94** (2), 494–510.

Bushby, S.T. (1990a) Testing conformance to energy management and control system communication protocols – Part 1: test architecture. *ASHRAE Transactions*, **96** (1), 1127–1133.

Bushby, S.T. (1990b) Testing conformance to energy management and control system communication protocols – Part 2: test suite generation. *ASHRAE Transactions*, **96** (1), 1134–1141.

Bynum, H.D. (1991) Plan and specification documentation of a direct digital control/building management system. *ASHRAE Transactions*, **97** (1), 773–779.

CIBSE (1985) *Applications Manual – Automatic Controls and Their Implications for Systems Design.* Chartered Institution of Building Services Engineers, London.

Clark, R.J., Ghandi, T., Hanus, S. (1991) DDC vs. pneumatics: choosing HVAC controls for today's buildings. *Consulting and Specifying Engineer*, **9** (8), 84–88.

Clemens, R.B. (1996) Control options for various humidification technologies. *ASHRAE Transactions*, **102** (2), 607–612.

Cole, J.P., Holness, G.V.R. (1989) Use of programmable controllers for HVAC control and facilities monitoring systems. *ASHRAE Transactions*, **95** (1), 492–497.

Englander, S.L., Norford, L.K. (1992) Saving fan energy in VAV systems – Part 2: supply fan control for static pressure minimisation using DDC zone feedback. *ASHRAE Transactions*, **98** (1), 19–32.

Fisk, D.J. (1981) *Thermal Control of Buildings.* Applied Science, London.

Hartman, T.B. (1990) Employing EMS to test short-term energy effectiveness of control systems. *ASHRAE Transactions*, **96** (1), 1113–1116.

Hartman, T.B. (1993) *Direct Digital Controls for HVAC Systems.* McGraw-Hill, New York.

Honda, Y., Inoue, M., Ito, Y., Sato, T. (1993) Integrated network architecture for heating, refrigerating and air conditioning. *ASHRAE Transactions*, **99** (2), 230–236.

Janu, G.J., Wenger, J.D., Nesler, C.G. (1995) Strategies for outdoor airflow control from a systems perspective. *ASHRAE Transactions*, **101** (2), 631–643.

Kärki, S.H., Lappalainen, V.E. (1994) A new emulator and a method for using it to evaluate BEMS. *ASHRAE Transactions*, **100** (1), 1494–1503.

Kelly, G.E., May, W.B. (1990) The concept of an emulator/tester for building energy management system performance evaluation. *ASHRAE Transactions*, **96** (1), 1117–1126.

Kelly, G.E., May, W.B., Kao, J.Y. (1994) Using emulators to evaluate the performance of building energy management systems. *ASHRAE Transactions*, **100** (1), 1482–1493.

Kintner-Meyer, M., Emery, A.F. (1995) Optimal control of an HVAC system using cold storage and building thermal capacity. *Energy and Buildings*, **23**, 19–31.

Lewis, M.A. (1990) Microprocessor control of centrifugal chillers – new choices. *ASHRAE Transactions*, **96** (2), 800–805.

Li, H., Ganesh, C., Munoz, D.R. (1996) Optimal control of duct pressure in HVAC systems. *ASHRAE Transactions*, **102** (2), 170–174.

Lowry, G. (1996) Survey of building and energy management systems user perceptions. *Building Services Engineering Research and Technology*, **17** (4), 199–202.

Lute, P.J., van Paassen, D.H.C. (1990) Integrated control system for low energy buildings. *ASHRAE Transactions*, **96** (2), 889–895.

Mamula, L.J., Cuk, D., Kramer, B. (1989) Microprocessor-based control of a close tolerance temperature industrial building. *CLIMA 2000 – Second World Congress on Heating, Ventilation, Refrigeration and Air Conditioning*, 338–345.

Mumma, S.A., Tsai, Y.-T. (1991) Direct digital control documentation employing algorithms in matrix format. *ASHRAE Transactions*, **97** (1), 780–790.

Mutammara, A.W., Hittle, D.C. (1990) Energy effects of various control strategies for variable-air-volume systems. *ASHRAE Transactions*, **96** (1), 98–102.

Newman, H.M. (1994) *Direct Digital Control of Building Systems.* John Wiley, New York.

Norford, L.K., Rabi, A. (1987) Energy management systems as diagnostic tools for building managers and energy auditors. *ASHRAE Transactions*, **93** (2), 2360–2375.

Payne, P.P. (1988) What distributed microcontrollers bring to the building management system. *ASHRAE Transactions*, **94** (1), 1503–1513.

Peitsman, H.C., Wang, S., Kärki, S.H. *et al.* (1994) The reproducibility of test on energy management and control systems using building emulators. *ASHRAE Transactions*, **100** (1), 1455–1464.

Peterson, K.W., Sosoka, J.R. (1990) Control strategies utilising direct digital control. *Energy Engineering*, **87** (4), 30–35.

Shepherd, K.J., Levermore, G.J., Letherman, K.M., Karayiannis, T.G. (1993) Analysis of VAV networks for design and control. *CLIMA 2000 Conference*, London.

Smith, R.B. (1990) Importance of flow transmitter selection for return fan control in VAV systems. *ASHRAE Transactions*, **96** (1), 1218–1223.

Sosoka, J.R., Peterson, K.W. (1988) Building a control system from the bottom up using application-specific controllers. *ASHRAE Transactions*, **94** (1), 1521–1529.

Tung, D.S.L., Deng, S. (1997) Variable air volume air conditioning system under reduced static pressure control. *Building Services Engineering Research and Technology*, **18** (2), 77–83.

Visier, J.C., Vaezi-Nejad, H., Jandon, M., Henry, C. (1994) Methodology for assessing the quality of building energy management systems. *ASHRAE Transactions*, **100** (1), 1474–1481.

Wang, S., Haves, P., Nusgens, P. (1994) Design, construction, and commissioning of building emulators for EMCS applications. *ASHRAE Transactions*, **100** (1), 1465–1473.

Warren, B.F., Harper, N.C. (1991) Demand controlled ventilation by room CO_2 concentration: a comparison of simulated energy savings in an auditorium space. *Energy and Buildings*, **17**, 87–96.

Wong, A.K.H., James, R.W. (1990) Multiple liquid chillers: intelligent control. *Building Services Engineering Research and Technology*, **11** (4), 125–128.

Zaheer-uddin, M., Rink, R.E., Gourishankar, V.G. (1990) Heuristic control profiles for integrated boilers. *ASHRAE Transactions*, **96** (2), 205–211.

2 Device technology

2.1 Sensors

A precise definition of a *sensor* (often called a *detector* in HVAC control) has always been elusive but for practical purposes a sensor can be thought of as a device which converts a physical property (e.g. temperature) or quantity (e.g. flow rate) into a conveniently measurable effect or signal (e.g. current, voltage or number). For HVAC control, sensor groups of interest are as follows:

- temperature (and comfort),
- pressure,
- flow,
- humidity (and enthalpy), and
- indoor air quality.

Most sensors consist of two 'components' – a *transducer* which converts the raw measured signal into a 'convenient' signal (usually electrical), and the associated *signal conditioning* which ensures that the raw transducer signal is converted to a scaleable electrical signal which can be calibrated with the raw measured signal (ideally resulting in a linear relationship between the two – Figure 2.1).

Recent and future generations of sensors additionally have 'intelligence' – a built-in microprocessor enabling additional data to be reported besides that measured. Comfort and enthalpy 'sensors' are examples of this category.

Sensor specification

A sensor can usually be fully specified with reference to at least 12 performance, practical and economic factors. Those concerned with performance are the following:

1 *Range*. The range of the measured variable for which the following characteristics are maintained at stated values.
2 *Accuracy*. The degree to which the measured output compares with some known benchmark – given for example as a *point accuracy* (e.g. ±0.1K at 25°C) or percentage of full-scale deflection (FSD) (e.g. ±1%FSD).
3 *Repeatability*. The ability of the sensor to reproduce consistently the same output from the same measured value.
4 *Sensitivity*. The smallest detectable change in measured value that results in an output change by the sensor.

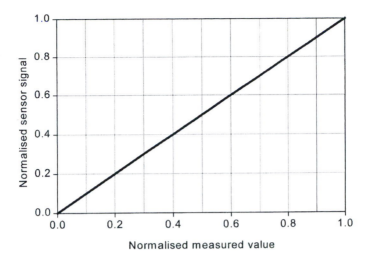

Figure 2.1 Ideal sensor response.

5   *Drift.* The degree to which the sensor fails to give a consistent perfor-
    mance throughout stated life.
6   *Linearity.* The closeness to linear proportionality between the output
    and the measured value across the range.
7   *Response time.* The rate of response with respect to time of the output
    following an input change (often expressed as a *time constant*).

The practical and economic considerations are as follows:

8   *Cost.*
9   *Maintenance.* Any special maintenance and recalibration requirements.
10 *Compatibility.* Compatibility and interchangeability with other compo-
    nents and standards.
11 *Environment.* The ability to withstand harsh environments.
12 *Interference.* Susceptibility to ambient 'noise' (e.g. Goff, 1966), such as
    radio frequencies.

We will now take a brief look at some of the various sensor types. Newman
(1994) gives a detailed treatment of this upon which the following summary
is partly based.

Temperature sensors

Resistance temperature detectors

Resistance temperature detectors (RTDs) make use of the variation in
electrical resistance of a metal with temperature as described by a polynomial of the form

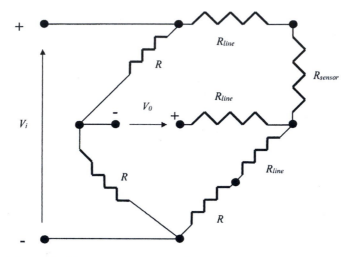

Figure 2.2 Three-wire Wheatstone bridge network.

$$R = R_{(ref)}\left(1 + a_1\theta + a_2\theta^2 + \ldots\right) \tag{2.1}$$

The reference resistance, $R_{(ref)}$, for RTDs is usually given at 0°C (e.g. 100Ω at 0°C), and a_1 and a_2 are constants.

Most commonly, the metal element is platinum. There are two main forms. The wound bobbin type consists of a wire wound around a spindle and has a high time constant. More recently, thin-film RTDs have been developed, consisting of a very thin film of metal on an insulating substrate which is laser-etched to form the necessary pattern to achieve the desired length and shape of resistive element. This type is low-cost and occupies very small space but is susceptible to a self-heating effect due to its resistance to the current passing through it.

In common with most other sensors whose transduced output is an electrical resistance, RTD measurements are conducted in a Wheatstone bridge network. Care should be taken to compensate for voltage losses in connecting leads since, in many distributed applications, signal leads may run into tens or even hundreds of metres. Figure 2.2 shows a bridge network for a *three-wire RTD*, in which the balanced network condition (reference condition) fully compensates for the resistances in the RTD connecting leads (R_{line}), though the compensation is less than perfect at other conditions (Newman, 1994).

Typical applications are as room and duct/pipe immersion temperature sensors.

Thermistors

These are similar in principle to RTDs but based on the temperature–resistance relationship of a semiconductor (typically metallic oxide which is

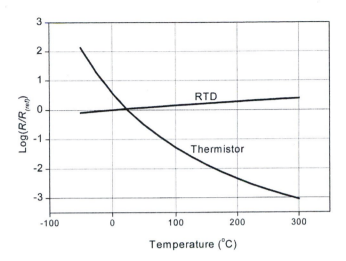

Figure 2.3 Typical comparative sensitivity of a thermistor and RTD. (© H. M. Newman, 1994.)

baked in a powdered form onto a substrate). Thermistors have a negative temperature coefficient of resistance whilst RTDs have a positive temperature coefficient of resistance. For a thermistor, the temperature–resistance relationship can be expressed as

$$R = R_{(ref)} \exp\left[\beta\left(\frac{1}{\theta} - \frac{1}{\theta_{(ref)}}\right)\right] \tag{2.2}$$

The reference resistance for a thermistor is usually given at 25°C and β is a constant.

The thermistor is highly sensitive when compared with the RTD (Figure 2.3). Whilst an RTD might change by a fraction of an ohm per Kelvin temperature difference a thermistor might change by 100Ω or so. This makes them very suitable for temperature measurement where the operating range of temperature is low (e.g. chilled water systems, close control applications).

Typical applications are in close control air conditioning and chilled water plant control.

Thermocouples

Thermocouples make use of the current flowing in a circuit consisting of two dissimilar metals which are joined at differing temperatures – a reference temperature and the measured temperature (the 'Seebeck effect'). These are robust and suitable for very high temperatures but are not very sensitive.

Typical applications are for temperature measurement in combustion and other high-temperature systems.

Figure 2.4 Comfort 'sensor' and controller. (Courtesy: Landis & Staefa Ltd.)

Integrated-circuit temperature sensors

These are fairly recent (and currently ongoing) development, consisting of semiconductor diode junctions which display a linear current–voltage characteristic with temperature. These devices are inexpensive but are fragile and range-limited. Applications are restricted to relatively low temperature measurement (e.g. room and in-duct temperatures).

Comfort 'sensors'

It is of course not possible to measure comfort as might be expressed by, for example, Fanger's *predicted mean vote* (ASHRAE, 1995). However, an instrument capable of measuring room air temperature, mean radiant temperature, humidity and air velocity can, when assumptions about clothing levels and metabolic activity are made, give an indication of comfort conditions. Such instruments are available (Figure 2.4). They function along with an application-specific controller which determines the required value of a degree of freedom – that controlled condition which is physically capable of being changed by the plant (usually room air temperature), based on measurements of all other comfort-dependent parameters (except *clothing level* and *metabolic values* which form control set points). These 'sensors' require on-board processing and are intrinsically 'intelligent'. See also Sugawara & Hara (1990).

The difficulty with these instruments lies in locating a suitable site for the sensor. Ideally, the sensor should be at the centre of the room but this is rarely practicable.

Pressure sensors

Pressure measurement is predominantly achieved through the measurement of the displacement of a diaphragm – caused by the pressure effect acting on one side of the diaphragm in relation to some reference condi-

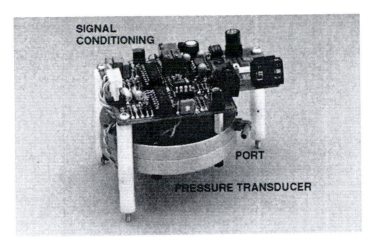

Figure 2.5 Inductive pressure sensor for low-pressure applications. (Courtesy: CMR Controls Ltd.)

tion on the other side. For absolute pressure measurement the reference pressure will be atmospheric. For differential pressure measurement, high- and low-pressure ports are connected across the diaphragm. Most pressure sensors are therefore based on *displacement transducers*. A summary of some leading examples follows.

Capacitive

These have a variable gap between the two plates of a capacitor in which one plate is connected to the diaphragm, the capacitance varying as the pressure changes. These sensors have low sensitivity and are restricted to low-pressure applications such as ventilation filter differential pressure measurement and VAV system fan control.

Inductive

These are based on a linear voltage differential transformer (LVDT) – the diaphragm forms a moving ferrite core within a dual-wound tube. The first winding has a voltage applied across it and the core provides a linked magnetic field between the two. As the core moves with changing pressure, the two coils are increasingly linked, increasing the voltage across the secondary coil in proportion to the applied pressure. These instruments have much in common with the capacitive type in that they are compact (see Figure 2.5) and restricted to the relatively low-pressure applications of ventilation systems.

Strain gauges

These are based on the change in electrical resistance of a body as it deforms under pressure (for example, as a wire is stretched under pressure its cross-sectional area reduces, so increasing its electrical resistance). These are suitable for high-pressure applications but have low sensitivity. There are few applications in mainstream HVAC systems.

Piezoelectric

These make use of a property that certain electrically conductive naturally occurring materials such as crystalline quartz have, in that a change in electrical characteristic occurs as the material is bodily deformed due to an applied load or pressure. These are very sensitive and can cope with rapidly fluctuating conditions and a wide pressure range, making them suitable for sound and vibration measurement applications.

Potentiometric

This is simply a three-terminal resistor with a variable centre node. The centre node forms a 'slider' whose position is changed with displacement. This is a long-established instrument which has a wide range and is inexpensive, but suffers from large physical size and is prone to wear.

Flow sensors

Many flow measurements are derived from pressure measurements.

Pitot tubes

These are essentially for in-duct ventilation applications and are based on two open-ended tubes; one mounted facing the air stream and the other perpendicular to it. The difference in pressure measured between the two tubes amounts to the in-duct velocity-pressure which can be readily converted to air velocity since, based on the Bernoulli equation,

$$c_a = \sqrt{\frac{2p_v}{\rho_a}} \tag{2.3}$$

where c_a and p_v are the velocity and velocity-pressure of air respectively (ms^{-1}, Nm^{-2}) and ρ_a is air density (kgm^{-3}).

There are a number of variants, mostly based on arrays of samplers so that an effective mean velocity across the duct can be obtained. One such example is the Veloprobe (Figure 2.6). This method of flow measurement is very common and highly robust. Accuracy depends very much on the number of sampling measurements taken across the duct and on the instrument used for differential pressure measurement. Whilst unwieldy, a liquid manometer will always be the most robust and accurate means for obtaining the latter. Griggs et al. (1990) discuss the placement of airflow sensors for control.

Orifice plates

These are based on achieving a pressure difference across the pipe or duct by throttling the flow through an orifice and measuring the pressure

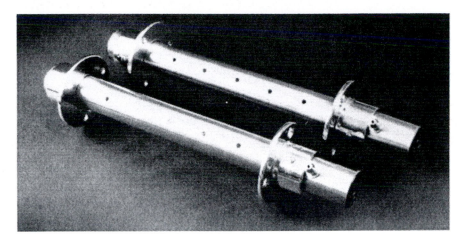

Figure 2.6 The Veloprobe. (Courtesy: CMR Controls Ltd.)

difference. They are very simple but prone to wear when used with liquids, especially those carrying small particles of dirt. The flow rate can be related to the square root of the pressure difference and machining dimensions are critical (BS1042, 1989). These have been used extensively in the past – mainly for piped liquids.

Venturi meter

This is similar in principle to the orifice plate except that a gradual reduction in pipe or duct bore forms a throat section, instead of an abrupt orifice, and there is a gradual enlargement to full bore downstream. Thus, pressure loss at the contraction is almost entirely regained in the enlargement which gives a major advantage over the orifice plate, and they are not prone to wear. However, venturi meters tend to be physically large and expensive.

Hot-wire anemometer

This is essentially for ventilation air flow measurements. The hot-wire anemometer is sensitive and can detect flow at very low velocities making it suitable for free air movement measurements as well as in-duct flow. For in-duct applications, the hot-wire anemometer is less robust than pitot methods. An element is heated and the current needed to maintain it at a constant temperature is measured. Thus the device is based on the cooling effect of the air velocity passing over the heated element through the surface heat transfer coefficient. To compensate for the variability of the air temperature passing over the sensing element, a temperature sensor is included so that a constant element–air temperature difference is maintained which then isolates air velocity as the only determinant to cooling effect and electrical resistance.

Turbine flow meters

In these, a turbine paddle blade rotates at the centre of the pipe or duct at a speed proportional to the velocity of flow. These are mainly used in liquid-in-pipe applications and are susceptible to wear and clogging and not suitable for dirty flow streams. Until relatively recently, these meters were the predominant choice for heat meter applications in hot water and chilled water systems.

Vortex-shedding meters

These are based on the generation and dissipation of vortices or eddies caused as fluid impinges on a bluff body. The frequency of pressure fluctuations caused by these vortices is proportional to fluid flow velocity. Pressure frequency measurement is therefore the basis of these devices. They are suitable for liquids and can be very accurate, but complicated signal conditioning makes them costly.

Electromagnetic flow meters

A magnetic field is set up across the flowing fluid using a coil around the pipe/duct through which an alternating current passes. Provided that the fluid is electrically conducting the magnetic field is cut at a rate which is proportional to the velocity of flow. These devices offer no resistance to the flowing fluid which offers a major advantage over other methods such as the turbine meter which they appear to be gradually replacing. They are also suitable for dirty liquids and slurries.

Ultrasonic flow meters

These are based either on the Doppler effect (i.e. the measurement of sound reflection from particles in the flowing stream), or on the time of passage of a sound wave across two diametrically opposed points on the pipe wall. In the former case, the fluid needs to contain some impurities – essentially dirt – in order to perform. The second type is restricted by the type of piping material, but both types are essentially non-invasive and portable. Accuracy is questionable.

Humidity sensors

Humidity sensors fall into more or less four categories: *hygrometers, psychrometers, electronic humidity sensors* and *dew-point sensors*. Humidity measurement has long been problematical because electromechanical hygrometers have suffered from serious non-linearity and a tendency to drift, whilst the more recently developed electronic sensors are prone to contamination in the air stream being measured and many are not suitable for sustained operation in conditions at, or close to, saturation. Nevertheless, certain electronic sensors are steadily improving in terms of accuracy, long-term stability and tolerance to contaminants.

Hurley & Hasegawa (1985) give a very comprehensive overview of a number of methods of direct and indirect humidity measurement for HVAC applications. A brief summary is given below.

The hygrometer

The earliest-developed method, this uses a naturally occurring material which changes dimensionally with moisture absorption and desorption (e.g. hair, wood, plastic). They are highly non-linear and prone to drift, requiring periodic recalibration. With the advent of electronic methods, the hygrometer is more or less becoming extinct.

The psychrometer

Placing a distilled-water-wetted wick around a conventional temperature sensor (e.g. RTD) results in the temperature of wet-bulb depression which can be related to relative humidity using properties of humid air. This is actually a very robust and potentially accurate method but the difficulty is that the air passing the wick must be maintained at a high enough velocity ($\geq 5ms^{-1}$) in order to maintain effective evaporation from the wick and, of course, the wick must be maintained in a wetted condition. Hence this method is rarely found outside the laboratory or for use in periodic field measurements.

Resistive and capacitive sensors

Examples of these include the *ion-exchange resin* type, the *Jason hygrometer*, *carbon-film sensor*, the *thin-film polymer sensor* and the *Dunmore hygrometer*.

The ion-exchange, Jason and carbon-film devices are all based on the doping of a substrate and the measurement of the electrical resistivity resulting from changes in moisture content in the permeable substrate. The ion-exchange type uses a chemically treated polystyrene with electrodes connected to the treated region; the Jason device uses gold doping onto oxidised layers on aluminium, the doped regions being responsive to moisture; and the carbon-film device uses a mixture of cellulose and carbon deposited on a polystyrene strip, the cellulose changing dimensionally with moisture content and causing a shift in carbon particles and, correspondingly, a change in electrical resistance. The Dunmore sensor uses a thin film of lithium chloride solution connected to a pair of electrodes.

Most of the polymer sensors make use of changes in electrical capacitance due to the polymer acting as the dielectric in a capacitor and its responsiveness to changes in moisture content. These are probably the most widespread of commercially available electronic methods today and current devices have improved sensitivity (2% relative humidity). Another advantage is that this sensor tends to have better immunity to contaminated air streams than the other electronic types, but this general family of sensors all suffer from impairment and in some cases irreparable damage if operated for significant periods at humidities greater than 90% – a frequent requirement for in-duct measurement and fresh air measurement. One solution is to heat the air sample prior to passing over the sensor transducer and to correct the resulting reading with reference to actual temperature conditions which can be achieved with the addition of some local sensor 'intelligence'.

Dew-point sensors

There are two commercially available types: the *chilled mirror sensor* and the *saturated salt solution* device. The latter consists of an RTD surrounded by a wick which is impregnated with lithium chloride solution. A variable heating element adjusts the temperature of the wick until a stable condition is arrived at and the power from the heating element just balances with the heat of evaporation from the wick. At this point, the wick and surrounding air are both at an equilibrium vapour pressure and at a temperature measured by the RTD equal to the dew-point temperature of the air.

The chilled mirror device has long been a 'benchmark' for dew-point and humidity measurement and is capable of detecting dew-point with an uncertainty of under 1K. A mirror is maintained at a uniform temperature by a miniature thermoelectric heat pump. The temperature is varied until dew starts to form on its surface – as measured by a photometer. A local microprocessor can be used to calculate the humidity if required using properties of moist air.

Figure 2.7 gives a schematic of the arrangement. Again, though as accurate a method as exists, the chilled mirror requires frequent cleaning in order to eliminate the effects of contaminants in the air stream. However it

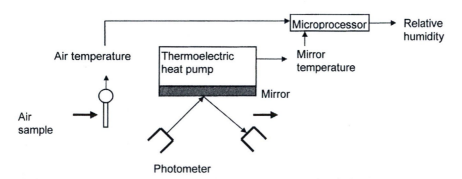

Figure 2.7 Schematic of the chilled mirror dew-point sensor. (Hurley and Hasegawa, 1985.)

is not as susceptible to operation at high relative humidities as many of the other electronic devices are. These combinations make it a good choice for close control air conditioning applications.

Tests have been conducted on three of the humidity sensors described above by Thomas (1992) (see also Huang, 1991). The chilled mirror dew-point sensor was compared with a polymer dielectric (capacitive) sensor and an ion-exchange resin type sensor in laboratory conditions and Thomas concluded that the chilled mirror device was twice as accurate as the capacitive and ion-exchange sensors and the only instrument which compared favourably with manufacturer's claims. The tests also revealed that the capacitive sensor's accuracy was influenced by errors caused by the exposure of its signal conditioning electronics to the humid air stream itself; when the electronics were placed outside the measured air stream into 'normal' humidity conditions the accuracy improved considerably.

Enthalpy sensors

These are a special case of the humidity sensor. As a thermodynamic property, enthalpy cannot be directly measured but can be expressed from a knowledge of two properties as far as air is concerned: dry-bulb temperature and moisture content. In turn, moisture content can be determined from a knowledge of the dry-bulb temperature and some humidity property such as wet-bulb temperature, dew-point temperature or relative humidity itself. Practical air enthalpy 'sensors' therefore consist of a dry-bulb temperature sensor (RTD or thermistor), a humidity sensor (in most practical devices, this is usually of the polymer dielectric (capacitive) variety) and a microprocessor which calculates the corresponding enthalpy using properties of humid air. For example, in a reasonable approximation to properties of humid air at a dry-bulb temperature θ_{air} (in degrees Celsius) and relative humidity ϕ (as a fraction), the air moisture content g is given by (ASHRAE, 1995)

$$g = \frac{0.623\phi p_{vs}}{(1013 - \phi p_{vs})} \tag{2.4}$$

in which the saturation vapour pressure (mbar) is given by

$$\log_{10}(p_{vs}) = \frac{(8.2857\theta_{air} + 186.45)}{(237.3 + \theta_{air})} \tag{2.5}$$

and the moist air enthalpy h_{air} (kJkg^{-1}) is

$$h_{air} = \theta_{air} + g(2501 + 1.805\theta_{air}) \tag{2.6}$$

As such, the air enthalpy 'sensor' is an example of an intelligent sensor and will of course be subject to the same problems associated with contaminated air streams and operation at high humidity as the humidity sensor itself is.

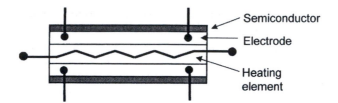

Figure 2.8 Typical non-selective air quality sensor.

Indoor air quality (IAQ) sensors

Historically, CO_2 sensors have been used in air-quality-critical applications. These can work adequately in many circumstances since a reasonably proportional relationship exists between occupancy and CO_2 concentration. The problem with CO_2 sensing alone is that it is assumed that other constituent gases and vapours present in the space are insignificant. However, these gases and vapours tend to be those predominantly responsible for odours and contribute to challenging indoor air quality in practical cases far more than CO_2 alone can. Examples include ammonium hydroxide, fatty acids, methane (mainly from humans) and formaldehyde and other new building material emissions, kitchen emissions, and so on. A comprehensive IAQ sensor should therefore be capable of measuring all of these odour-producing gases more or less collectively.

One such sensor, described by Thompson (1991), *non-selectively* measures the concentration of all oxidisable gases collectively. It consists of a tube coated by a thin-film semiconductor (zinc dioxide), a pair of electrodes and a miniature heating element within the tube. Maintained at a constant temperature the semiconductor adsorbs the gases which results in the release of electrons which in turn alters the resistance across the two electrodes; hence a signal. The process is bidirectional. Though the predominant human emission CO_2 is not measured, many of the gases that are measured are essentially from human sources and so CO_2 is indirectly accommodated. Figure 2.8 gives a schematic of the arrangement.

The instrument is sensitive and has a reasonably fast response time making it quite suitable for IAQ applications in comfort ventilation and air conditioning applications.

A/D and D/A converters

Analogue signals from sensors (and any other sources of this type) require to be converted to binary digital numbers for onward processing at the control level and beyond. This is carried out first by *multiplexing* the incoming signals (a process in which each of a number of incoming signals to the local controller or outstation are addressed *in turn*) and then applying the analogue signal in the form of a dc voltage to an *analogue-to-digital* (A/D) *converter* (Newman, 1994).

A/D converters can be 8, 10 or 12 bits (binary digits) in resolution. Since the maximum value a binary number can take is $2^n - 1$ (where n is the number of bits), then it follows that in an 8-bit A/D converter there will be 255 divisions of the incoming analogue signal whilst a 12-bit A/D converter offers 4095 divisions. For example, if a 0–10V(dc) incoming analogue signal represents air flow rate in the corresponding range 0–3000Ls^{-1}, an 8-bit A/D converter will represent the air flow measurement signal with a resolution of $3000 \div 255 = 11.76$Ls^{-1} (to 0.39%). A 12-bit A/D converter will represent the same measurement with a resolution of $3000 \div 4095 = 0.73$Ls^{-1} (to 0.024%).

The most common A/D converter works by comparing the applied input voltage to a balancing voltage which is built up in equal increments. Each time an increment is added to the balancing voltage, a digital number is incremented by the numerical equivalent based on the measurement range until the two voltages balance within the available resolution. So in the 8-bit A/D example mentioned above, the balancing voltage is incremented by $10 \div 255$ (0.0392V(dc)) and the corresponding digital number is incremented by 11.76 until the balancing voltage reaches a point where it is within 0.0392V(dc) of the applied voltage signal at which point the corresponding digital number represents the conversion.

Clearly, the resolution of the A/D converter is crucial in reconstructing a representative numerical signal and the 8-bit A/D converter is adequate for few practical applications. At stake is not so much cost as conversion time; though in the relatively slow-response HVAC system, speed of A/D signal conversion is conveniently of little importance compared with, for example, the time constant of the sensors themselves. There is little reason therefore why the 12-bit A/D converter should not be regarded as a standard for HVAC applications.

The *digital-to-analogue* (D/A) *converter* performs the same role as the A/D converter in reverse, leading to analogue signals for positioning control elements based on binary digital numbers representing the value of the control signal calculated by the controller. The same 'staircase' resolution applies though this tends not to be as critical on the output side as illustrated in the following example.

Example 2.1

A temperature sensor has a range of -50 to $+50°C$ with a point accuracy of $\pm 0.1K$. The A/D and D/A converters have 12-bit resolution. The temperature value at a particular time is 21.3°C in relation to a set point of 22°C and the control action consists of a proportional gain (see section 2.3) of $0.2K^{-1}$, which is applied to the control error. The actuator/positioning device has a 'hysteresis' (see section 2.2) error of 0.5% of range and the controller output at zero error (steady state) is 0.

Find the maximum compound uncertainty due to error and signal conversion resolution for this instantaneous signal transaction.

Solution

For a signal range of 0–1, the error-free situation will give a control element position of

$$\text{signal} = 0.2(22 - 21.3) + 0(\text{steady-state output}) = 0.14$$

(i.e. controller proportional gain multiplied by error).

For the actual situation, the 12-bit A/D converter will have a resolution of $2^{12} - 1 = 4095$ resolution parts. Therefore

$$\text{A/D resolution} = \frac{\text{sensor range}}{\text{A/D resolution}} = \frac{100\text{K}}{4095} = 0.0244\text{K}$$

The controller output at maximum error will be

$$0.2[22 - (21.3 + 0.1 + 0.0244)] = 0.1151$$

(which includes the sensor error and the A/D converter resolution). The D/A converter resolution will be $1 \div 4095 = 0.000\,244$ (as in the case of the A/D converter). Hence the control signal with maximum error and resolution uncertainties will be

$$0.1151 - 0.000\,244 = 0.114\,86$$

Deducting the positioning device hysteresis error,

$$0.114\,86 - 0.005 = 0.109\,86$$

Therefore, the maximum uncertainty for this signal transaction as a percentage of the error-free control value will be

$$\frac{0.14 - 0.109\,86}{0.14} \times 100 = 21.5\%$$

This is clearly a significant result. The breakdown by category is:

- Detector   66.4%
- A/D converter   16.2%
- D/A converter   0.8%
- Hysteresis   16.6%

Conclusion

Whilst the D/A and A/D converters have the same resolution, the effect on overall signal error is dramatically different. A much more significant effect occurs on the A/D converter which is due to the scaling effect of the controller gain though, as is generally the case in practice, the error of the sensor will often dominate overall system error.

2.2 Control elements

Control valves

Linearity

Valves are used extensively in HVAC applications for flow isolation, flow regulation and flow control. For control applications, control valves are used to regulate *heat transfer through flow* rather than to regulate flow as such. The key concern for both design and commissioning is to ensure that the valve achieves a linear response such that a proportional relationship exists between valve position and heat transfer at the plant. Failure to achieve something reasonably close to such a relationship will mean that the plant, if under well tuned control, will at best drift in and out of regions of good control performance and bad control performance and, if badly tuned, control response may be unstable in some regions of operation. A fuller discussion of this is given in Example 9.3.

When steam is the heating medium this is quite easy since, due to the reasonably constant steam latent heat across a fairly wide range of pressures, heat transfer ends up being directly proportional to the flow rate of steam delivered. Therefore, a valve designed to achieve a *linear* relationship between position and flow rate works well.

With hot water heating (or chilled water cooling), a logarithmic relationship between heat emission and flow rate exists in heat transfer equipment. For example, the emission of a heating coil follows, approximately, a logarithmic mean temperature difference

$$q \propto \frac{[(\theta_{wi} - \theta_{ao}) - (\theta_{wo} - \theta_{ai})]}{\ln[(\theta_{wi} - \theta_{ao})/(\theta_{wo} - \theta_{ai})]} \tag{2.7}$$

whereas hot water radiators are expected to follow the exponential characteristic

$$q \propto \left[\frac{(\theta_{wi} + \theta_{wo})}{2} - \theta_r \right]^{1.3} \tag{2.8}$$

Here q = heat emission, θ_{wi} = inlet water temperature, θ_{wo} = outlet water temperature, θ_{ai} = inlet air temperature, θ_{ao} = outlet air temperature and θ_r = room temperature. The control valve needs to compensate for this non-linearity in its relationship between position and flow rate if linear control is to be achieved.

Therefore, the stem position and flow rate relationship for an HVAC control valve, termed the *valve characteristic*, is an important consideration; for steam heating valves, a linear characteristic should be specified, and for hot water and chilled water valves, a logarithmic characteristic (often referred to as an 'equal-percentage' characteristic) should be specified for reasons which will become clear later.

A control valve has four distinct component features. The *port* through which the flowing fluid passes is simply an orifice with a machined seat onto

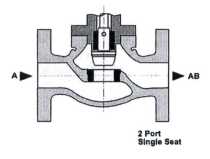

**2 Port
Single Seat**

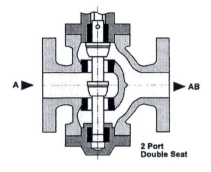

**2 Port
Double Seat**

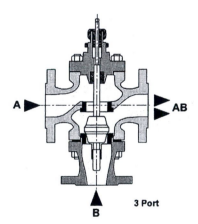

3 Port

Figure 2.9 Typical control valve cut-aways. (© Caradon Trend Ltd.)

which the valve *plug* closes. The plug is connected to a *stem* or *spindle* which
in turn connects the plug to the outside world; in the case of a control valve
this connects to a positioning device or actuator. To achieve both a leak-
tight valve body as well as freedom of movement for the valve stem, a *packing
gland* is formed at the head of the valve. The valve characteristic is achieved
by machining the plug and port (mainly the former) to a specific profile
shape (Figure 2.9).

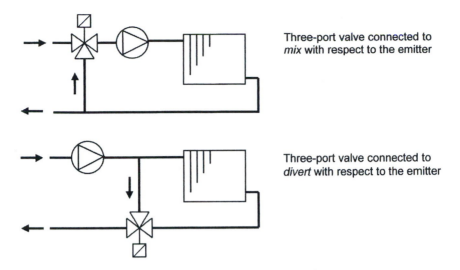

Three-port valve connected to *mix* with respect to the emitter

Three-port valve connected to *divert* with respect to the emitter

Figure 2.10 Three-port valve connections.

Three-port valves

Three-port valves are used extensively in HVAC applications. Mostly, these valves are designed as mixing valves having two inlet ports and one outlet such that, with correct connection arrangements, variable-flow constant-temperature control or variable-temperature constant-flow control can be realised with respect to emission (CIBSE, 1985). However, diverting valves are also possible, having one inlet port and two outlets.

There are two strategies for the use of a three-port control valve:

- Constant flow rate, variable flow temperature with respect to the emitter (mixing at the emitter).
- Constant flow temperature, variable flow rate with respect to the emitter (diverting at the emitter).

Figure 2.10 shows how a three-port mixing valve can be connected to achieve these objectives (note that a diverting valve with correct connection details could achieve exactly the same outcomes).

The inlet ports of these valves are characterised to compensate for the non-linearity of the emission system whilst the bypass port is also characterised to ensure that the combined flow rate handled by the valve is (ideally) constant irrespective of the valve position, as is the overall pressure drop across the valve. This is in fact why three-port valves are used; in the combined (pumped) part of the circuit in which they are connected, the circulating pump should experience a constant flow rate at constant pressure drop irrespective of the prevailing valve position. In practice however these conditions are not precisely achievable (see Underwood & Edge, 1995), but good design practice must seek to ensure that they hold as far as is possible if overall circuit flow conditions are to remain stable.

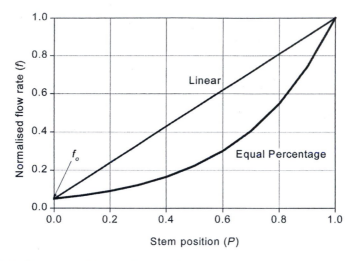

Figure 2.11 Control valve inherent characteristics.

Two-port control valves

Two-port valves are increasing in popularity especially in large networks where there is a large number of zone control valves (e.g. fan coil air conditioning systems) – see Lau (1996a,b). Flow stability in the main circuit is achieved either by using variable-speed pumping in a manner analogous to fan control in variable-air-volume air conditioning (see section 1.2), or by using a differential pressure regulator across the primary circuit. A two-port valve has to stroke against the pressure drop across the seat (more so than a three-port valve which gets some 'assistance' from the bypass port), as well as the friction between the stem and packing gland. If the pressure drop is large, the positioning device will accordingly be large and costly. Often, a better solution is to use a double seat valve in which the pressure drop across the upper seat tends to drive the plug to the open position whilst the opposite happens at the lower seat which results in a lower net force requirement acting on the stem. However, double seat valves do not give a tight closure – resulting in a very small leakage or *let-by* at the nominally closed position. Both seating arrangements can be seen in Figure 2.9.

Matching valve to system

(a) Inherent valve characteristic The characteristic of a control valve is expressed in two ways. The *inherent characteristic* expresses its idealised performance in the absence of accompanying system pressure fluctuations as might for example be obtainable under test conditions (BS5793, 1989). We will consider this first. The two inherent characteristics of interest for HVAC control valves are illustrated in Figure 2.11.

Let the inherent characteristic, γ, express the normalised flow rate (the flow rate expressed as a fraction of design-rated flow rate for the valve at the fully open position) as a function of the stem position (expressed usually as a fraction), P. We now define a flow rate at nominal closure, f_o, usually referred to as the *let-by* of the valve. This is the closure flow rate arising from the need to maintain a small clearance between plug and port thereby preventing the plug from 'sticking' to its seat and requiring an inordinate force by the positioning device to lift it. The let-by for most control valves, though dependent on valve differential pressure, is very low (usually less than 1% of the design-rated flow rate).

For the linear case, it is thus clear that the inherent characteristic can be expressed as

$$\gamma = f_o + P(1 - f_o) \tag{2.9}$$

For the equal-percentage case, equal increments of valve position will produce equal ratios of flow. Thus for an equal-percentage valve under constant-pressure conditions,

$$\frac{d\gamma}{\gamma} \propto dP$$

Introducing proportionality constant c,

$$\frac{d\gamma}{\gamma} = c\,dP$$

For a general solution with allowance for let-by,

$$\int_{f_o}^{f_o \leq \gamma \leq 1} \frac{d\gamma}{\gamma} = c \int_0^{0 \leq P \leq 1} dP$$

from which

$$\ln(\gamma) - \ln(f_o) = cP$$

From the condition $P = 1$ implies $\gamma = 1$, we get

$$c = -\ln(f_o) \qquad \text{and} \qquad \gamma = \exp[(1 - P)\ln(f_o)]$$

which reduces to

$$\gamma = f_0^{(1-P)} \tag{2.10}$$

(b) Installed characteristic and valve authority In practice, the very action of the valve in controlling flow results in pressure changes in the controlled circuit so that the inherent characteristic is not physically realisable and an *installed characteristic* which includes system effects becomes applicable.

The installed characteristic can be derived from the inherent characteristic and a new term which links valve and system, the *valve authority N*, defined as follows,

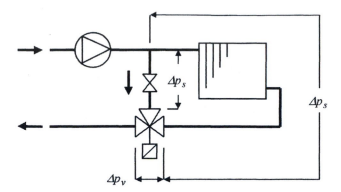

Figure 2.12 Relevant pressure drops for assessing three-port valve authority, N.

$$N = \frac{\Delta p_v}{\Delta p_v + \Delta p_s} \qquad\qquad (2.11)$$

where Δp_v and Δp_s are, respectively, the pressure drop across the fully open valve and controlled circuit at design conditions.

For two-port control valve applications, the design circuit pressure drop is the pressure drop across the entire circuit since this will be, in effect, what the valve will be working against. However, care has to be taken in choosing the correct controlled circuit pressure drop in the case of three-port valve applications. Figure 2.12 clarifies this for the 'diverting' case. For 'mixing', the relevant circuit pressure drop will be that on the heating *source* side, between the control valve connection on flow and the bypass connection on return.

The valve authority is usually fixed at some suitable value; the higher the value, the closer to a linear response the heat exchange process will get but this will be at the expense of a high valve pressure drop which presents other problems such as a high positioning device power requirement, wear at the valve plug and port, turbulence noise and a high pumping power requirement. CIBSE (1985) recommend practical choices of N of 0.3 for mixing-connected circuits and 0.5 for diverting circuits as a compromise between reasonable linearity and moderate valve pressure drop.

An installed characteristic can now be derived. Assuming that the pressure developments around the circuit can be related to the square of the volume flow rate, then, letting v refer to volume flow rate and the prime notation refer to design reference conditions,

$$\frac{\Delta p_s}{\Delta p_s'} = \left(\frac{v}{v'}\right)^2 \qquad \text{for the system}$$

and

$$\frac{\Delta p_v}{\Delta p_v'} = \left(\frac{v}{\gamma v'}\right)^2 \qquad \text{for the valve}$$

It is required that the pressure drop across the valve and connecting circuit path remains constant under all conditions if the valve is to perform correctly, so that

$$\Delta p_v + \Delta p_s = \Delta p_v' + \Delta p_s' = \Delta p_s'\left(\frac{v}{v'}\right)^2 + \Delta p_v'\left(\frac{v}{\gamma v'}\right)^2 \qquad (2.12)$$

Noting that $v/v' = \psi$ which is the *installed characteristic* and, from equation (2.11),

$$\Delta p_s' = \Delta p_v'\left(\frac{1 - N}{N}\right)$$

then by substitution of these two expressions in equation (2.12) and re-arranging we have

$$\psi = \frac{\gamma}{\sqrt{[\gamma^2(1 - N) + N]}} \qquad (2.13)$$

which defines the installed characteristic of a valve in terms of its inherent characteristic and the valve authority.

The significance of valve authority and inherent characteristic on the matching of a valve to a system is illustrated in the following example.

Example 2.2

A 100kW hot water heating coil receives hot water at 120°C (flow) and 90°C (return) whilst heating air from −3°C (inlet) to 20°C (discharge) at design conditions. Using a *logarithmic mean temperature difference* for the coil, compare the required water flow characteristic for the coil as a function of heat emission, with the installed characteristic of both linear and equal-percentage valves at $N = 0.2$, 0.5 and 0.7. Take the let-by for both valves to be 1%.

Solution

At design,

$$m_w = \frac{q}{c_{pw}(\theta_{wi} - \theta_{wo})} = \frac{100}{4.2 \times (120 - 90)} = 0.794 \text{kgs}^{-1}$$

A part-load logarithmic mean temperature difference for the coil expressed as a fraction of the design logarithmic mean temperature difference is, using prime notation for the design case,

$$\frac{q}{q'} = \frac{\Delta\theta_{LM}}{\Delta\theta_{LM}'}$$

and

Table E2.2.1 Coil flow characteristic

q/q'	θ_{ai}	θ_{wo}	m_w	m_w/m'_w
1.0	−3.0	90.0	0.794	1.000
0.8	1.6	59.9	0.317	0.400
0.6	6.2	35.8	0.170	0.214
0.4	10.8	20.4	0.096	0.121
0.2	15.4	16.0	0.046	0.058
0.0	20.0	20.0	0.000	0.000

$$\Delta\theta'_{LM} = \frac{(\theta'_{wi} - \theta'_{ao}) - (\theta'_{wo} - \theta'_{ai})}{\ln[(\theta'_{wi} - \theta'_{ao})/(\theta'_{wo} - \theta'_{ai})]} = \frac{(120 - 20) - (90 - (-3))}{\ln[(120 - 20)/(90 - (-3))]} = 96.5K$$

So

$$\frac{q}{100} = \frac{\Delta\theta_{LM}}{96.5}$$

and at $q/100 = 0.8$ (i.e. 80kW) the part-load inlet air temperature is

$$\theta_{ai} = 20 - 0.8 \times (20 - (-3)) = 1.6°C$$

Therefore

$$0.8 = \frac{(120 - 20) - (\theta_{wo} - 1.6)}{96.5 \times \ln[(120 - 20)/(\theta_{wo} - 1.6)]}$$

$$77.2\ln\left[\frac{100}{(\theta_{wo} - 1.6)}\right] = 101.6 - \theta_{wo}$$

Thus by successive approximation or graphical means, $\theta_{wo} = 59.9°C$. Therefore

$$m_w = \frac{80}{4.2 \times (120 - 59.9)} = 0.317\text{kgs}^{-1} \qquad \text{(at 80kW)}$$

In a similar manner, the required control flow characteristic for the coil, m_w/m'_w, can be found – these are given in Table E2.2.1.

The installed valve characteristics are now calculated. First, the inherent valve characteristic is determined using equation (2.9) for the linear valve type, and equation (2.10) for the equal-percentage valve type, noting that the let-by in both cases is 0.01. Then the installed characteristics for both valves are determined using equation (2.13) in which N takes the various values given (0.2, 0.5 and 0.7). The results are given in Table E2.2.2.

The results are summarised in Figure E2.2.1 for the linear valve and Figure E2.2.2 for the equal-percentage valve.

Conclusion

The linear valve inadequately matches the required flow characteristic irrespective of the value of valve authority, N, chosen. The equal-percentage

Table E2.2.2 Comparative installed valve characteristics

P	Inherent		Installed					
	Linear	Eq. %	Linear			Equal Percentage		
			N = 0.2	N = 0.5	N = 0.7	N = 0.2	N = 0.5	N = 0.7
0.0	0.010	0.010	0.022	0.014	0.012	0.022	0.014	0.012
0.2	0.208	0.025	0.429	0.288	0.246	0.056	0.035	0.030
0.4	0.406	0.063	0.705	0.532	0.469	0.140	0.089	0.075
0.6	0.604	0.158	0.861	0.731	0.671	0.337	0.221	0.188
0.8	0.802	0.398	0.949	0.885	0.849	0.696	0.523	0.460
1.0	1.000	1.000	1.000	1.000	1.000	1.000	1.000	1.000

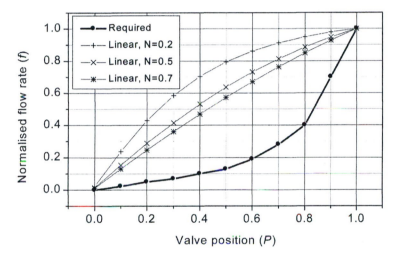

Figure E2.2.1 Installed valve performance – linear.

valve gives a good match with the required flow characteristic at $N = 0.5$–0.7, though the higher value of N will result in a high valve pressure drop requirement. In practice, $N = 0.5$ gives a good compromise between characteristic and pressure drop. Clearly therefore, besides ensuring an appropriate valve characteristic, a control valve must have a significant pressure drop in relation to the controlled circuit if it is to do its job correctly.

In practice, linear valves tend not to be used with hot/chilled water heating and cooling systems, unless the temperature drop/rise is low, but are used for steam heating applications. For situations where a very low water temperature rise/drop is intended a linear valve will usually make a better choice than an equal-percentage type (indeed some manufacturers offer an inherently near-linear valve for chilled water applications in which the design water temperature rise tends to be quite low).

Early work on HVAC control valve performance was due to Hamilton *et al.* (1974) as part of an investigation into a discharge air temperature control system of a heating coil. Linear and equal-percentage two-port pneumatically actuated control valves were compared in an otherwise con-

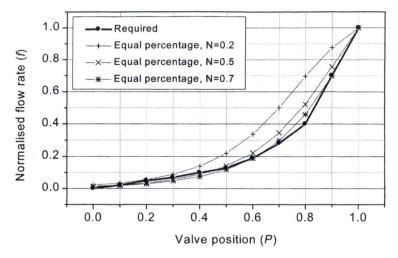

Figure E2.2.2 Installed valve performance – equal percentage.

ventional duct-mounted hot water heating coil system. This work identified poor control with the linear valve at part-load when compared with the equal-percentage valve, though at nearly full load, control with the linear valve was marginally better than with the equal-percentage valve. See also Siemers (1990).

In a departure from what has become established UK practice, Clark & Borresen (1984) considered the performance of coil control in a configuration similar to that investigated by Hamilton, but in which a three-port control valve was connected to mix with respect to the load – thereby achieving variable water flow temperature at constant flow rate. The work concluded that the combination of valve pressure drop and the characteristics of both the inlet and bypass ports were critical. Valves using equal-percentage (inlet) with equal-percentage (bypass) characteristics were found to be unsuitable for this application. For slow-responding systems, valves using high authority and equal-percentage (inlet) with linear (bypass) characteristics should be used, whilst for fast-responding systems, a high authority and linear (inlet) with linear (bypass) characteristics should be used.

(c) Rangeability In practice, machining tolerances of the valve plug and seat mean that the characteristic of the valve is not fully maintained across the full range of stem positions. The rangeability represents the ratio of maximum to minimum flow for which the characteristic is maintained. For a good valve, this will be 50:1 or better. Generally, the practical consequence of this is that the specified characteristic falls off at very low flow rates, when the plug is close to its seat and is, therefore, not frequently of much concern unless the system is seriously oversized.

Table E2.3.1 Valve F_c and sizes (Courtesy: Satchwell Control Systems Ltd (MZ range))

Valve size (mm)	F_c $(m^3h^{-1}bar^{-1/2})$	Positioning device	
		Type 1: Max. Δp_v (bar)	Type 2: Max. Δp_v (bar)
25	8.0	8.5	4.2
32	12.0	5.5	2.7
40	20.0	3.5	1.7
50	32.0	2.2	1.1
65	63.0	1.4	0.8
80	80.0	1.0	0.5

Control valve sizing

Manufacturers give values of a *flow coefficient* (F_c) for valves which can be related to the valve pressure drop and flow rate by

$$v = F_c\sqrt{\frac{\Delta p_v}{s_w}} \qquad (2.14)$$

Whilst there is some variation among manufacturers, F_c is usually defined as the flow rate (v) passed by the valve (m^3h^{-1}) at a pressure drop of 1bar. s_w is the specific gravity of the fluid (1.0 for water). Hence for a chosen value of N, the required valve pressure drop and flow rate will be known from system design data from which F_c can be obtained and matched to manufacturer's data at the various valve sizes available. Example 2.3 illustrates the method.

Example 2.3

A manufacturer offers a range of modulating control valves (Table E2.3.1). A circuit is designed to handle $16Ls^{-1}$ of water and has a design pressure drop of $150kNm^{-2}$. Specify a suitable valve.

Solution

For practical applications, a target valve authority of 0.5 is acceptable. Hence, at $N = 0.5$:

$$N = 0.5 = \frac{\Delta p_v}{\Delta p_v + \Delta p_s}$$

$$\Delta p_v = \Delta p_s = 150kNm^{-2} = 1.5bar$$

$$v = 16Ls^{-1} = \frac{16 \times 3600}{10^3} = 57.6m^3h^{-1}$$

Therefore

$$F_c = \frac{v}{\sqrt{\Delta p_v}} = \frac{57.6}{\sqrt{1.5}} = 47.0\,\mathrm{m}^3\mathrm{h}^{-1}\mathrm{bar}^{\frac{1}{2}n}$$

The choice therefore lies between a 50mm valve ($F_c = 32.0$) and a 65mm valve ($F_c = 63.0$).

If we choose the 50mm valve, the actual valve pressure drop and valve authority will be

$$\Delta p_v = \left(\frac{v}{F_c}\right)^2 = \left(\frac{57.6}{32.0}\right)^2 = 3.24\,\mathrm{bar}$$

and

$$N = \frac{\Delta p_v}{\Delta p_v + \Delta p_s} = \frac{3.24}{3.24 + 1.5} = 0.68$$

The valve pressure drop is high and an overall pump pressure head of $1.5 + 3.24 = 4.74\mathrm{bar}$ ($474\mathrm{kNm}^{-2}$) would be required which seems to be very excessive; but the valve authority is fine (see Figure E2.2.1 at $N = 0.7$). The real problem however is the differential pressure across which the positioning device has to stroke; from Table E2.3.1 we find that neither of the positioning devices will function at this high pressure drop (maximum 2.2bar for type 1 positioning device). Therefore a 50mm valve will not work.

If we choose the 65mm valve, the actual valve pressure drop and authority will be

$$\Delta p_v = \left(\frac{57.6}{63.0}\right)^2 = 0.84\,\mathrm{bar}$$

and

$$N = \frac{0.84}{0.84 + 1.5} = 0.36$$

Now the overall pump head will need to be $0.84 + 1.5 = 2.34\mathrm{bar}$ ($234\mathrm{kNm}^{-2}$). The valve authority is less than ideal; lying somewhere between the $N = 0.2$ and $N = 0.5$ curves in Figure E2.2.1, but is adequate. The type 1 positioning device with a maximum differential pressure of 1.4bar can be used. Hence the 65mm valve with type 1 positioner forms our choice.

Control dampers

Damper applications and formats

Dampers tend to be used in control application for the direct control of flow, and therefore a linear installed characteristic is sought. Common control damper applications are as follows.

* Mixing dampers; for the control of variable fresh and recirculating air in central air handling plant.

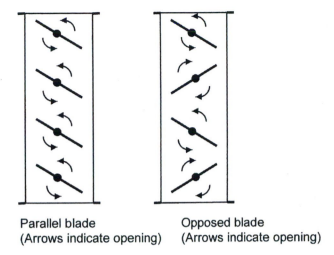

Parallel blade   Opposed blade
(Arrows indicate opening)   (Arrows indicate opening)

Figure 2.13 Blade attitude in parallel and opposed blade dampers.

- Face and bypass dampers; for capacity control of ventilation heat recovery devices (mainly plate heat exchangers) and, occasionally, cooling coils.
- Variable air volume system terminal control.

In all but the last of these applications, control dampers are of rectangular cross-section with multiple blades for larger dimensions (300mm × 300mm and above typically) and a single blade ('butterfly' damper) for smaller dimensions.

There are two main choices of damper, each choice affecting the way in which the damper blades rotate to closure. In the *opposed* blade format, each damper blade in a multi-blade assembly rotates in the opposite direction to its neighbours, and in the *parallel* blade format, all damper blades rotate in the same direction (Figure 2.13). These considerations clearly do not apply to single-blade dampers which form the majority of the circular-duct-mounted variable-air-volume control dampers. Another possibility for circular duct applications is the *iris* damper which presents an orifice of variable diameter to the air stream but these are not commonly used in control applications.

The spindle or stem for each blade passes outside the duct wall where it connects to a linkage mechanism which, in turn, connects to the positioning device (Figure 2.14).

Matching damper and system

Many of the principles established for control valves apply equally to control dampers; most particularly the damper authority (equation (2.11)) and the installed characteristic (equation (2.13)). The system pressure

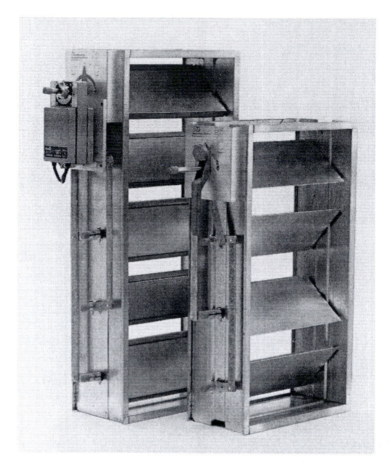

Figure 2.14 Control dampers showing linkages and (rear) positioning device. (Courtesy: MAP – Motorised Air Products Ltd.)

drop is assessed for the controlled part of the duct network. For the special case of mixing dampers, Δp_s will be the pressure drop between ambient intake and the mix point for the fresh air damper, and between the recirculating connection and ambient exhaust for the exhaust damper. For the recirculating damper however, Δp_s is the difference between the static pressure at the recirculation connection on the exhaust side (positive on fan discharge) and the mix point on the supply side (negative at fan suction). Hence for mixing damper assemblies, the recirculating damper is sized to a higher pressure drop than its fresh and exhaust counterparts. Many air handling plant manufacturers offer mixing damper sets included within the overall packaged plant and sized to some convenient dimensions rather than with reference to these pressure losses. This results in many instances of non-linearity or poor performance in the control of these dampers.

Flat plate with half-round   Crimped-edge with half-square

Figure 2.15 Typical flat-plate and crimped-edge damper blade profiles. (Legg, 1986.)

Table 2.1 Damper constants a and b (Legg, 1986)

Damper format	No. of blades	Blade shape	a	b
Parallel	4	Flat plate	−2.027	0.0828
	4	Crimped-edge	−1.350	0.0745
	2	Flat plate	−1.980	0.0876
	2	Crimped-edge	−1.510	0.0842
Opposed	4	Flat plate	−1.020	0.1010
	4	Crimped-edge	−1.250	0.1090
	2	Flat plate	−2.040	0.1100
	2	Crimped-edge	−1.510	0.1050
Butterfly	1	Flat plate	−2.110	0.1087

Unlike control valves, a damper can be specified with reference to an infinite range of sizes. At the same size and flow rate (in the case of a control damper, 'size' is the face dimensions or diameter, if circular), parallel and opposed blade dampers have different installed characteristics. In general for a given application, an opposed blade damper will achieve its near-linear installed characteristics at a lower pressure drop than a parallel blade damper though will, as a consequence, be larger. This may be an important consideration where the pressure drop across the controlled duct system is high – a condition which however turns out to be quite rare in most HVAC control damper applications.

Control damper blade rotation is, nominally, 0° fully open, 90° fully closed. A control damper can be conveniently ranged to required flow requirements by fixing the blade 'start angle', α', at some predetermined value other than 0°. For dampers with a low start angle, the inherent characteristic can be expressed as a function of the blade angle of rotation α (Legg, 1986),

$$\gamma = \exp[b(\alpha' - \alpha)/2] \tag{2.15}$$

where b is an empirical constant which depends on blade profile, blade format and number of blades.

The most common blade shapes in HVAC applications are flat-plate and crimped-edge types (Figure 2.15), though more complex profiles such as wedge and aerofoil sections are possible.

The pressure drop across a damper is, in common with other flow-resistance elements,

$$\Delta p_d = k_{da} \frac{\rho_a c_a^2}{2} \tag{2.16}$$

in which the damper pressure loss coefficient, k_{da}, is blade-angle-dependent and can be determined from the following, which is based on the same source as equation (2.15):

$$k_{da} = \exp(a + b\alpha) \tag{2.17}$$

where a is an empirical constant which depends on blade profile, etc. Table 2.1 gives typical values of constants a and b for a variety of flat and crimped-edge blade profiles.

Because we have the practical flexibility to specify the damper cross-sectional dimensions freely, we are not so constrained to the choice of a critical N as we found to be the case with control valves. Instead, when sizing a damper, Legg suggests that the inherent characteristic be calculated (from equation (2.15)) at a value corresponding to a blade angle of 50% of the overall travel at which point we assume that the installed characteristic, ψ, is 0.5. We can therefore calculate the required damper authority by rearranging equation (2.13), and putting $\psi = 0.5$,

$$N = \frac{3\gamma^2}{1 - \gamma^2} \tag{2.18}$$

Hence, to size a damper, the required inherent characteristic is obtained from equation (2.15) for the appropriate damper format and geometry. From this the required damper authority is calculated from equation (2.18) which from equation (2.11) leads to the required damper pressure drop at the fully open position. With the damper pressure loss factor at this position determined from equation (2.17), we obtain the required damper face velocity, c_a, from equation (2.16) and, hence, the required damper cross-sectional area, A_d, since

$$A_d = \frac{v_a}{c_a} \tag{2.19}$$

where v_a is the volume flow rate of air to be handled by the damper and will be known from design information.

The sizing procedure and matching of damper to system will now be explored in an example.

Example 2.4

The flow of air in a duct is to be regulated. The duct handles $3\text{m}^3\text{s}^{-1}$ at design and the corresponding pressure drop is 120Nm^{-2}. A four-blade flat-plate parallel or opposed blade damper could be used, each with a start angle of 10°.

Determine an appropriate size for the parallel and opposed blade dampers and compare their installed characteristics with ideal conditions.

Solution

Sizing the damper for a mid-range installed characteristic of 0.5, the mid-range blade angle will be

$$10° + \frac{90° + 10°}{2} = 55°$$

For the parallel blade choice, the damper constants from Table 2.1 are $a = -2.027$ and $b = 0.0828$ whilst for the opposed blade choice the constants are $a = -1.020$ and $b = 0.101$. Using equation (2.15), the inherent characteristic of the two damper choices will therefore be

$$\gamma_{parallel} = \exp[b(a' - a)/2] = \exp[0.0828(10 - 50)/2] = 0.191$$

$$\gamma_{opposed} = \exp[0.101(10 - 50)/2] = 0.133$$

Hence, the corresponding damper authorities will be

$$N_{parallel} = \frac{3\gamma^2}{1 - \gamma^2} = \frac{3 \times 0.191^2}{1 - 0.191^2} = 0.114$$

$$N_{opposed} = \frac{3 \times 0.133^2}{1 - 0.133^2} = 0.054$$

Correspondingly, the required pressure drop across the damper choices at design conditions and based on equation (2.11) will be

$$\Delta p_{d,parallel} = \frac{N_{parallel}\Delta p_s}{1 - N_{parallel}} = \frac{0.114 \times 120}{1 - 0.114} = 15.4 \text{Nm}^{-2}$$

$$\Delta p_{d,opposed} = \frac{0.054 \times 120}{1 - 0.054} = 6.85 \text{Nm}^{-2}$$

Using equation (2.17) the required pressure drop coefficient at the fully open position for both dampers will be

$$k_{da',parallel} = \exp(a + ba') = \exp(-2.027 + 0.0828 \times 10) = 0.302$$

$$k_{da',opposed} = \exp(-1.02 + 0.101 \times 10) = 0.990$$

Taking the density air to be 1.2kgm^{-3}, the required face velocities to maintain the required conditions will from equation (2.16) be

$$c_{a,parallel} = \sqrt{\frac{2\Delta p_{d,parallel}}{k_{da',parallel}\rho_{air}}} = \sqrt{\frac{2 \times 15.4}{0.302 \times 1.2}} = 9.22 \text{ms}^{-1}$$

$$c_{a,opposed} = \sqrt{\frac{2 \times 6.85}{0.99 \times 1.2}} = 3.4 \text{ms}^{-1}$$

From which the required cross-sectional areas of the two damper choices will be (equation (2.19))

$$A_{d,parallel} = \frac{v_a}{c_a} = \frac{3.0}{9.22} = 0.325 \text{m}^2 \qquad (570\text{mm} \times 570\text{mm if square aspect used})$$

$$A_{d,opposed} = \frac{3.0}{3.4} = 0.882 \text{m}^2 \qquad (905\text{mm} \times 905\text{mm if square aspect used})$$

Table E2.4.1 Comparative installed damper characteristics

α	Ideal ψ	$\gamma_{parrallel}$	$\gamma_{opposed}$	$\psi_{parallel}$	$\psi_{opposed}$
10	1.000	1.000	1.000	1.000	1.000
20	0.875	0.661	0.604	0.933	0.956
30	0.750	0.437	0.364	0.821	0.860
40	0.625	0.289	0.220	0.667	0.696
50	0.500	0.191	0.133	0.499	0.500
60	0.375	0.126	0.080	0.352	0.327
70	0.250	0.083	0.048	0.240	0.203
80	0.125	0.055	0.029	0.161	0.124
90	0.000	0.036	0.018	0.106	0.077

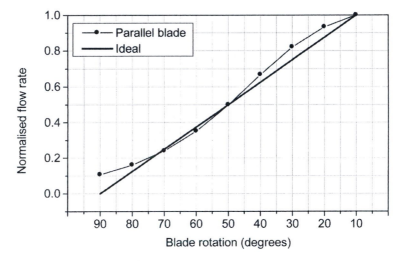

Figure E2.4.1 Parallel blade damper response across range.

We see from this that the opposed blade damper offers a lower pressure drop (though both damper pressure drop results are moderate in relation to the system pressure drop). However, the opposed blade damper is well in excess of twice the size of the parallel type.

We now look at the installed characteristics. γ and ψ can be calculated across the full range of damper travel (using equations (2.15) and (2.13)) for both damper types, at their respective damper authorities. These can then be compared with an ideal linear performance (in which the flow fraction in the range 0–1 is related to blade angle in the range 10–90°) in order to establish which of the two dampers gives the best full range response. These calculations are summarised in Table E2.4.1. Figures E2.4.1 and E2.4.2 give summaries of the comparative full range performance of the two selected dampers.

Conclusions

The parallel blade damper is the smaller of the two, but offers a higher resistance to air flow than the opposed blade damper. This may have

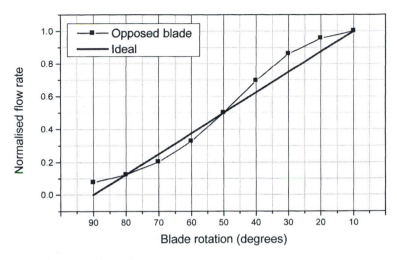

Figure E2.4.2 Opposed blade damper response across range.

implications in retrofit situations where excess fan power may be limited, and in energy-conscious design applications where the increased fan power requirement over installation life may be significant. In terms of full range response however, the parallel blade damper gives a marginally closer to ideal performance than the opposed blade damper, though it does exhibit a higher closure leakage rate than the opposed blade damper – where tight shut-off is required a separate two-position damper will need to be used.

Variable-speed drives

Drive types

Variable-speed drives and variable-capacity control for fans, pumps and refrigeration system compressors are increasing in popularity mainly due to improving economics of the technology involved as well as the scope for energy efficiency through their use.

Variable-speed drives (VSDs) of the *electronic* and *electromagnetic* types form the most common applications because they easily retrofit and also lead to high part-load efficiency savings.

Other mechanical and hydraulic variable-capacity controls are machine-specific electromechanical or pneumatic actuator-driven methods which include *belt-tension changers, hydraulic fluid couplings, sliding disc* (for centrifugal fans), *variable inlet guide vanes* (for centrifugal fans and compressors) and *variable blade pitch adjustment* (for axial fans). These methods have received some attention in the past but today the only consistently used method among these is the use of variable inlet guide vanes in very large-scale refrigeration machines employing centrifugal compressors.

Variable-capacity controls for HVAC fans (e.g. Kradow & Lin, 1995), pumps (e.g. Rishel, 1991) and compressors (e.g. Deng & James, 1994;

Wong & James, 1988) are achieved through the use of variable-speed drives of which there are five well-established types:

- Multiple-speed motors
- Variable-voltage motors
- Variable-frequency motors
- Switched reluctance drives
- Eddy current coupling

Concentrating on these types, ETSU (1991) give an overview of the technology, application and economics of the various methods from which the following is a summary.

Multiple-speed motors These are essentially restricted to discrete *stepwise control* in up to four stages (25%, 50%, 75%, rated speed) but are capable of achieving high motor efficiency across the range of speeds (88–90%). The most common is the 'tapped' single-wound motor with two speed settings though up to four speeds are possible with a dual winding (e.g. an eight-pole dual-wound motor running at 3000rpm (rated), 1500rpm, 1000rpm and 750rpm).

Variable voltage motors These are suitable only for specially adapted ac induction motors servicing loads which exhibit a relationship in which the torque varies as the square of the speed (which is the case for many HVAC fan and pump applications). The AC voltage and frequency are adjusted in proportion. This method tends not to be economic for motor sizes in excess of 75kW and most favourable applications are where the speed range is restricted to 50% of rated.

Variable frequency motors These are becoming the most common type of variable-capacity control for HVAC fans and pumps and consist of rectifier and inverter circuits. The rectifier converts the incoming ac power cycle to dc and the inverter reconstructs this as variable-frequency ac output by switching the dc current on and off at a variable rate. There are several switching methods but the most common is *pulse width modulation* (PWM) which exhibits minimal waveform distortion in the output current (Figure 2.16).

The speed range with this method is good – down to about 10% of rated speed – and the motor efficiency remains flat at, typically, 82–86% across range. However, they tend to generate harmonics which may not be a problem with a small number of drives in one building, but may cause problems where a large number of variable-speed drives are to be used unless harmonic filtering is adopted.

Switched reluctance drives These employ a specially developed eight-pole motor in which the poles are switched on and off in rapid succession to achieve accurate control down to 1% of rated speed. Excellent motor efficiency across range is maintained (typically 86–91%), but the method is

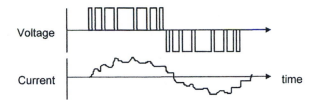

Figure 2.16 PWM variable-frequency drive characteristic.

restricted to motors of up to about 50kW in rating.

Eddy current coupling Like the switched reluctance drive, this is an electromagnetic method. A fixed-speed motor drives a magnetic drum which surrounds a poled rotor which is attached to the drive. A control signal generated from a shaft tachometer adjusts an excitation current which is fed to the coupling thereby achieving controlled speed. The method suffers from low part-load motor efficiency (40% at 50% capacity and below) and has a cut-off speed of about 10% of rated. However the main merit of this system is that it generates no harmonics.

Characteristics

Most VSD applications in HVAC control envisage the fan or pump controlled from a static pressure sensor such that static pressure is maintained as throttling occurs within the system. The fan speed controller's job is therefore to achieve variable flow conditions within the system at constant pressure. These applications are common for variable-air-volume (VAV) control and the emerging variable-speed pumping where two-port control valves are in use at terminals. From the fan/pump laws, we know that, at constant system pressure, the fan power will vary as the flow rate v and as the cube of the fan speed w, hence

$$w \propto v^{1/3} \tag{2.20}$$

It is not desirable to modulate a system to zero, or practically zero, flow in most cases. Certainly, VAV systems usually cut off at about 25% of design flow rate, partly to ensure adequate minimum fresh air throughput and partly to preserve reasonable patterns of room air distribution. Also, many VSDs themselves have a cut-off speed, typically 10%. For the VAV case, equation (2.20) therefore suggests a minimum fan speed of 63% of rated speed at 25% flow cut-off. Figure 2.17 shows that, in fact, over this fairly narrow range of speed control a reasonably linear response by the VSD might be expected.

Actuators

The actuator is, essentially, the positioning device in a control system. In the case of electromechanical control, most actuators consist of a servo

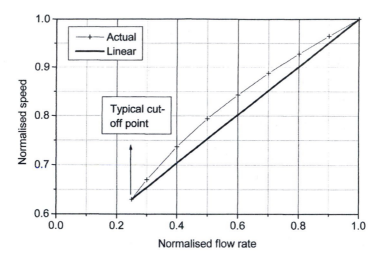

Figure 2.17 VSD characteristic at constant system pressure.

motor geared for high torque at low speed. Several performance issues are noteworthy here. The two forms of action are *rotary* and *linear.* The rotary actuator drives the stem directly and tends mainly to be used for very large positioning loads. In the linear case, the rotary output of the actuator is transferred to linear motion using a gear/linkage mechanism. Most actuators stroke through 90° or less commonly 180° whilst linear strokes are up to 38mm typically. Figure 2.18 shows a common commercially available valve actuator.

Inputs There is typically a 0–5V(dc) or 0–10V(dc) signal from the D/A converter. The power supply will usually be 24V(ac) or 230V(ac) for all but the very large actuators.

Set-up Three physical settings are possible: *span, start* and *action* (direction). These are illustrated in Figure 2.19 for a 0–10V(dc) positioning signal.

Cycle times These tend to be slow; for example 80–150s (rotary), or 2–9smm^{-1} for linear travel. Hence the typical electromechanical actuator will have a time constant of about one minute.

Power failure It is necessary to specify whether the actuator needs to be self-correcting in the event of power failure. A return spring ensures this. The return spring can be loaded to cause the actuator to close or open in the event of power failure – an important consideration in some situations such as heating control valves during especially cold weather.

Hysteresis An error, δ, results mainly (though not exclusively) from slack in the actuator linkage mechanism and any associated freedom in

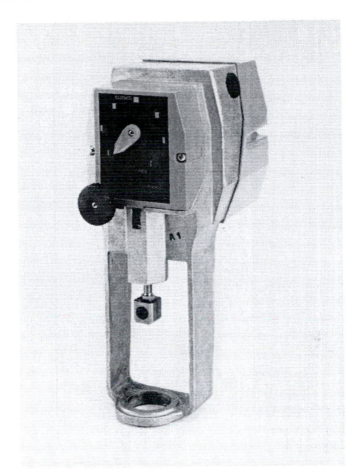

Figure 2.18 Typical valve actuator. (© Caradon Trend Ltd.)

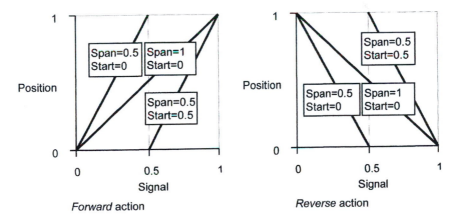

Figure 2.19 Actuator set-up.

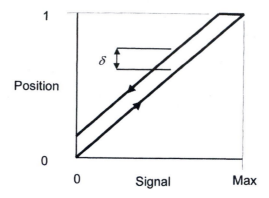

Figure 2.20 Hysteresis positioning error.

the final control element (e.g. the packing gland of a valve). The effect of this is illustrated in Figure 2.20 and manifests itself upon a reversal in the direction of the control signal; there is a pause in stem movement as the slack is taken up. This can be significant (e.g. >2% of device output range) in systems which contain mechanical linkages and in pneumatic systems and tends to increase with age due to wear among components.

Two-motor actuators In very small applications *two-motor* actuators are often used by control equipment suppliers consisting of two miniature unidirectional motors one of which drives to open whilst the other drives to closed (Hartman, 1993). These are of lower cost than the conventional bidirectional servo motor which is a major advantage in applications where there is a large number of small terminal control systems (e.g. VAV terminal dampers, fan coil system valves, etc.). They can be driven by a solid-state Triac which eliminates any mechanical device in the switching of the motor and can utilise direct digital outputs from the control system (two such signal points are needed; one for increase, the other for decrease). Hence these actuators are sometimes called *floating-point control actuators*.

2.3 PID (proportional/integral/derivative) controllers

Much of the emphasis in later chapters of this book is on the required characteristics of the controller to achieve a quality, stable, system response. However, it is appropriate to mention some of the principles at this point.

Proportional control

Early HVAC analogue controller were mostly of the simple proportional type in which the output of the controller, u, is related to the prevailing control error, ε, 'magnified' by a proportional gain, K_c,

$$u = K_c \varepsilon + u_o \qquad\qquad (2.21)$$

As control heads to a stationary condition (steady state), the signal settles at u_o. In many situations this will take the value of 50% of the control signal range which ensures that control actions take place either side of the set point. The snag with purely proportional control is that, though it works well in many instances, it can lead to *offset* – a sustained error at steady-state conditions. The causes and conditions for offset are explored at length in section 4.1.

Integral control

The control signal is proportional to the integral of the error with respect to time, t, in this mode:

$$u = u_o + \frac{K_c}{i_t} \int_{t=0}^{t=\infty} \varepsilon dt \qquad\qquad (2.22)$$

where i_t is the *integral action time* or *reset time*. Often, K_c/i_t appears as a single constant in practical controller set-ups, called the *integral gain.*

In this mode of control, we see that a period of time will be required following some disturbance before the error integral accumulates as increasing signal, u. In other words, integral control action tends to be slow in response. On the other hand, the integration of error with respect to time can only conceivably settle at a stationary value when the error is zero, hence there can be no offset with this method. Because many practical HVAC control applications are themselves very slow in response (e.g. room heating), some control system manufacturers were prompted to develop a purely integral controller on the grounds that its slowness in response was insignificant when compared with system effects but an essentially offset-free response prevailed. However, these controllers have been more or less replaced by two- or three-term controllers.

Proportional plus integral control

In practice, the integral control mode tends to be combined with the proportional mode in order to produce offset-free control. Hence the two-term or PI controller which is now very common in HVAC control (see e.g. Mehta, 1984):

$$u = u_o + K_c\left(\varepsilon + \frac{1}{i_t} \int_{t=0}^{t=\infty} \varepsilon dt \right) \qquad\qquad (2.23)$$

Note that the integral influence in this controller is inversely proportional to the integral action time, and $i_t = \infty$ produces zero integral action. Well tuned, this mode of control is offset-free though the slowness of response due to integral action prevails.

Proportional plus integral plus derivative control

Adding a derivative term to the PI controller results in a three-term proportional plus integral plus derivative (PID) controller. The derivative term responds quickly to sudden error changes but does not act upon very slow changes in error. Hence the derivative term can be thought of as adding compensation for the slowness in response resulting from integral action:

$$u = u_o + K_c \left(\varepsilon + \frac{1}{i_t} \int_{t=0}^{t=\infty} \varepsilon \mathrm{d}t + d_t \frac{\mathrm{d}\varepsilon}{\mathrm{d}t} \right) \qquad (2.24)$$

where d_t is the *derivative action time* or *rate action time* and clearly determines the magnitude of the derivative influence. The product $K_c d_t$ is referred to as *derivative gain* by some specialists. With programmable digital controllers, this type of controller is a standard today.

Proportional plus derivative control

Though very rarely used in practice, PD control can be useful in situations where abrupt errors occur, such as a rapidly switching heat gain source:

$$u = u_o + K_c \left(\varepsilon + d_t \frac{\mathrm{d}\varepsilon}{\mathrm{d}t} \right) \qquad (2.25)$$

Note that this combination of control action does not eliminate offset unless a high enough proportional gain can be used without fear of instability – a rare luxury which is why PD is rarely used in practice today.

In one variant for both the PID and the PD case, the derivative becomes the feedback variable value, instead of the error value, preventing the derivative term from acting on set point changes which can in some instances lead to reduced stability.

Figure 2.21 gives a comparison of the various conventional control actions, based on a unit step change (i.e. a sudden change of unity magnitude) in set point at $t = 0$, and underlines the various characteristics discussed above.

In practice, the controller is crucial since it represents that part of a control system which offers *degrees of freedom*, or the ability to adjust system response through the various settings we have considered here: proportional gain, integral and derivative action times. The modern *universal controller* will offer PID control as standard. All control parameters are adjustable from a keypad such as the 800 series controller from Eurotherm (Figure 2.22).

Thus satisfactory control system performance is as much about controller set-up as anything else. We will therefore go on to consider the influence of the parameters behind this at various stages in later chapters. Before we can do this however, we need to know something of how to represent control

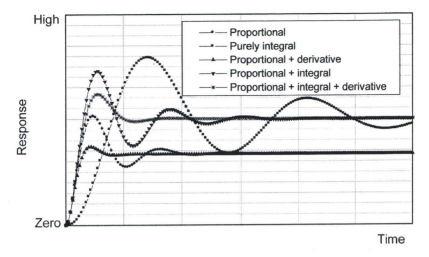

Figure 2.21 Typical controller responses to a step change.

Figure 2.22 Typical practical PID controller. (Courtesy: Eurotherm Controls.)

systems for analytical treatment. Hence, in Chapter 3 we will take a look at
system modelling, and in Chapter 4 we will look at how we can apply system
models to the crucial bottom line in control – how to identify system
stability.

References

ASHRAE (1995) *Handbook of Fundamentals.* American Society of Heating, Refrigerating and Air Conditioning Engineers, Atlanta, GA.

BS1042 (1989) *Methods of Measurement of Fluid Flow in Closed Conduits*, Part 1 – *Pressure Differential.* British Standards Institute, London.

BS5793 (1989) *Industrial Process Control Valves*, parts 1 and 2. British Standards Institute, London.

CIBSE (1985) *Applications Manual – Automatic Controls and Their Implications for System Design.* Chartered Institution of Building Services Engineers, London.

Clark, D.R., Borresen, B.A. (1984) Dynamics of a heating coil control loop. *Proceedings of the Symposium on the Performance of HVAC Systems and Controls*, BRE, Garston, 57–73.

Deng, S.M., James, R.W. (1994) Feedforward control for a DX air cooling plant: model simulation of a new system. *Building Services Engineering Research and Technology*, **15** (1), 31–34.

ETSU (1991) *Best Practice Programme: Good Practice Guide 14 – Retrofitting AC Variable Speed Drives.* Energy Efficiency Office, Harwell.

Goff, K.W. (1966) Estimating characteristics and effects of noisy signals. *Journal of the Instrument Society of America*, **I 45–49** (II 44–54), 45–49.

Griggs, E.I., Swim, W.B., Yoon, H.G. (1990) Duct velocity profiles and the placement of air control sensors. *ASHRAE Transactions*, **96** (1), 523–541.

Hamilton, D.C., Leonard, R.G., Pearson, J.T. (1974) Dynamic response characteristics of a discharge air temperature control system at near full and part heating load. *ASHRAE Transactions*, **80** (1), 181–194.

Hartman, T.B. (1993) *Direct Digital Controls for HVAC Systems.* McGraw-Hill, New York.

Huang, P.H. (1991) Humidity measurements and calibration standards. *ASHRAE Transactions*, **97** (2), 298–304.

Hurley, C.W., Hasegawa, S. (1985) Humidity sensors for HVAC applications. *Proceedings of the CIB Conference*, Trondheim, 173–188.

Krakow, K.I., Lin, S. (1995) PI control of fan speed to maintain constant discharge pressure. *ASHRAE Transactions*, **101** (2), 398–407.

Lau, K.K. (1996a) Differential pressure bypass chilled water systems: capacity ratios between on–off and modulating units. *Building Services Engineering Research and Technology*, **17** (4), 177–183.

Lau, K.K. (1996b) Differential pressure bypass chilled/heating water systems: criterion and procedure for balancing. *Building Services Engineering Research and Technology*, **17** (4), 185–189.

Legg, R.C. (1986) Characteristics of single and multi-blade dampers for ducted air systems. *Building Services Engineering Research and Technology*, **7** (4), 129–145.

Mehta, D.P. (1984) Dynamic performance of PI controllers: experimental validation. *ASHRAE Transactions*, **93** (2b), 1775–1793.

Newman, H.M. (1994) *Direct Digital Control of Building Systems.* John Wiley, New York.

Rishel, J.B. (1991) Control of variable-speed pumps on hot- and chilled-water systems. *ASHRAE Transactions*, **97** (1), 746–750.

Siemers, H. (1990) Selecting valves by choosing the optimum flow characteristic. *Proceedings of the British Hydraulic Research Group 3rd. Int. Conference on Valves and Actuators*, Oxford, 97–128.

Sugawara, S., Hara, M. (1990) A new air conditioning environmental control method. *Energy and Buildings*, **14**, 183–187.

Thomas, R.M. (1992) Humidity sensors in heating, ventilating and air conditioning. *ASHRAE Transactions*, **7** (3), 529–539.

Thompson, D. (1991) Automatic control of air quality. *Australian Journal of Refrigeration, Air Conditioning and Heating*, **45** (9), 32–35.

Underwood, C.P., Edge, J.S. (1995) Flow characteristics in circuits using three-port modulating control valves. *Building Services Engineering Research and Technology*, **16** (3), 127–132.

Wong, A.K.H., James, R.W. (1988) Capacity control of a refrigeration system using a variable speed compressor. *Building Services Engineering Research and Technology*, **9** (2), 63–68.

3 System modelling

3.1 Simple room modelling for control analysis

The essence of modelling for control system analysis is the description of the *dynamic* characteristics of systems and plant, usually over comparatively short periods of time. Time, t, forms the independent variable for such investigations, and system *stability* and *response quality* form the outcomes.

When analysing the stability and response of control systems, interest is sometimes restricted to system response over a short-time period immediately following some disturbance. This short term response might be all that is needed to judge whether a control system will behave satisfactorily.

From a thermal modelling viewpoint, this is clearly a special case, since the room fabric does not influence dynamic thermal response for periods much shorter than, say, one hour in practice. Thus the short-term response of a control system to some disturbance can often be analysed by simply considering the room air volume (together with the plant and control components) and neglecting the thermal response of the fabric (Fisk, 1981; see also Athienitis *et al.*, 1990).

A very simple energy balance for the room air can thus be written:

$$C_r \frac{d\theta_r}{dt} = q_{plant} - \Sigma(AU_i)(\theta_r - \bar{\theta}_f) - \frac{n_v V_r}{3} \times (\theta_r - \theta_o) + q_{gain} \qquad (3.1)$$

where C_r = room air thermal capacity (product of volume, density and specific heat capacity) (JK^{-1}), θ_r = room air temperature (°C), q_{plant} = energy input from plant (W), $\bar{\theta}_f$ = the surface temperature of the room fabric (°C), $\Sigma(AU_i)$ = area-integrated fabric surface U-value, $n_v V_r/3$ = ventilation coefficient (WK^{-1}) in which n_v, V_r are the ventilation air change rate (h^{-1}) and room volume (m^3) respectively, θ_o = external temperature (°C) and q_{gain} = coincident room sensible heat gain (W).

The left-hand side (LHS) of equation (3.1) represents the rate of change of energy in room air which must balance with the difference between energy received from plant and energy lost by fabric and infiltration, expressed on the right-hand side (RHS). Note that, in the steady state, the LHS is zero and the above becomes a simple heat-loss expression.

This is a linear differential equation since it contains no functions of variables such as exponentiation, or products of two or more variables such as flow rate and temperature. This means that it can be easily solved using, for example, Laplace transforms, whilst non-linear differential equations require special consideration which we shall give later. This therefore represents the main advantage of using such a room energy model – it is simple and convenient to apply. In return there are restrictions:

- The equation is strictly limited to short-term analyses (of minutes rather than hours).
- As a consequence of the above, it is unsuitable for analysing control of high-thermal-capacity systems (e.g. high-water-content heating, embedded panel systems, etc.).

If we now multiply out the brackets in equation (3.1), substitute for $C_r = V_r \rho_a c_{pa}$ (where V_r, ρ_a, c_{pa} are room volume, density and specific heat capacity of air respectively), and express each term as a *deviation* rather than as an absolute value, then constant terms will disappear and we get

$$V_r \rho_a c_{pa} \frac{\mathrm{d}}{\mathrm{d}t}(\delta \theta_r) = \delta q_{plant} - \left(\Sigma(AU_i) + \frac{n_v V_r}{3} \right) \delta \theta_r \qquad (3.2)$$

where the δ notation implies deviations in the variables from some known steady-state values. It is important to recognise that in much of transform modelling applied to control system investigations, and certainly all such cases contained in this book, the variables concerned are deviation variables rather than absolute values. There is great convenience in this as we shall see in the following.

From here on, we shall drop the δ notation for convenience – as long as we remember what our transform variables actually represent.

By considering the deviation variables, constants (i.e. parameters having nil deviation at least in the short term for which our analysis is applicable) $\bar{\theta}_f$, θ_o, q_{gain} disappear from the equation. This is very useful – it means that our equation has initial conditions of zero (i.e. zero deviation at the initial steady state) and is a great convenience when dealing with Laplace transforms (Table 3.1).

With the equation in this form, we can either go on to solve it as an 'open-loop' problem, or perform a Laplace transformation and combine it with other components to make up a system (i.e. a *block diagram*). The latter can be accomplished by substituting $\mathrm{d}\theta_r/\mathrm{d}t = \theta_r(s)s$,

$$V_r \rho_a c_{pa} \theta_r(s)s = q_{plant}(s) - \left(\Sigma(AU_i) + \frac{n_v V_r}{3} \right) \theta_r(s) \qquad (3.3)$$

which reduces to

$$\theta_r(s) \left(\frac{V_r \rho_a c_{pa}}{\Sigma(AU_i) + n_v V_r/3} s + 1 \right) = \frac{q_{plant}(s)}{\Sigma(AU_i) + n_v V_r/3} \qquad (3.4)$$

or

$$\frac{\theta_r(s)}{q_{plant}(s)} = \frac{K_r}{\tau_r s + 1} \qquad (3.5)$$

This is the room air capacity *transfer function* in which

$$\tau_r = \frac{V_r \rho_a c_{pa}}{\Sigma(AU_i) + n_v V_r/3} \qquad (3.6)$$

Table 3.1 Some common Laplace transforms and their time functions

s-Transform	Time function
Integrator	
$\dfrac{a}{s}$	at
Simple lag	
$\dfrac{1}{s + a}$	$\exp(-at)$
First order + integrator	
$\dfrac{a}{s(s + a)}$	$a[1 - \exp(-at)]$
Second order	
$\dfrac{b - a}{(s + a)(s + b)}$	$\exp(-at) - \exp(-bt)$
Second order + integrator	
$\dfrac{ab}{s(s + a)(s + b)}$	$\dfrac{b}{(a - b)}\exp(-at) - \dfrac{a}{(a - b)}\exp(-bt)$

is the room temperature *time constant* to fluctuations in plant output, and

$$K_r = \frac{1}{\Sigma(AU_i) + n_v V_r / 3} \tag{3.7}$$

is the room temperature *gain* to fluctuations in plant output.

This is pretty much all there is to deriving a linear transfer function. Naturally, some are more complex than this case, but the procedural principles do not differ: write down the governing differential equation(s) in a convenient linear form; express the variables as deviations such that all other non-deviant terms become constants and disappear; finally, express in 'time constant' form. Of course few problems present us with a convenient set of linear differential equations and we shall see how to deal with these cases in section 3.3. Meanwhile, we will pay further consideration to the linear case.

Example 3.1

A warehouse space measures 10.0m × 5.0m × 5.0m high and is heated by a thermostatically controlled two-stage indirect gas-fired heater of negligible thermal capacity. High fire position has an output of 10.0kW and switches on at (or below) 19°C. Low fire position has an output of 5.0kW,

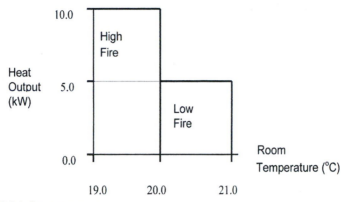

Figure E3.1.1 Firing sequence.

switches on at 20°C and switches off at 21°C as shown in Figure E3.1.1. The space has a natural ventilation air change rate of $1.0\,h^{-1}$ and fabric surface heat transfer coefficient of $8.33\,Wm^{-2}K^{-1}$.

Establish the room air capacity temperature time constant and gain, and determine the heat-up and cool-down transient over a heating cycle assuming that the warehouse temperature at the start of the cycle is 19°C.

Solution

We have

$$\Sigma(AU_i) = 8.33 \times (2 \times 10 \times 5 + 2 \times 10 \times 5 + 2 \times 5 \times 5) = 2082.5\,WK^{-1}$$

$$\frac{n_vV_r}{3} = \frac{1.0 \times (10 \times 5 \times 5)}{3} = 83.33\,WK^{-1}$$

Therefore

$$K_r = \frac{1}{2082.5 + 83.33} = 4.617 \times 10^{-4}\,WK^{-1}$$

$$\tau_r = \frac{1.2 \times 1025.0 \times (10 \times 5 \times 5)}{2082.5 + 83.33} = 142.0s$$

The air capacity model for the warehouse will therefore be

$$\frac{\theta_r(s)}{q_{plant}(s)} = \frac{4.617 \times 10^{-4}}{(142.0s + 1)} \qquad\qquad (E3.1.1)$$

Now, to solve this, we note that the space is subject to a step change in plant capacity as it switches on, of 10 000W. We can tackle input disturbances of a stepwise nature in Laplace problems using the *Heaviside unit step function* (e.g. Stroud, 1973). For example, we note the following Laplace transformations of some typical step functions (Figure E3.1.2):

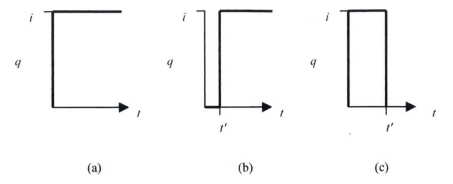

Figure E3.1.2 Typical Heaviside step functions.

a $\quad q(s) = \dfrac{i(s)}{s}$

b $\quad q(s) = \dfrac{i(s)\exp(-t's)}{s}$

c $\quad q(s) = \dfrac{i(s)}{s} - \dfrac{i(s)\exp(-t's)}{s}$

Thus, returning to our problem, the Laplace transformation of the input function $q_{plant}(s)$ will be

$$q_{plant}(s) = \frac{10^4}{s} - \frac{5\times 10^3 \exp(-t's)}{s} - \frac{5\times 10^3 \exp(-t''s)}{s}$$

where t' and t'' are the event times when the room temperature has reached 20°C and 21°C respectively.

Now, returning to our room energy model for this problem (equation (E3.1.1)), and multiplying both sides by the input function, $q_{plant}(s)$, we get our expression in terms of the room temperature:

$$\theta_r(s) = \frac{4.617}{s(142.0s + 1)} - \frac{2.309 \exp(-t's)}{s(142.0s + 1)} - \frac{2.309 \exp(-t''s)}{s(142.0s + 1)} \qquad \text{(E3.1.2)}$$

A solution to this giving us a time-domain expression for the room temperature is obtained by inverse transformation. Inverse transforms can be obtained from standard tables of transforms (see e.g. Leigh, 1992; Houpis & Lamont, 1992). Brief examples were given in Table 3.1.

Taking each term in our expression (equation (E3.1.2)) in turn, and noting that $a = 1/\tau_r$ when referring to Table 3.1,

$\theta_r(t) = 4.617A$ $\qquad\qquad\qquad$ $(0 \le t \le t')$ $\qquad\qquad\qquad$ (E3.1.3)

$\theta_r(t) = 4.617A - 2.309B$ $\qquad\qquad$ $(t' < t \le t'')$ $\qquad\qquad$ (E3.1.4)

$\theta_r(t) = 4.617A - 2.309(B + C)$ $\quad$ $(t > t'')$ $\qquad\qquad\qquad$ (E3.1.5)

Table E3.1.1 Results throughout one cycle

t (seconds)	t' (seconds)	t" (seconds)	Result (°C)
0.00	—	—	19.00
10.00	—	—	19.32
20.00	—	—	19.75
34.66	0.00	—	20.00
94.66	60.00	—	20.45
194.66	160.00	—	20.88
239.66	205.00	0.00	21.00
264.66	230.00	25.00	20.68
317.66	283.00	78.00	20.15
337.66	303.00	98.00	20.00

where

$$A = 1 - \exp\left(\frac{-t}{142}\right), \qquad B = 1 - \exp\left(\frac{-t'}{142}\right), \qquad C = 1 - \exp\left(\frac{-t''}{142}\right)$$

The heater will cause the space to rise from 19°C to 20°C before switching to the lower (5000W) fire position which will cause the rate of room temperature rise to slow down. At 21°C the heater will switch off and the room temperature will decay. We can calculate several random points throughout this cycle, noting that since our model refers to deviation variables, we must add the results to the initial condition of 19°C (Table E.3.1.1); ... and so on. Graphically, the complete cycle is shown in Figure E3.1.3.

The result in Example 3.1 is a classical thermostat switching sequence. We can see the initial rapid rate of temperature rise on high fire position, followed by a period of lower rate of rise when at the low fire position. We can also appreciate from this result why on:off control systems can sometimes be approximated by a 'saw-tooth' function made up of a series of straight lines.

There are many problems which cannot be solved using the simple room air capacity model – in particular those cases where the response of the plant may not be considered instantaneous, or where there is radiant heat exchange with the building fabric. Such cases require the room envelope to play a part in the model. We will consider this in the next section.

3.2 Accounting for fabric in room modelling

Elemental fabric layers

Consider the case of a thin, uniform layer of construction material, so thin in fact that the temperature throughout its body can be assumed constant. Let us suppose it is a wall (though for that matter, it could be any construction element) and is therefore bounded by a room temperature (θ_r) on one

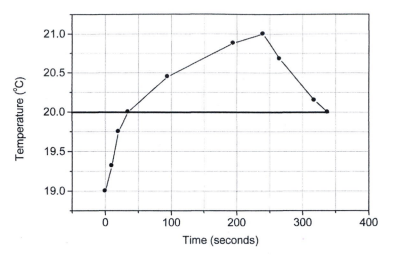

Figure E3.1.3 Temperature transient.

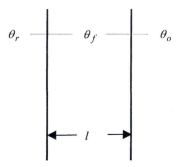

Figure 3.1 Single-layer building element.

side and an external temperature (θ_o) on the other (Figure 3.1). A transfer function representation for this can be built up as follows.

An energy balance for this element produces the following:

$$l\rho_f c_{pf}\frac{\mathrm{d}\theta_f}{\mathrm{d}t} = u_i(\theta_r - \theta_f) - u_o(\theta_f - \theta_o) \tag{3.8}$$

where ρ_f = density of the material ($\mathrm{kgm^{-3}}$), c_{pf} = specific heat of the material ($\mathrm{Jkg^{-1}K^{-1}}$), u_i = combined convection/radiation inside surface heat transfer coefficient ($\mathrm{Wm^{-2}K^{-1}}$) and u_o = combined material and external surface heat transfer coefficient ($\mathrm{Wm^{-2}K^{-1}}$).

Treating the variables as deviations from some known initial conditions, we make our usual substitution (s for the d/dt operator), obtaining the following:

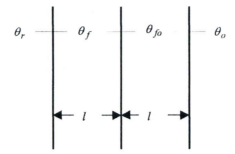

Figure 3.2 Two-layer building element.

$$\theta_f(s) = \frac{K_{\theta_r}}{(\tau_f s + 1)} \theta_r(s) + \frac{K_{\theta_o}}{(\tau_f s + 1)} \theta_o(s) \tag{3.9}$$

where

- the wall time constant(s)

$$\tau_f = \frac{l \rho_f c_{pf}}{(u_i + u_o)}$$

- the wall/room temperature gain

$$K_{\theta_r} = \frac{u_i}{(u_i + u_o)}$$

- the wall/external temperature gain

$$K_{\theta_o} = \frac{u_o}{(u_i + u_o)}$$

Clearly, if the external temperature (which might be the interface temperature of this layer of material and the next layer, or the outside temperature) is constant, then the above reduces to a first-order lag relating the wall temperature to the room temperature.

Unfortunately, many construction elements are not so thin as to allow us to make such a uniform body temperature assumption. We can address this by splitting our material layer up into a number of layers or slices of equal thickness, l.

A two-layer element can be envisaged as Figure 3.2. An energy balance can be set up for this case as follows:

$$l\rho_f c_{pf} \frac{d\theta_f}{dt} = u_i(\theta_r - \theta_f) - u_f(\theta_f - \theta_{fo}) \tag{3.10}$$

$$l\rho_f c_{pf} \frac{d\theta_{fo}}{dt} = u_f(\theta_f - \theta_{fo}) - u_o(\theta_{fo} - \theta_o) \tag{3.11}$$

where u_f is the heat transfer coefficient due to the layer thermal conductivity ($\mathrm{Wm^{-2}K^{-1}}$). Taking Laplace transforms and combining these two expressions leads us to the following second-order representation for a uniform building element which relates the two boundary temperatures to the inside surface layer temperature:

$$\theta_f(s) = \frac{As + B}{Cs^2 + Ds + E}\theta_r(s) + \frac{F}{Cs^2 + Ds + E}\theta_o(s) \tag{3.12}$$

where

$$A = l\rho_f c_{pf} u_i \qquad\qquad B = (u_f u_i + u_o u_i)$$

$$C = (l\rho_f c_{pf})^2 \qquad\qquad D = l\rho_f c_{pf}(2u_f + u_i + u_o)$$

$$E = (u_f u_i + u_i u_o + u_f u_o) \qquad F = u_f u_o$$

In a similar manner, third- and fourth-order expressions for the more rigorous three-layer and four-layer element descriptions can be arrived at. The following expression is obtained for a three-layer element:

$$\theta_f(s) = \frac{A's^2 + B's + C'}{D's^3 + E's^2 + F's + G'}\theta_r(s) + \frac{H'}{D's^3 + E's^2 + F's + G'}\theta_o(s) \tag{3.13}$$

where

$$A' = u_i(l\rho_f c_{pf})^2 \qquad\qquad B' = l\rho_f c_{pf}(3u_i u_f + u_i u_o)$$

$$C' = (u_i u_f^2 + 2u_i u_f u_o) \qquad\qquad D' = (l\rho_f c_{pf})^3$$

$$E' = (l\rho_f c_{pf})^2(4u_f + u_i + u_o) \qquad F' = l\rho_f c_{pf}(3u_i u_f + u_i u_o + 3u_f^2 + 3u_f u_o)$$

$$G' = (u_i u_f^2 + 2u_i u_f u_o + u_o u_f^2) \qquad H' = u_o u_f^2$$

For a four-layer element:

$$\theta_f(s) = \frac{A''s^3 + B''s^2 + C''s + D''}{E''s^4 + F''s^3 + G''s^2 + H''s + I''}\theta_r(s)$$
$$+ \frac{J''}{E''s^4 + F''s^3 + G''s^2 + H''s + I''}\theta_o(s) \tag{3.14}$$

where

$$A'' = u_i(l\rho_f c_{pf})^3 \qquad\qquad\qquad B'' = 6(l\rho_f c_{pf})^2(u_i u_f)$$

$$C'' = l\rho_f c_{pf}(8u_i u_f^2 + 2u_i u_f u_o) \qquad D'' = (u_i u_f^3 + 3u_i u_o u_f^2)$$

$$E'' = (l\rho_f c_{pf})^4 \qquad\qquad\qquad F'' = (l\rho_f c_{pf})^3(u_i + 7u_f)$$

$$G'' = (l\rho_f c_{pf})^2(6u_i u_f + 13u_f^2 + 2u_f u_o)$$

$$H'' = l\rho_f c_{pf}(8u_i u_f^2 + 2u_i u_f u_o + 6u_f^3 + 4u_o u_f^2)$$

$$I'' = (u_i u_f^3 + 3u_i u_o u_f^2 + u_o u_f^3) \qquad\qquad J'' = u_o u_f^3$$

We must ask ourselves which of the above should be used for a given modelling situation. Clearly, we want an accurate description of the heat transfer dynamics of our problem but we do not want to introduce any unnecessary complication. We will attempt to answer this question with the following worked example which looks at predictions from the various element models for several common construction materials.

Example 3.2

To establish which of the various building element transfer functions gives the 'best' performance, we will compare them for a variety of common building materials of varying thermal capacity and thermal conductivity with reference to a unit step change in room air temperature but constant external temperature. The material data used are given in Table E3.2.1 (CIBSE, 1986). Note that since a standard thickness for brick or block is 112 mm, all material nominal thicknesses will be fixed at this value for comparison.

For the purpose of this exercise, standard *combined* (radiant/convective) surface heat transfer coefficients will be taken for inside and outside surfaces of $8.33 \mathrm{Wm}^{-2}\mathrm{K}^{-1}$ and $18.18 \mathrm{Wm}^{-2}\mathrm{K}^{-1}$ repectively (CIBSE, 1986) for a wall of high surface emissivities and 'normal' exposure.

Solution

First, note that the elemental layer thicknesses will be 0.112m, 0.056m, 0.037m and 0.028m for first-order, second-order, third-order and fourth-order transfer functions respectively. We also note that all but the outermost elemental layer u-values are calculated from $u_f = k_f / l$ (where k_f is the material thermal conductivity). The outermost elemental layer u-value will be

$$u_o = \left(\frac{l}{k_f} + \frac{1}{h_o} \right)^{-1}$$

where h_o is the external surface heat transfer coefficient.

In order to avoid some inconveniently large values for certain coefficients, the time base has been fixed in hours and energy units in kJ.

We now calculate the coefficients of the various transfer functions based on equations (3.9), (3.12), (3.13) and (3.14) for first-order to fourth-order transfer functions noting that the external temperature deviation (and, hence, $\theta_o(s)$) is zero. The results are summarised below.

Table E3.2.1 Material properties

Material	Density (kgm^{-3})	Conductivity (Wm^{-1}K^{-1})	Specific heat capacity (Jkg^{-1}K^{-1})
Brick	1700	0.840	800
Light concrete block	600	0.190	1000
Mineral wool	30	0.035	1000

- First order

$$\frac{\theta_f(s)}{\theta_r(s)} = \frac{0.611}{(3.10s + 1)} \qquad \text{(brick)}$$

$$\frac{\theta_f(s)}{\theta_r(s)} = \frac{0.843}{(1.89s + 1)} \qquad \text{(block)}$$

$$\frac{\theta_f(s)}{\theta_r(s)} = \frac{0.964}{(0.108s + 1)} \qquad \text{(insulation)}$$

- Second order

$$\frac{\theta_f(s)}{\theta_r(s)} = \frac{(0.557s + 0.611)}{\left(1.41s^2 + 3.11s + 1\right)} \qquad \text{(brick)}$$

$$\frac{\theta_f(s)}{\theta_r(s)} = \frac{(1.26s + 0.843)}{\left(1.41s^2 + 2.72s + 1\right)} \qquad \text{(block)}$$

$$\frac{\theta_f(s)}{\theta_r(s)} = \frac{(0.366s + 0.964)}{\left(0.021s^2 + 0.447s + 1\right)} \qquad \text{(insulation)}$$

- Third order

$$\frac{\theta_f(s)}{\theta_r(s)} = \frac{\left(0.127s^2 + 0.697s + 0.611\right)}{\left(0.215s^3 + 1.65s^2 + 3.07s + 1\right)} \qquad \text{(brick)}$$

$$\frac{\theta_f(s)}{\theta_r(s)} = \frac{\left(0.493s^2 + 1.52s + 0.843\right)}{\left(0.368s^3 + 1.93s^2 + 2.84s + 1\right)} \qquad \text{(block)}$$

$$\frac{\theta_f(s)}{\theta_r(s)} = \frac{\left(0.031s^2 + 0.403s + 0.964\right)}{\left(0.001s^3 + 0.050s^2 + 0.477s + 1\right)} \qquad \text{(insulation)}$$

- Fourth order

$$\frac{\theta_f(s)}{\theta_r(s)} = \frac{\left(0.013s^3 + 0.214s^2 + 0.884s + 0.611\right)}{\left(0.016s^4 + 0.329s^3 + 1.98s^2 + 3.61s + 1\right)} \qquad \text{(brick)}$$

$$\frac{\theta_f(s)}{\theta_r(s)} = \frac{\left(0.086s^3 + 0.752s^2 + 1.72s + 0.843\right)}{\left(0.048s^4 + 0.578s^3 + 2.23s^2 + 3.04s + 1\right)} \qquad \text{(block)}$$

$$\frac{\theta_f(s)}{\theta_r(s)} = \frac{\left(0.002s^3 + 0.053s^2 + 0.467s + 0.964\right)}{\left(4.6 \times 10^{-5}s^4 + 0.003s^3 + 0.073s^2 + 0.536s + 1\right)} \qquad \text{(insulation)}$$

We can now calculate the unit step response of each transfer function. This is not so difficult for the first-order cases since we recall that for a unit step in $\theta_r(s)$ at $t = 0$ we can obtain the transient from

$$\theta_f(t) = \theta_f(0) + K_{\theta_f} \times \left[1 - \exp\left(\frac{-t}{\tau_f}\right)\right]$$

where K_{θ_f} and τ_f are the gain and time constant.

However, for the higher-order cases this will prove to be more complicated. A convenient alternative is to simulate the various transfer functions for a unit step change. This has been carried out here using MATLAB/ Simulink (Simulink-2, 1996) according to the block diagram of Figure E3.2.1. The block diagram has been generated to enable all transfer function types to generate results alongside one another for a given material type. Shown here are the transfer functions of the brick case (the use of multiplex (Mux) component allows us to plot all transfer function results on common axes).

The results are expressed in Figures E3.2.2, E3.2.3 and E3.2.4 for brick, concrete block and mineral wool respectively. From the figures, it is evident that a first-order transfer function representation for common building materials will result in some loss of accuracy when compared with higher-order expressions. Equally, there seems little to be gained in terms of accuracy in moving from the second-order representation to third- or fourth-order transfer functions given the substantial increases in complexity of the resulting building envelope model that will result.

In conclusion therefore, most practical building materials can be expressed with adequate accuracy using a second-order transfer function. We will stick to this level of detail as far as building fabric is concerned from this point on – for an alternative simplified treatment of building response, refer to Crabb *et al.* (1987).

Composite construction elements

Of course, with the exception of single glazing and certain partition types, few building envelopes are made up of a series of single-layer constructions. In practice, walls, floors and roofs are a combination of several layers of

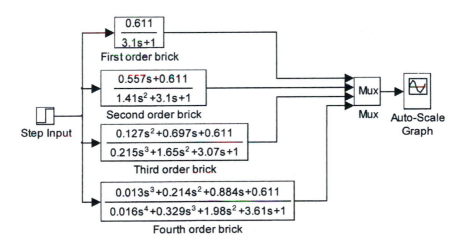

Figure E3.2.1 MATLAB/Simulink block diagram.

differing materials and constructions of up to four such layers are common-place. A composite construction element model therefore needs to be 'chained' from individual element transfer functions in order to arrive at a composite construction representation. For a three-layer composite, using second-order element transfer functions and assuming exposure to an external temperature, Figure 3.3 gives the resulting general model.

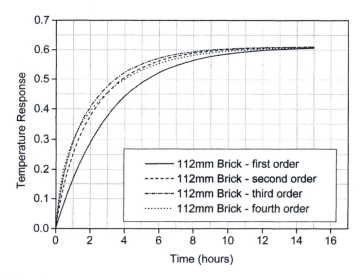

Figure E3.2.2 Response of brick.

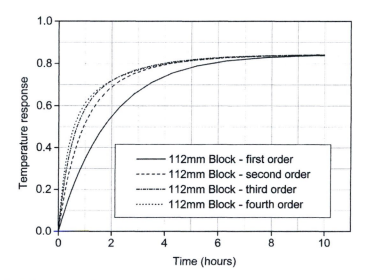

Figure E3.2.3 Response of block.

Example 3.3

Using data from Example 3.2 for concrete block and brick, investigate the response in inside surface temperature to a unit step excitation of 5K in internal room air temperature for a wall consisting of 112 mm concrete block and 112 mm brick when (a) the concrete block forms the inner layer of material, and (b) the brick forms the inner layer of material.

Solution

First, we determine the coefficients for the composite block diagram model of the two-layer wall. Figure E3.3.1 shows the resulting block diagram model for the case with the brick on the inside. Simulating this arrangement using Simulink for the two cases produces the responses in Figure E3.3.2.

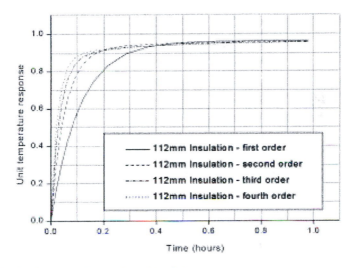

Figure E3.2.4 Response of mineral wool insulation.

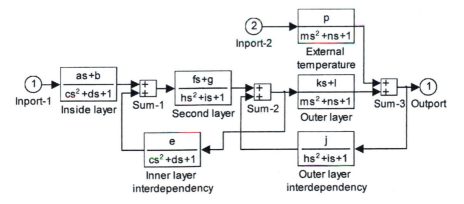

Figure 3.3 Model of a three-layer composite construction.

The result of Figure E3.3.2 shows that the brick-inner construction approaches asymptotic conditions after about 24 hours whilst the block-inner construction takes around 18 hours.

Another way of expressing this is by the *indicative time constant* which assumes that the response can be approximated by a first-order lag plus a dead-time component – a form of model which we will use throughout this book. The form of model is given by equation (3.15) in its *s*-domain form and equation (3.16) in its step response *t*-domain form and we note that it

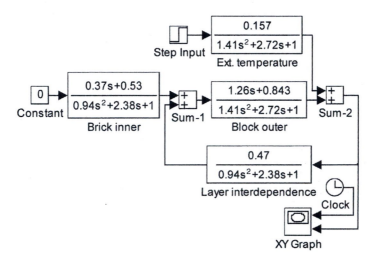

Figure E3.3.1 Composite construction model.

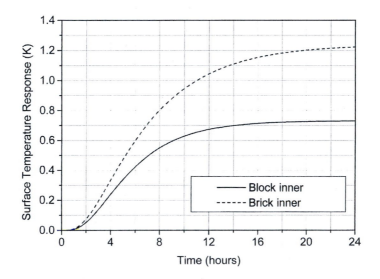

Figure E3.3.2 Comparative surface temperature responses.

is identical to the form of expression used in Example 3.1 to deal with time-delayed switching of a heater:

$$\theta_f(s) = \frac{\exp(-t_{df}s)}{(\tau_f s + 1)}\theta_r(s) \tag{3.15}$$

and

$$\theta_f(t) = \theta_r(0) + \Delta\theta_r(t) \times \{1 - \exp[(t_{df} - t)/\tau_f]\} \tag{3.16}$$

where t_{df} is the dead time, τ_f the wall time constant and $\Delta\theta_r(t)$ the *apparent overall* response in $\theta_r(t)$.

This is in fact a reasonable alternative to a second-order model – representing something in between first order and second order but having the advantage of being much easier to deal with than the latter. Indeed for many heat exchange systems, this is a very good approximation to dynamic response.

With such an approximation the indicative time constant represents the time taken for the step response to reach 63.21% of its apparent overall change (you will be able to verify this for yourself by substituting $t = t_{df} + \tau_f$ in the above expression). On this basis, the indicative time constants for the two-layer wall would be about 6.5 hours when the block layer is on the inside, and 7.5 hours when the brick layer is on the inside. Thus the same physical wall but with material layers juxtaposed will have exactly the same steady-state heat transfer characteristics but quite different dynamic response – a situation of great significance in passive designs especially where incoming solar radiation plays a major part in room heat exchange.

3.3 Heat emitter and non-linear modelling

Modelling of buildings coupled with plant for control system analysis has received some attention, mostly recently. Refer for example to Athienitis *et al.* (1990), Zaheer-uddin (1993), Athienitis (1993) or Zaheer-uddin & Goh (1991). In this section we will consider a simple paradigm for the coupling of an emitter and a space.

Development of an emitter model

The heat emitter (or for that matter heat sink – such as chilled beam or ceiling) is coupled to the room thermal environment through the room air temperature and, if the emission is in part radiant, through the various room surface temperatures. Most practical hot water emitters will be partly or fully characterised by natural convection for which the output can be expressed as follows (see for example McIntyre, 1986):

$$q_{plantc} = E(\bar{\theta}_w - \theta_r)^{n_e} \tag{3.17}$$

where E = emission constant (typically WK^{-n_e}), $\bar{\theta}_w$ = mean inlet and outlet water temperature (°C), θ_r = room air temperature (°C) and n_e = an emission index.

Where the emission is in part realised in the form of long-wave radiation between the heat emitter and room surfaces, then it is appropriate to include a separate radiant term. This might be based for practical purposes on the mean room surface temperature and an overall radiant emission constant (essentially, a combination of the Stefan–Boltzmann constant, surface emissivities and unity 'view' factor for all room surfaces combined):

$$q_{planter} = E_c\left(\bar{\theta}_w - \theta_r\right)^{n_c} + E_r\left(\bar{\theta}_w^4 - \bar{\theta}_f^4\right) \tag{3.18}$$

(in which there are now independent emission constants for the radiant and convective portions of heat emission).

Consider now the room emitter as a set of smaller emitters connected in series such that the water temperature in any of these emitters is uniform throughout and equal to the outlet water temperature. Thus we can replace the mean water temperature with the outlet water temperature and express a general *lumped capacity* water-based model as follows:

$$C_e\frac{\mathrm{d}\theta_{wo}}{\mathrm{d}t} = m_w c_{pw}(\theta_{wi} - \theta_{wo}) - E_c(\theta_{wo} - \theta_r)^{n_c} - E_r\left(\theta_{wo}^4 - \bar{\theta}_f^4\right) \tag{3.19}$$

where, for any consistent system of units, the heat emitter overall thermal capacity

$$C_e = V_m c_{pm}\rho_m + V_w c_{pw}\rho_w$$

m_w = the mass flow rate of water, V_m, c_{pm}, ρ_m = the volume, specific heat capacity and density of the heat emitter material and V_m, c_{pw}, ρ_w = the volume, specific heat capacity and density of water held in the heat emitter. (The temperatures in the radiant term must be expressed in the absolute Kelvin scale.)

Dealing with non-linearities

Equation (3.19) is unusual in relation to what we have seen so far in that it is a non-linear ordinary differential equation. All previous ordinary differential equations we have encountered have been linear. This equation is non-linear on three counts. Firstly, we note that the temperature difference in the convective term is raised to a power, n_c. Secondly, the emitter outlet temperature and mean room surface temperature in the radiant term are raised to powers. Thirdly, and somewhat less obviously, if control over our heat emitter is to be by varying the water flow rate, m_w, and the inlet water temperature to the emitter also varies (due for example to boiler cycling or to the influence of an upstream series-connected emitter), then the product $m_w\theta_{wi}$ also forces a non-linearity.

Modelling in the s-domain cannot cope directly with non-linearities – in essence, we end up with non-constant gains and time 'constants' in the resulting transfer functions which do not yield solutions. We can get round this at least as far as the application of our model to relatively minor deviations around some notional operating point is concerned, with the use of a Taylor-series expansion of the model about the initial steady-state operating point.

Neglecting second- and higher-order terms in the Taylor-series expansion, for a non-linear function $f(x)$,

$$f(x) \cong f(x)_{ss} + (x - x_{ss})\frac{\mathrm{d}f(x)_{ss}}{\mathrm{d}x_{ss}} \tag{3.20}$$

(The ss subscript refers to the initial steady-state conditions.) For a function of several variables, $f(x, y, \ldots)$,

$$f(x, y, \ldots) \cong f(x, y, \ldots)_{ss} + (x - x_{ss})\frac{\partial f(x, y)_{ss}}{\partial x_{ss}} + (y - y_{ss})\frac{\partial f(x, y)ss}{\partial y_{ss}} + \ldots$$
$$\tag{3.21}$$

. . . and so on.

Since the scope of interest for many control stability and response analyses lies within such relatively minor operating deviations, this method works quite well. However, if a more substantial range of operating conditions is of interest, then non-linear methods will need to be resorted to (see example 9.3).

For a more substantial treatment on the subject of linearising heat exchange models, see for example Enns (1962).

Development of a linearised heat emitter model

We now return to our heavily non-linear heat emitter model of equation (3.19). Linearising term by term, starting with the water heat balance term,

$$f(m_w, \theta_{wi}, \theta_{wo}) = m_w c_{pw}(\theta_{wi} - \theta_{wo})$$

inlet water term

$$\left.\frac{\partial f(m_w, \theta_{wi}, \theta_{wo})}{\partial m_{wss}}\right|_{m_w c_{pw}\theta_w} = c_{pw}\theta_{wiss}$$

and

$$\left.\frac{\partial f(m_w, \theta_{wi}, \theta_{wo})}{\partial \theta_{wiss}}\right|_{m_w c_{pw}\theta_{wi}} = c_{pw}m_{wss}$$

outlet water term

$$\left.\frac{\partial f(m_w, \theta_{wi}, \theta_{wo})}{\partial m_{wss}}\right|_{m_w c_{pw}\theta_{wo}} = -c_{pw}\theta_{woss}$$

and

$$\frac{\partial f(m_w, \theta_{wi}, \theta_{wo})}{\partial \theta_{woss}}\bigg|_{m_w c_{pw} \theta_{wo}} = -c_{pw} m_{wss}$$

Applying equation (3.21), the linearised water heat balance term reduces to

$$m_w c_{pw}(\theta_{wi} - \theta_{wo}) \cong c_{pw}(\theta_{wiss} - \theta_{woss})(m_w - m_{wss}) + m_{wss} c_{pw}(\theta_{wi} - \theta_{wo}) \tag{3.22}$$

Now for the convective term,

$$f(\theta_{wo}, \theta_r) = E_c(\theta_{wo} - \theta_r)^{n_c}$$

we get

$$\frac{\partial f(\theta_{wo}, \theta_r)}{\partial \theta_{woss}} = n_c E_c(\theta_{woss} - \theta_{rss})^{n_c-1}$$

and

$$\frac{\partial f(\theta_{wo}, \theta_r)}{\partial \theta_{rss}} = -n_c E_c(\theta_{woss} - \theta_{rss})^{n_c-1}$$

which leads to the linearised convective term

$$E_c(\theta_{wo} - \theta_r)^{n_c} \cong E_c(\theta_{woss} - \theta_{rss})^{n_c} + n_c E_c(\theta_{woss} - \theta_{rss})^{n_c-1}(\theta_{wo} - \theta_{woss})$$
$$-n_c E_c(\theta_{woss} - \theta_{rss})^{n_c-1}(\theta_r - \theta_{rss}) \tag{3.23}$$

Finally, for the radiant term,

$$f(\theta_{wo}, \overline{\theta}_f) = E_r(\theta_{wo}^4 - \overline{\theta}_f^4)$$

we get

$$\frac{\partial f(\theta_{wo}, \overline{\theta}_f)}{\partial \theta_{woss}} = 4E_r \theta_{woss}^3$$

and

$$\frac{\partial f(\theta_{wo}, \overline{\theta}_f)}{\partial \overline{\theta}_{fss}} = -4\overline{\theta}_f^3$$

resulting in the linearised radiant term

$$E_r(\theta_{wo}^4 - \overline{\theta}_f^4) \cong E_r(\theta_{woss}^4 - \overline{\theta}_{fss}^4) + 4E_r \theta_{woss}^3(\theta_{wo} - \theta_{woss}) - 4E_r \overline{\theta}_{fss}^3(\overline{\theta}_f - \overline{\theta}_{fss}) \tag{3.24}$$

We now substitute equations (3.22), (3.23) and (3.24) into equation (3.19), dropping all steady-state terms and taking Laplace transforms in the process, to produce the following:

$$\theta_{wo}(s) = \frac{K_{m_w}}{(\tau_e s + 1)} m_w(s) + \frac{K_{\theta_{wi}}}{(\tau_e s + 1)} \theta_i(s) + \frac{K_{\theta_r}}{(\tau_e s + 1)} \theta_r(s) + \frac{K_{\overline{\theta}_f}}{(\tau_e s + 1)} \overline{\theta}_f(s) \tag{3.25}$$

where

$$\tau_e = \frac{C_e}{m_{wss}c_{pw} + n_c E_c(\theta_{woss} - \theta_{rss})^{n_c-1} + 4E_r\bar{\theta}_{woss}^3} = \frac{C_e}{T_e}$$

$$K_{m_w} = \frac{c_{pw}(\theta_{wiss} - \theta_{oss})n}{T_e} \qquad K_{\theta_{wi}} = \frac{m_{wss}c_{pw}}{T_e}$$

$$K_{\theta_r} = \frac{n_c E_c(\theta_{woss} - \theta_{rss})^{n_c-1}}{T_e} \qquad K_{\bar{\theta}_f} = \frac{4E_r\bar{\theta}_{fss}^3}{T_e}$$

which represents a generalised s-domain model for a room heat emitter. There is not a lot we can do with this model until we couple it with a room. We will now therefore revisit the simple room model in section 3.1 and develop it further so that it can be coupled with the emitter to form a comprehensive room heat exchange model suitable for control system studies.

3.4 A coupled room and emitter model

Returning to equation (3.1) which was developed for instantaneous plant input, q_{plant}, we now need to make substitutions for the plant term. It is possible to envisage three general cases for room–plant interaction as far as thermal control is concerned:

1   *Case 1.* The room receives a mechanically delivered air supply at a variable or constant air mass flow rate and a variable or constant supply temperature.
2   *Case 2.* The room is naturally ventilated containing a heat emitter which may exhibit a variable proportion of convective and radiant emission (note that this includes the possibility of chilled surface cooling).
3   *Case 3.* This is a combination of case 1 and case 2.

Consider *case 1*. Using a simple sensible energy balance for the plant influence,

$$q_{plant} = m_a c_{pa}(\theta_{ai} - \theta_r) \tag{3.26}$$

which can be linearised to account for non-linearity caused by the products of mass flow rate and temperatures (both of which may vary). Using the method set out in section 3.3, the linearised form of q_{plant} will be

$$m_a c_{pa}(\theta_{ai} - \theta_r) \cong (\theta_{aiss} - \theta_{rss})m_a c_{pa} + m_{ass}c_{pa}(\theta_{ai} - \theta_r) \tag{3.27}$$

Substituting in equation (3.1) and retaining the external temperature since, with fabric coupling, this now becomes significant, we can express the following comprehensive room model for case 1:

$$\theta_{r(s)} = \frac{K_{m_a}}{(\tau_r s + 1)}m_a(s) + \frac{K_{\theta_{ai}}}{(\tau_r s + 1)}\theta_{ai(s)} + \frac{K_{\bar{\theta}_f}}{(\tau_r s + 1)}\bar{\theta}_f(s) + \frac{K_{\theta_o}}{(\tau_r s + 1)}\theta_o(s) \tag{3.28}$$

where

$$\tau_r = \frac{C_r}{m_a c_{pa} + \Sigma(AU_i) + n_v V_r/3} = \frac{C_r}{T_r}$$

$$K_{m_a} = \frac{c_{pa}(\theta_{ass} - \theta_{rss})}{T_r} \qquad K_{\theta_{ai}} = \frac{m_a c_{pa}}{T_r}$$

$$K_{\bar{\theta}_f} = \frac{\Sigma(AU_i)}{T_r} \qquad K_{\theta_o} = \frac{n_v V_r}{3T_r}$$

Fabric temperature in the above comes from the mean of the surface temperatures predicted by the composite construction model of Figure 3.3 (there will be as many of these as there are fabric elements making up the room envelope).

The situation of *case 2* is a little more tricky since we now have a radiant term which interacts directly with the fabric rather than the room air model. We can deal with the convection term of the heat emitter by substituting equation (3.23) into the plant term of equation (3.1), which results in the following room model for case 2 assuming one emitter (or one equivalent emitter) in the room:

$$\theta_r(s) = \frac{K_{\theta_{wo}}}{(\tau_r s + 1)}\theta_{wo}(s) + \frac{K_{\bar{\theta}_f}}{(\tau_r s + 1)}\bar{\theta}_f(s) + \frac{K_{\theta_o}}{(\tau_r s + 1)}\theta_o(s) \qquad (3.29)$$

where

$$\tau_r = \frac{C_r}{n_c E_c(\theta_{woss} - \theta_{rss})^{n_c - 1} + \Sigma(AU_i) + n_v V_r/3} = \frac{C_r}{T_r}$$

$$K_{\theta_{wo}} = \frac{n_c E_c(\theta_{woss} - \theta_{rss})^{n_c - 1}}{T_r}$$

$$K_{\bar{\theta}_f} = \frac{\Sigma(AU_i)}{T_r} \qquad K_{\theta_o} = \frac{n_v V_r}{3T_r}$$

The water outlet temperature term in the above comes from the emitter model of equation (3.25). To incorporate the radiant component of the heat emitter, we need to modify the inside layer transfer function of all room fabric elements (i.e. equation (3.12) where it occupies the inside layer position of Figure 3.3) as follows:

$$\theta_f(s) = \frac{As + B}{Cs^2 + Ds + E}\theta_r(s) + \frac{Fs + G}{Cs^2 + Ds + E}\theta_{wo}(s) + \frac{H}{Cs^2 + Ds + E}\theta_o(s) \qquad (3.30)$$

where

$$A = lp_f c_{pf} u_i \qquad\qquad B = (u_i u_f + u_o u_i)$$

$$C = (lp_f c_{pf})^2 \qquad\qquad D = lp_f c_{pf}(2u_f + u_i + u_o + R_f)$$

$$E = u_f u_i + u_i u_o + u_f u_o + R_f(u_f + u_o) \qquad F = R_r lp_f c_{pf}$$

$$G = R_r(u_f + u_o) \qquad\qquad H = u_f u_o$$

and radiant coefficients

$$R_f = 4E_r\theta_{woss}^3 \qquad R_r = 4E_r\overline{\theta}_{fss}^3$$

Case 3 can be developed by combining the results of case 1 and case 2 essentially by merging equations (3.28) and (3.29). Equation (3.28) applies but with the addition of the $\theta_{wo}(s)$ term from equation (3.29). Correspondingly, the convective emission term in the time constant and gain terms of equation (3.29) is added appropriately to equation (3.28).

This is now a good place to set up another example in order to see how we might combine all these equations to form a coupled room and emitter model.

Example 3.4

A room measures 12m × 7m × 3m (high) and has 45m^2 of external wall which consists of 112mm light concrete block (inner), 50mm mineral wool insulation and 112mm of brick (outer). The floor, ceiling and partition walls are constructed from a low-thermal-capacity material and can be assumed to play no part in the dynamic characteristics of the space.

Using the data given, derive a coupled heat emitter and room model for both natural convector space heating and underfloor space heating and simulate the response in room air temperature to disturbances in heating water flow rate, flow temperature and external air temperature.

Data

The data for the materials of the room are listed in Table E3.4.1. For the room, $\Sigma\,(AU_i) = 468.0\,\mathrm{kJh^{-1}K^{-1}}$ and $n_v = 0.5\mathrm{h^{-1n}}$.

Natural convector heating consists of 24m of mild steel tube, 35.9mm inside diameter, 3.25mm wall thickness. There are 9448 aluminium fins of surface dimensions 100mm × 100 mm and thickness 0.35mm. The densities of mild steel and aluminium are respectively 7860kgm^{-3} and 2710kgm^{-3} and corresponding specific heat capacities 0.42kJkg^{-1}K^{-1} and 0.913kJkg^{-1}K^{-1}. For the natural convectors take $n_c = 1.25$ and $E_c = 108.0\mathrm{kJh^{-1}K^{-n_c}}$. The steady-state heating outlet water temperature and room air temperature can be taken as 70°C and 20°C respectively.

Underfloor heating consists of 22mm embedded tubes at 100mm centres laid in 12m lengths across the 7m dimension of the floor. The tubes have

Table E3.4.1 Data for the room

Construction	Density (kgm^{-3})	Spec. heat cap. (kJkg^{-1}K^{-1})	U-values (kJh^{-1}m^{-2}K^{-1})		
			Inner	Material	Outer
Brick	1700	0.80	5.04	54.00	29.59
Block	600	1.00	30.00	12.22	12.22
Insulation	30	1.00	12.22	5.04	5.04

negligible wall thickness. The screed topping consists of 75mm of fine sand and cement screed with a density and specific heat capacity of $1200 \, \mathrm{kgm}^{-3}$ and $0.84 \mathrm{kJkg}^{-1}\mathrm{K}^{-1}$ respectively. For the underfloor heating take $n_c = 1.00$, $E_c = 144.0 \mathrm{kJh}^{-1}\mathrm{K}^{-n_c}$ and $E_r = 4.68 \times 10^{-7} \mathrm{kJh}^{-1}\mathrm{K}^{-4}$.

The steady-state heating outlet water temperature and room air and surface temperatures can be taken as 25°C and 20°C respectively. The design heating water flow rate, specific heat capacity and density for both cases can be taken as $342.8 \mathrm{kgh}^{-1}$, $4.2 \mathrm{kJkg}^{-1}\mathrm{K}^{-1}$ and $1000 \mathrm{kgm}^{-3}$, and the design heating water temperature difference is $10 \, \mathrm{K}$ in both cases.

Solution

There are several stages to this problem.

a   Derive the fabric model for the convective heating case.
b   Derive the room transfer function for the convective heating case.
c   Calculate the thermal capacity of the natural convector case and, hence, its transfer function.
d   Form a block diagram model for the natural convector case.
e   Modify the fabric model for the radiant effect of the underfloor heating case.
f   Derive the room transfer function for the underfloor heating case.
g   Calculate the thermal capacity of the underfloor heating and, hence, its transfer function.
h   Form a block diagram model of this second case.
i   Obtain and compare results.

We will now work steadily through stages a–i in turn.

Stage a

For the fabric model (natural convector heating case), we require to express and link transfer functions for the three-layer wall. The transfer functions are based on equation (3.12) and the linking of the three will be in accordance with Figure 3.3.

For the inside layer (concrete block), the coefficients of transfer function are calculated as follows (a timebase of hours has been chosen for convenience):

$$A = l\rho_f c_{pf} u_i = 0.056 \times 600.0 \times 1.00 \times 30.0 = 1008.0$$

$$B = (u_f u_i + u_o u_i) = 2 \times (u_i u_f) \quad \text{(since } u_o = u_f \text{ for this case)}$$
$$= 2 \times (12.22 \times 30.0) = 732.9$$

$$C = (l\rho_f c_{pf})^2 = (0.056 \times 600.0 \times 1.00)^2 = 1129.0$$

$$D = l\rho_f c_{pf}(2u_f + u_i + u_o)$$
$$= 0.056 \times 600.0 \times 1.00 \times (2 \times 12.22 + 30.0 + 12.22) = 2239.3$$

$$E = (u_f u_i + u_i u_o + u_f u_o) = (2 \times 12.22 \times 30.0 + 12.22^2) = 882.1$$

$$F = u_f u_o = 12.22^2 = 149.2$$

The transfer function for the inside layer will therefore be

$$\theta_{fi}(s) = \frac{1.143s + 0.831}{1.280s^2 + 2.539s + 1}\theta_r(s) + \frac{0.169}{1.280s^2 + 2.539s + 1}\theta_{fm}(s) \qquad (E3.4.1)$$

where $\theta_{fm}(s)$ is the temperature of the middle layer of material at the layer/layer interface.

Similarly, transfer functions can be expressed for the middle and outer layers of material forming the wall. For the middle layer, which is the insulation,

$$\theta_{fm}(s) = \frac{0.062s + 0.829}{0.0038s^2 + 0.138s + 1}\theta_{fi}(s) + \frac{0.71}{0.0038s^2 + 0.138s + 1}\theta_{wo}(s) \qquad (E3.4.2)$$

And, for the outer material layer,

$$\theta_{fo}(s) = \frac{0.190s + 0.209}{2.873s^2 + 5.380s + 1}\theta_{fm}(s) + \frac{0.791}{2.873s^2 + 5.380s + 1}\theta_o(s) \qquad (E3.4.3)$$

Stage b

We now turn to the room transfer function for the natural convection heating case. Equation (3.29) applies with the above fabric transfer functions, since there is no radiant emission with this case.

First, the room time constant (note that the multiplying factor of 3.6 in the ventilation term converts this result from basic units of WK^{-1} to our more manageable units $kJh^{-1}K^{-1}$):

$$\tau_r = \frac{C_r}{T_r} = \frac{V_r \rho_a c_{pa}}{n_c E_c(\theta_{woss} - \theta_{rss})^{n_c - 1} + \Sigma(AU_i) + n_v V_r/3}$$

$$= \frac{12 \times 7 \times 3 \times 1.2 \times 1.025}{1.25 \times 108.0(70 - 20)^{1.25-1} + 468 + 0.5 \times 252 \times 3.6/3}$$

$$\tau_r = \frac{310.0}{978.2} = 0.317 \text{ hours}$$

Now the various gains,

$$K_{\theta_{wo}} = \frac{n_c E_c(\theta_{woss} - \theta_{rss})^{n_c - 1}}{T_r} = \frac{1.25 \times 108.0(70 - 20)^{1.25-1}}{978.2} = 0.367$$

$$K_{\bar{\theta}_f} = \frac{\Sigma(AU_i)}{T_r} = \frac{468}{978.2} = 0.478 \qquad K_{\theta_o} = \frac{n_v V_r}{3T_r} \times 3.6 = 151.2$$

The room transfer function will therefore be

$$\theta_r(s) = \frac{0.367}{(0.317s + 1)}\theta_{wo}(s) + \frac{0.478}{(0.317s + 1)}\bar{\theta}_{fi}(s) + \frac{0.155}{(0.317s + 1)}\theta_o(s) \qquad (E3.4.4)$$

Stage c

Now we calculate the emitter thermal capacity and the emitter transfer function from equation (3.25). For the thermal capacity, we require the sum of thermal capacities for the water content and heat transfer surfaces which consist of steel tubing and aluminium fins:

$$C_e = C_w + C_{tube} + C_{fin}$$

with

$$C_w = V_w \rho_w c_{pw} = 24 \times \frac{\pi \times 0.0359^2}{4} \times 1000 \times 4.2 = 102.0 \text{kJK}^{-1}$$

$$C_{tube} = (V \rho c_p)_{tube} = 24 \times 0.03915\pi \times 7860 \times 0.42 = 31.7 \text{kJK}^{-1}$$

$$C_{fin} = (V \rho c_p)_{fin} = 9448 \times 0.00035 \times \left(0.1^2 - \frac{\pi \times 0.03915^2}{4}\right) \times 2710 \times 0.913$$

$$= 72.0 \text{kJK}^{-1}$$

Though the water thermal capacity is dominant, the aluminium fins end up having a much higher thermal capacity than the steel tubing, due to the large overall surface of aluminium.

Noting that the radiant term in the denominator of the time constant of equation (3.25) will be zero in this case, the time constant for the natural convector will be

$$\tau_e = \frac{C_e}{T_e} = \frac{C_e}{m_{wss}c_{pw} + n_c E_c (\theta_{woss} - \theta_{rss})^{n_c - 1}}$$

$$= \frac{102.0 + 31.7 + 72.0}{342.8 \times 4.2 + 1.25 \times 108.0(70 - 20)^{1.25 - 1}}$$

$$\tau_e = \frac{205.7}{1798.8} = 0.114 \text{ hours}$$

The gains, again noting that there will be no $\bar{\theta}_f(s)$ term in this purely convective case, will be

$$K_{m_w} = \frac{c_{pw}(\theta_{wiss} - \theta_{woss})}{T_e} = \frac{4.2 \times 10}{1798.8} = 0.023$$

$$K_{\theta_{wi}} = \frac{m_{wss}c_{pw}}{T_e} = \frac{342.8 \times 4.2}{1798.8} = 0.800$$

$$K_{\theta_r} = \frac{n_c E_e(\theta_{woss} - \theta_{rss})^{n_c - 1}}{T_e} = \frac{1.25 \times 108.0 \times (70 - 20)^{1.25 - 1}}{1798.8} = 0.200$$

All this leads to the emitter transfer function for this case:

$$\theta_{wo}(s) = \frac{0.023}{(0.114s + 1)} m_w(s) + \frac{0.800}{(0.114s + 1)} \theta_{wi}(s) + \frac{0.200}{(0.114s + 1)} \theta_r(s) \qquad \text{(E3.4.5)}$$

Stage d

We are now in a position to construct a block diagram model of·our coupled room and emitter. Equations (E3.4.1)–(E3.4.3), which form the fabric transfer functions, can be coupled together as in Figure 3.3. The remaining equations – (E3.4.4) for the room and (E3.4.5) for the emitter – are 'connected' so that appropriate outputs meet appropriate inputs. That leaves three variables, θ_o, m_w, θ_{wi}, which are inputs to our block diagram model, which is therefore said to be an *open-loop system*. Figure E3.4.1 gives the resulting block diagram model for the natural convector heating case.

Stage e

We now move on to the underfloor heating case. The fabric model used in the previous case needs to be modified slightly to enable the radiant emission component of the underfloor heating to be taken into account. This can be done by applying equation (3.30) which will affect the inside layer of the wall construction only – the transfer functions for the other two layers remain unchanged. We note that coefficients A, B, C for equation (3.30) remain as before and H takes on the value assigned to F in the previous case.

The remaining coefficients are as follows (noting that the radiant coefficients require to be calculated using the absolute temperature scale):

$$R_f = 4E_r\theta_{woss}^3 = 4 \times 4.68 \times 10^{-7} \times (273.15 + 25)^3 = 50.0$$
$$R_r = 4E_r\overline{\theta}_{fss}^3 = 4 \times 4.68 \times 10^{-7} \times (273.15 + 20)^3 = 47.2$$

and

$$D = lp_f c_{pf}(2u_f + u_i + u_o + R_f)$$
$$= 0.056 \times 600 \times 1.00 \times (2 \times 12.22 + 30.0 + 12.22 + 50.0) = 3919.3$$

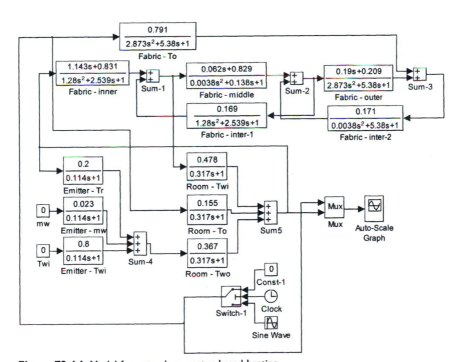

Figure E3.4.1 Model for natural convector-based heating.

$$E = u_f u_i + u_i u_o + u_f u_o + R_f(u_f + u_o)$$
$$= 2 \times 12.22 \times 30.0 + 12.22^2 + 50.0 \times (2 \times 12.22) = 2103.6$$

$$F = R_r \times l\rho_f c_{pf} = 47.2 \times 0.056 \times 600 \times 1.00 = 1585.9$$
$$G = R_r(u_f + u_o) = 47.2 \times (2 \times 12.22) = 1153.1$$

Thus the new inside surface layer transfer function is

$$\theta_{fi}(s) = \frac{0.479s + 0.348}{0.537s^2 + 1.863s + 1}\theta_r(s) + \frac{0.754s + 0.548}{0.537s^2 + 1.863s + 1}\theta_{wo}(s)$$
$$+ \frac{0.071}{0.537s^2 + 1.862s + 1}\theta_{fm}(s) \qquad \text{(E3.4.6)}$$

Stage f

The coefficients of the room transfer function will also differ from those used in the previous case since the convective heat transfer coefficient and index have changed:

$$n_c E_c(\theta_{woss} - \theta_{rss})^{n_c-1} = E_c \qquad \text{for this case since } n_c = 1 \text{ and } E_c = 144.0$$

Therefore the new room time constant and gains are calculated to give

$$\tau_r = 0.406\text{h}, \qquad K_{\theta_{wo}} = 0.189, \qquad K_{\bar{\theta}_f} = 0.613 \qquad \text{and} \qquad K_{\theta_f} = 0.198$$

Hence the room transfer function for the underfloor heating case is

$$\theta_r(s) = \frac{0.189}{(0.406s + 1)}\theta_{wo}(s) + \frac{0.613}{(0.406s + 1)}\theta_{fi}(s) + \frac{0.198}{(0.406s + 1)}\theta_o(s) \qquad \text{(E3.4.7)}$$

Stage g

The emitter model is given by equation (3.25) but this time we need to include the radiant transfer function term and the appropriate radiant term in the denominator of the time constant. We first need to calculate the thermal capacity for this case (this will be high due to the sand/cement screed):

$$C_e = C_w + C_{screed}$$

and there is $12 \times 7/0.1 = 840\,\text{m}$ of 22 mm tubing

$$C_w = V_w \rho_w c_{pw} = 840 \times \frac{\pi \times 0.022^2}{4} \times 1000 \times 4.2 = 1341.1\,\text{kJK}^{-1}$$

$$C_{screed} = (V\rho c_p)_{screed}$$
$$= \left[(12 \times 7 \times 0.075) - \left(840 \times \frac{\pi \times 0.022^2}{4} \times 1200 \times 0.84\right)\right] = 6028.5\,\text{kJK}^{-1}$$

The time constant for the underfloor heating is

$$\tau_e = \frac{C_e}{T_e} = \frac{C_e}{m_{wss}c_{pw} + n_c E_c(\theta_{woss} - \theta_{rss})^{n_c-1} + 4E_r\bar{\theta}_{woss}^3} = \frac{1341.1 + 6028.5}{(342.8 \times 4.2 + 144.0 + 50.0)}$$

$$\tau_e = 4.51\,\text{hours}$$

The various gains are

$$K_{m_w} = 0.026, \qquad K_{\theta_{wi}} = 0.881, \qquad K_{\theta_r} = 0.088 \qquad \text{and} \qquad K_{\bar{\theta}_f} = 0.031$$

Hence the transfer function for the underfloor heating is as follows:

$$\theta_{wo}(s) = \frac{0.026}{(4.51s + 1)} m_w(s) + \frac{0.881}{(4.51s + 1)} \theta_{wi}(s)$$

$$+ \frac{0.088}{(4.51s + 1)} \theta_r(s) + \frac{0.031}{(4.51s + 1)} \theta_{fi}(s) \qquad \text{(E3.4.8)}$$

Stage h

The block diagram model for underfloor heating is built up in exactly the same way as before. In fact the only difference (apart from the transfer function coefficients) is that there are two new transfer function terms – one for the emitter and one for the inside wall layer – which account for the presence of radiant heat exchange. Figure E3.4.2 gives the coupled model for this case.

Stage i

Results from these models are now generated using Simulink. These have been generated using arbitrary disturbances in heating water flow rate, heating water inlet temperature and external temperature. The distur-

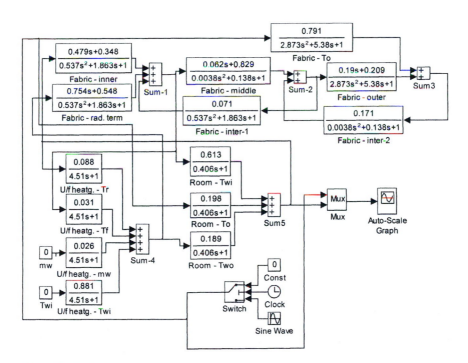

Figure E3.4.2 Model for underfloor heating.

bances are, notionally, of a form that might take place in practice – sinusoidal variations have been used with amplitudes of 171.4kgh^{-1} in water flow rate (equivalent to 50% of the design water flow rate), 10K in inlet water temperature and 5K in external temperature. A single sinusoidal cycle has been applied in each case with a cycle time of one hour for each of the heating disturbances and 24 hours for the external temperature. Note that the models shown in Figures E3.4.1 and E3.4.2 are based on the external temperature disturbances. Hence the results in room temperature obtained represent *frequency responses* to a single cycle input disturbance. The results are plotted for both system in Figures E3.4.3, E3.4.4 and E3.4.5.

All results show the lagging effect of the room temperature response – in the case of water flow rate and temperature, it takes about three hours for conditions to reach steady state following the disturbance of one hour duration. However in the case of the external temperature variation, the response is far less damped and this is due to the relatively rapid influence of the natural ventilation which only meets with the room air capacity as it influences room air temperature. The particularly interesting feature of the heating responses is the dramatic effect of the damping due to the heating systems – clearly much higher with the underfloor heating than with natural convectors. This is one reason why underfloor heating is not suitable for highly responsive buildings.

So much for room-based heat emitters. Now we will move on and take a look at other examples of plant component models for dynamic analysis.

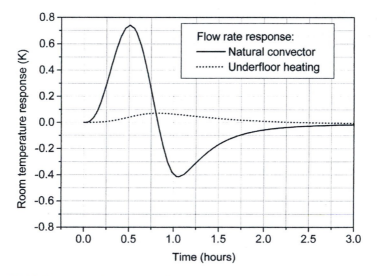

Figure E3.4.3 Response to heating water flow rate.

3.5 Component modelling

Coils

In-duct coils are thermally decoupled from the room and are, therefore, somewhat easier to deal with than for the case of convective/radiant heat emitters – case 1 in section 3.4 is applicable.

Heating coil modelling is relatively straightforward (see for example Shavit & Brandt, 1982; Gondal, 1987). Cooling coil modelling on the other

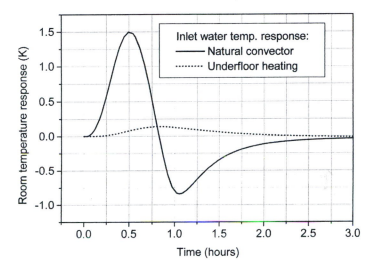

Figure E3.4.4 Response to heating water inlet temperature.

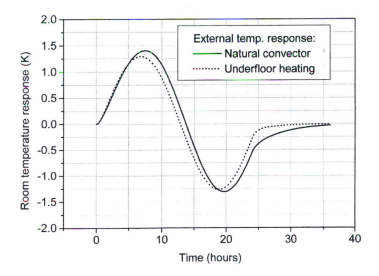

Figure E3.4.5 Response to external air temperature.

hand is much more complex, especially where sensible and latent cooling is concerned (for a detailed treatment see Yik et al., 1997).

For analysis restricted to control stability or operation within a finite band, Adams & Holmes (1977) have proposed a method for finding coil time constants for disturbances in both inlet water temperatures and water mass flow rate. For convenience, equations from which time constant information can be derived are presented here by curve-fitting to their results. A summary of the method is as follows.

1   *Step 1.* Establish the design coil water velocity, air velocity, length and number of tubes in the coil construction. In this context, the tube length is an effective length neglecting any sections across which no useful heat transfer takes place. The number of tubes is the number of parallel tubes connecting to each header.

2   *Step 2.*
    a   If control is to be by varying the inlet water flow rate, then the coil flow time constant, τ_{cm}, will be

$$\tau_{cm} = \frac{5.038 + 4.763L_t - L_t^2 + 1.102 \times 10^{-1}L_t^3 - 4.400 \times 10^{-3}L_t^4}{C_w} \quad (3.31)$$

where

$$C_w = a + bc_w + cc_w^2 + dc_w^3 + ec_w^4$$

L_t = the total length of each coil tube (m) and c_w = the velocity of water in each tube (ms^{-1}). And, in the following, c_a is the air side coil face velocity (ms^{-1}):

$a = 2.187 \times 10^{-1} + 2.090 \times 10^{-1}\, c_a - 5.710 \times 10^{-2}c_a^2 + 9.000 \times 10^{-3}c_a^3$
$\quad - 5.436 \times 10^{-4}c_a^4$

$b = 7.127 \times 10^{-1} + 8.099 \times 10^{-1}c_a - 2.723 \times 10^{-1}c_a^2 + 5.540 \times 10^{-2}c_a^3$
$\quad - 4.320 \times 10^{-3}c_a^4$

$c = -7.128 \times 10^{-1} - 5.550 \times 10^{-1}c_a + 1.452 \times 10^{-1}c_a^2 - 2.617 \times 10^{-2}c_a^3$
$\quad + 1.860 \times 10^{-3}c_a^4$

$d = 4.082 \times 10^{-1} + 1.688 \times 10^{-1}c_a - 2.400 \times 10^{-3}c_a^2 - 4.900 \times 10^{-3}c_a^3$
$\quad + 6.862 \times 10^{-4}c_a^4$

$e = -8.690 \times 10^{-2} - 1.150 \times 10^{-2}\, c_a - 1.330 \times 10^{-2}c_a^2 + 4.500 \times$
$\quad 10^{-3}c_a^3 - 4.563 \times 10^{-4}c_a^4$

    b   If control is to be by varying the inlet water temperature, then a coil water temperature time constant and dead time ($\tau_{c\theta}$ and $t_{dc\theta}$) are obtained:

$$\tau_{c\theta} = \frac{f + gL_t + hL_t + iL_t + jL_t}{C_w} \quad (3.32)$$

where, in the following, N_t is the number of parallel tubes connecting to each coil header:

$f = 6.255 + 1.975 \times 10^{-1}\, N_t - 9.200 \times 10^{-3}N_t^2 + 2.200 \times 10^{-4}N_t^3 -$
$\quad 1.980 \times 10^{-6}N_t^4$

$$g = 2.953 + 1.454 \times 10^{-1} \, N_t - 9.800 \times 10^{-3} N_t^2 + 2.974 \times 10^{-4} N_t^3 - 3.149 \times 10^{-6} N_t^4$$

$$h = -6.581 \times 10^{-1} - 4.330 \times 10^{-2} N_t + 3.300 \times 10^{-3} N_t^2 - 1.109 \times 10^{-4} N_t^3 + 1.250 \times 10^{-6} N_t^4$$

$$i = 7.320 \times 10^{-2} + 5.700 \times 10^{-3} N_t - 4.750 \times 10^{-4} N_t^2 + 1.653 \times 10^{-5} N_t^3 - 1.930 \times 10^{-7} N_t^4$$

$$j = -2.960 \times 10^{-3} - 2.624 \times 10^{-4} N_t + 2.296 \times 10^{-5} N_t^2 - 8.228 \times 10^{-7} N_t^3 + 9.795 \times 10^{-9} N_t^4$$

and

$$t_{dc\theta} = \frac{k + lL_t + mL_t^2 + nL_t^3 + pL_t^4}{c_w} \tag{3.33}$$

where

$$k = 1.757 \times 10^{-1} + 2.220 \times 10^{-2} \, N_t - 9.046 \times 10^{-4} N_t^2 + 1.923 \times 10^{-5} N_t^3 - 1.516 \times 10^{-7} N_t^4$$

$$l = 5.431 \times 10^{-1} + 7.310 \times 10^{-2} \, N_t - 3.600 \times 10^{-3} N_t^2 + 9.298 \times 10^{-5} N_t^3 - 9.048 \times 10^{-7} N_t^4$$

$$m = -8.390 \times 10^{-2} - 1.110 \times 10^{-2} \, N_t + 5.626 \times 10^{-4} N_t^2 - 1.527 \times 10^{-5} N_t^3 + 1.525 \times 10^{-7} N_t^4$$

$$n = 8.800 \times 10^{-3} + 1.200 \times 10^{-3} \, N_t - 6.124 \times 10^{-5} N_t^2 + 1.698 \times 10^{-6} N_t^3 - 1.715 \times 10^{-8} N_t^4$$

$$p = -2.941 \times 10^{-4} - 6.732 \times 10^{-5} \, N_t + 4.607 \times 10^{-6} N_t^2 - 1.506 \times 10^{-7} N_t^3 + 1.703 \times 10^{-9} N_t^4$$

3 *Step 3.*

a For control from the water flow rate, the coil model will be a single time constant (i.e. first-order lag), of the following form:

$$\frac{\theta_{ao}(s)}{m_w(s)} = \frac{K_{cm}}{(\tau_{cm}s + 1)} \tag{3.34}$$

where $\theta_{ao}(s)$ = coil outlet air temperature change (K), $m_w(s)$ = coil water flow rate change (typically kgs^{-1}) and K_{cm} = coil outlet air temperature/water flow rate gain (Kskg^{-1}). Note that the time-domain form of this for a unit step change in inlet water flow rate will be:

$$\theta_{ao}(t) = \theta_{ao}(0) + \Delta\theta_{ao}\left[1 - \exp\left(\frac{-t}{\tau_{cm}}\right)\right] \tag{3.35}$$

where $\Delta\theta_{ao}$ = the apparent overall change in θ_{ao}.

b For control from the inlet water temperature, a first-order lag plus dead-time model is formed:

$$\frac{\theta_{ao}(s)}{\theta_{wi}(s)} = \frac{K_{c\theta} \exp(-t_{dc\theta}s)}{(\tau_{c\theta}s + 1)} \tag{3.36}$$

where $\theta_{wi}(s)$ = coil inlet water temperature change (K) and $K_{c\theta}$ = coil outlet air temperature/inlet water temperature gain. And the time-domain form for a unit step change in inlet water temperature is

$$\theta_{ao}(t) = \theta_{ao}(0) + \Delta\theta_{ao}\left[1 - \exp\left(\frac{t_{dc\theta} - t}{\tau_{c\theta}}\right)\right] \qquad (t \geq t_{dc\theta}) \qquad (3.37)$$

Example 3.5

Derive transfer functions for a chilled water cooling coil of the following specification, for both (a) chilled water flow rate and (b) chilled water inlet temperature variations. Assume that the relationships between coil outlet air temperature and inlet water flow rate, and between outlet air temperature and inlet water temperature are linear across their full ranges.

Specification

- Sensible cooling capacity: 40kW
- Coil air volume flow rate: $2.0 \text{m}^3\text{s}^{-1}$
- Face air velocity: 2.5ms^{-1}
- Coil face aspect: square
- Rows: 3
- Tubes per header: 18
- Tube water velocity: 0.34ms^{-1}

Solution

(a) For the inlet water flow rate case, a first-order lag model will be obtained.

The coil aspect is square, and therefore for a face velocity of 2.5ms^{-1}, the coil width will be $\sqrt{2.0/2.5} = 0.894$m. Thus, the effective coil length $= 3$ rows $\times 0.894 = 2.68$m.

At this air velocity, the coefficients of equation (3.30) (for the coil time constant) will be

$$a = 0.5037, \qquad b = 1.7325, \qquad c = -1.5291, \qquad d = 0.7654, \qquad e = -0.1463$$

Using equation (3.30), the coil time constant will be

$$\tau_{cm} = \frac{5.038 + 4.7632 \times 2.68 - 2.68^2 + 0.1102 \times 2.68^3 - 0.0044 \times 2.68^4}{0.5037 + 1.7325 \times 0.34 - 1.5291 \times 0.34^2 + 0.7654 \times 0.34^3 - 0.1463 \times 0.34^4}$$
$$= \frac{12.519}{0.9441} = 13.26\text{s}$$

At the specified conditions and using standard air properties, the coil is capable of lowering the air temperature by

$$\Delta\theta_a = \frac{-q_{c-sensible}}{V_a \rho_a c_{pa}} = \frac{-40.0}{2.0 \times 1.2 \times 1.025} = -16.26\text{K}$$

(Note the negative sign – signifying removal of heat.)

Correspondingly, the chilled water flow rate capable of effecting this change in air temperature will be, again using standard water properties,

$$m_w = \frac{-q_{c-sensible}}{c_{pw}(\theta_{wi} - \theta_{wo})} = \frac{-40.0}{4.2 \times (-5.0)} = 1.905 \text{kgs}^{-1}$$

For a full range linear response therefore, the coil gain for this case is

$$\frac{\Delta\theta_a}{m_w} = \frac{-16.26}{1.905} = -8.54 \text{Kskg}^{-1}$$

and the coil transfer function will therefore be

$$\frac{\theta_{ao}(s)}{m_w(s)} = \frac{-8.54}{(13.26s + 1)}$$

The significance of the negative sign now becomes clearer – an increase in chilled water flow rate will result in a *fall* (i.e. negative change) in outlet air temperature.

(b)   For the inlet water temperature case, a first-order lag plus dead-time model will be obtained for which we can obtain the time constant and dead time from equations (3.32) and (3.33).

For 18 tubes per header, the coefficients of these equations are determined as follows:

$f = 7.9044$, $g = 3.7989$, $h = -0.8839$, $i = 0.0980$, $j = -0.0041$
$k = 0.3785$, $l = 1.1399$, $m = -0.1745$, $n = 0.1078$, $p = -0.00078$

The coil time constant for this case will be

$$\tau_{c\theta} = \frac{7.9044 + 3.7989 \times 2.68 - 0.8839 \times 2.68^2 + 0.098 \times 2.68^3 - 0.0041 \times 2.68^4}{0.9441}$$

$$= 14.21\text{s}$$

and the dead time

$$t_{dd\theta} = \frac{0.3785 + 1.1399 \times 2.68 - 0.1745 \times 2.68^2 + 0.1078 \times 2.68^3 - 0.00078 \times 2.68^4}{0.34}$$

$$= 12.40\text{s}$$

The gain for this case for linear performance across a full range of operation is obtained from a simple ratio of air and corresponding water temperature differences, i.e.

$$K_{c\theta} = \frac{-16.26}{-5.0} = 3.25$$

The transfer function for this case will therefore be

$$\frac{\theta_{ao}(s)}{\theta_{wi}(s)} = \frac{3.25 \exp(-12.4s)}{(14.21s + 1)}$$

Adams & Holmes' method is very convenient for coil dynamic model fitting, but in section 9.4 a non-linear coil model is developed in which the 'apparent' time constant is much higher than the values indicated in the above example. It is therefore suggested that the above data be used with some caution.

Sensors

In a control system, the sensor closes the control loop forming the feedback path. Certain types of sensor act rapidly which means that they can often be represented in control loop modelling as 'unity negative feedback' possessing unity gain. This effectively treats the sensor as if it acted instantaneously. Flow sensors often fall into this category, and it is sometimes possible to treat certain types of temperature sensors in this way provided that the flow velocity across the sensing element is high – as exists for instance in the case of duct and pipe immersion sensors. However, certain types of room temperature sensors usually exhibit significant time constants (CIBSE, 1985).

Adams & Holmes (1977) also give modelling parameters for several common types of temperature measuring device including the predominant types used in HVAC control – the resistance temperature detector (RTD) and the thermistor. Again, these data are expressed here in the form of the following curve fits which can be used to construct first-order lag plus dead-time models, based on the air velocity passing over the sensor element, c_a. The resulting models express the applied temperature, θ_a, and corresponding sensor output temperature signal, ϕ.

1 For an RTD:

time constant,

$$\tau_d = 198.9 - 95.4c_a + 23.16c_a^2 - 2.500c_a^3 + 9.700 \times 10^{-2}c_a^4 \tag{3.38}$$

dead time,

$$t_{dd} = 21.69 - 8.518c_a + 1.942c_a^2 - 1.974 \times 10^{-1}c_a^3 + 7.300 \times 10^{-3}c_a^4 \tag{3.39}$$

thus,

$$\frac{\theta_a(s)}{\phi(s)} = \frac{\exp(-\tau_{dd}\,s)}{(t_d s + 1)} \tag{3.40}$$

2 For a thermistor:

time constant,

$$\tau_d = 151.5 - 70.8c_a + 18.04c_a^2 - 2.023c_a^3 + 8.090 \times 10^{-2}c_a^4 \tag{3.41}$$

dead time,

$$t_{dd} = 19.37 - 8.930c_a + 2.426c_a^2 - 2.825 \times 10^{-1}c_a^3 + 1.170 \times 10^{-2}c_a^4 \tag{3.42}$$

equation (3.40) is once again applicable.

Note that, in the case of sensors, it is convenient for the output to be expressed in the same units as those of the set point. Control error is then simply the difference between these two values (the physical output of the above types will usually be a bridge resistance in reality). With this convention, the *gain of the sensor in the feedback path will always be unity.*

Valves and dampers

Most control valves and dampers themselves influence flow more or less instantaneously – but the driving servo motor or actuator can exhibit significant lag. Manufacturers will often give time constants for control valve and damper actuators. There are no general rules of thumb; electromechanical actuators can have time constants of many seconds to several minutes for larger valves, whilst the time constant of a direct-acting valve such as a thermostatic radiator valve can be as high as 20 minutes (Fisk, 1981).

Where such time constant information is available, a first-order lag model can usually be written for the valve or damper in which the gain will be the ratio of flow rate change to corresponding control signal change.

Transport lag

In most systems, especially distributed systems of the type found in HVAC applications, time delays between components (due for instance to the time taken for the flow of a fluid between two points) are likely to be found. A common example occurs when a control sensor is some distance from the corresponding point at which control action is to be exercised.

For example, consider the case in which an air temperature sensor is mounted in a duct some 20m away from a coil which is to be controlled using the signal from the air temperature sensor. If the air passes down the duct at a mean velocity of 4ms^{-1}, it will take 5s before the air temperature value leaving the coil will be sensed by the temperature sensor downstream. Such conditions in systems are sometimes called *distance–velocity lags* or *transport lags*. In fact, this type of lag is simply a pure time delay and can be represented with a transfer function of identical form to that used to express dead time when dealing with coil and sensor modelling in sections 3.5.1 and 3.5.2:

$$\frac{\theta(s)}{\theta'(s)} = \exp(-ls) \qquad\qquad (3.43)$$

where $\theta'(s)$, $\theta(s)$ = the upstream and downstream signals, respectively, and l = the time delay in appropriate units of time. The units of the upstream and downstream signals are immaterial, provided they are consistent.

Note that, in the form expressed above, we have a non-linear function of s for a transport lag which becomes unmanageable in many of the linear control system analysis methods we will be looking at later. The same applies to the dead-time components of the coil and sensor model forms described previously. A convenient way to deal with this is to express

$$\exp(-ls) = \frac{\exp(-ls/2)}{\exp(ls/2)}$$

and expand the resulting expression using a Taylor series expansion. In most cases, sufficient accuracy will result if the Taylor series is truncated after the first-order term which results in what is sometimes referred to as a first-order Padé approximation,

$$\exp(-ls) \cong \frac{(1 - ls/2)}{(1 + ls/2)} \tag{3.44}$$

PID controllers

With the current trend towards digital control and building management systems, full three-term control is becoming standard (earlier analogue controllers would usually be simple proportional or, less commonly, proportional plus integral). We will give extensive consideration to digital control of this type in Chapter 5.

For modelling purposes, the general continuous-time model for a three-term controller is

$$\frac{u(s)}{\varepsilon(s)} = K_c\left(1 + \frac{1}{i_t s} + d_t s\right) \tag{3.45}$$

where $u(s)$ = control signal, $\varepsilon(s)$ = control error, K_c = controller gain and i_t, d_t = integral and derivative times.

Limitations in linear modelling

We have made extensive use of the concept of the steady-state gain. We see that the gain of a component can be found by dividing the component steady-state output by the corresponding input that prompted it (Example 3.5). We have used it in such a way that implies that this gain value is constant across the full range of operating conditions of the component. In most practical situations this is not the case for the gain will usually vary across the entire operating range of the component.

Consider for example a control valve. The gain used as we have intended it implies that the flow produced varies linearly with the control signal applied. But we know this not to be the case. Most valves used for hot water heating and chilled water are characterised logarithmically to offset the non-linear performance of the coil or heat exchanger they are supposed to be controlling.

However, much control analysis seeks to investigate performance, not across the full range of operating conditions, but across a finite range – such as a unit change in set point or load. This is usually all that is necessary in order to judge performance of the loop since such changes in practice are unlikely to be large and sudden. Thus, for a finite range of operating conditions, the notion of a constant gain is a reasonable one and linear control theory holds reasonably well.

For more comprehensive problems, non-linear HVAC modelling has received widespread attention – mostly in the form of simulation studies of plant (see for instance Underwood, 1993; Novakovic & Grindal, 1993) or for the particular analysis of non-linear plant behaviour (e.g. Borresen, 1981; Thompson, 1981; Roberts & Oak, 1991). Recognising the limitations of purely theoretical modelling of non-linear plant behaviour, there has also been some work based on empirical model-fitting (Underwood & Crawford, 1991; Crawford *et al.*, 1991).

The purpose of modelling entire control loops is to enable the investigation of control system performance. This might involve the performance investigation of a new design proposal or the improvement of performance of an existing system. Usually, we can learn all we need to know from the short-term performance of the control loop. In control system design studies, two issues are of interest. These are:

- Loop stability
- Quality of response

Before we can assess and improve the quality of response of a control system, we must of course ensure a design that achieves stable operation under all conditions. We will now move on to Chapter 4 which will give detailed consideration to this and other aspects of control system design.

References

Adams, S., Holmes, M.J. (1977) *Determining Time Constants for Heating and Cooling Coils.* BSRIA Technical Note TN6/77, Building Services Engineering Research and Information Association, Bracknell.

Athienitis, A.K. (1993) A methodology for integrated building–HVAC system thermal analysis. *Building and Environment,* **28** (4), 483–496.

Athienitis, A.K., Stylianou, M., Shou, J. (1990) A methodology for building thermal dynamics studies and control applications. *ASHRAE Transactions,* **96** (2), 839–848.

Borresen, B.A. (1981) HVAC control process simulation. *ASHRAE Transactions,* **87** (2), 871–882.

CIBSE (1985) *Automatic Controls and Their Implications for System Design.* Chartered Institution of Building Services Engineers, London.

CIBSE (1986) *Guide Book A.* Chartered Institution of Building Services Engineers, London.

Crabb, J.A., Murdoch, N., Penman, J.M. (1987) A simplified thermal response model. *Building Services Engineering Research and Technology,* **8** (1), 13–19.

Crawford, R.R., Dykowski, R.G., Czajkowski, S.E. (1991) A separated linear least-squares modelling procedure for nonlinear HVAC components. *ASHRAE Transactions,* **97** (2), 11–18.

Enns, M. (1962) Comparison of dynamic models of a superheater. *ASME Transactions – Journal of Heat Transfer,* **84** (C4), 375–385.

Fisk, D.J. (1981) *Thermal Control of Buildings.* Applied Science, London.

Gondal, I.A. (1987) Linear analysis of an air temperature control loop. *ASHRAE Transactions*, **93** (2), 736–751.

Houpis, C.H., Lamont, G.B. (1992) *Digital Control Systems – Theory, Hardware, Software*. McGraw-Hill, New York.

Leigh, J.R. (1992) *Applied Digital Control – Theory, Design and Implementation*. Prentice-Hall, Englewood Cliffs, NJ.

McIntyre, D.A. (1986) Output of radiators at reduced flow rate. *Building Services Engineering Research and Technology*, **7** (2), 92–95.

Novakovic, V., Grindal, A. (1993) Designing the controllability of a HVAC plant by dynamic simulation. *Proceedings of the CLIMA 2000 Conference*, London.

Roberts, A.A., Oak, M.P. (1991) Nonlinear dynamics and control for thermal room models. *ASHRAE Transactions*, **97** (1), 722–728.

Shavit, G., Brandt, S.G. (1982) The dynamic performance of a discharge air temperature system with a PI controller. *ASHRAE Transactions*, **88** (2), 826–838.

Simulink-2 (1996) *SIMULINK 2 Dynamic System Simulation for MATLAB*. The Mathworks Inc., Natick, MA.

Stroud, K.A. (1973) *Laplace Transforms*. Stanley Thornes, London.

Thompson, J.G. (1981) The effect of room and control system dynamics on energy consumption. *ASHRAE Transactions*, **87** (2), 883–896.

Underwood, C.P. (1993) The application of modular simulation to evaluate HVAC control systems. *Proceedings of the CLIMA 2000 Conference*, London.

Underwood, D.M., Crawford, R.R. (1991) Dynamic nonlinear modelling of a hot-water-to-air heat exchanger for control applications. *ASHRAE Transactions*, **97** (1), 149–155.

Yik, F.W.H., Underwood, C.P., Chow, W.K. (1997) Chilled water cooling and dehumidifying coils with corrugated plate fins: modelling method. *Building Services Engineering Research and Technology*, **18** (1), 47–58.

Zaheer-uddin, M. (1993) Energy start–stop and fluid flow regulated control of multizone HVAC systems. *International Journal of Energy*, **18** (3), 289–302.

Zaheer-uddin, M., Goh, P.A. (1991) Transient response of a closed-loop VAV system. *ASHRAE Transactions*, **97** (2), 378–387.

4 System stability

4.1 Feedback control

Figure 4.1 represents the generalised feedback control system in *block diagram form*. We can now take a look at some of the basic terminology of feedback control.

Note that the plant in this network may contain a space. The *control element* is a positioning device such as a valve, damper or variable-speed drive.

There are two inputs shown, a *set point* or reference condition, *r*, and a *disturbance, d* (in practice there may be a number of disturbances present). The sensor generates a *feedback signal*, ϕ, which is subtracted from the set point value to produce an error signal, ε. This constitutes *negative feedback* therefore. The controller generates a *positioning signal, u*, and the control element results in a *manipulated variable, p*. Correspondingly, a *controlled variable, y*, is maintained.

Transfer functions G_c, G_v and G_p form the *forward path* of the block diagram model whilst the sensor transfer function G_d forms the *feedback path*. In some applications, the response of the sensor is fast compared with other components in the plant and it is then possible to set $G_d = 1$. In these circumstances, a plant with *unity negative feedback* exists.

Each block is represented by a transfer function. If for example each transfer function is expressed as a Laplace transformation, then each block represents the output Laplace variable, divided by the input Laplace variable. Therefore, the relationship between any variable in the system and any 'upstream' variable is simply obtained by multiplying the various intervening block transfer functions together. This in fact is the main motive for expressing dynamic systems in this way – it is flexible and enables the various relationships of interest to be generated by some simple algebraic operations.

Load disturbances

In practice the main (if not the only) job of the control system is to adjust for disturbances – termed *disturbance-rejection* control. In HVAC control, disturbances occur due to combinations of the following:

- *Climate* – temperature, solar, wind-speed, wind direction, humidity.
- *User* – occupancy metabolics, use of machinery and lighting systems.
- *Interaction* – other HVAC plant and subsystem influences when controlled elsewhere (for example, the cycling of on:off controlled boiler plant affecting the flow temperature to a heating coil will induce changes in heating coil load requiring local control action).

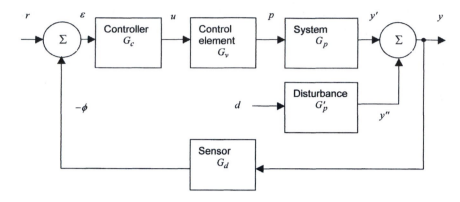

Figure 4.1 The generalised feedback control system.

Historically, the control engineer has often concentrated on set point control rather than on disturbance-rejection control. The argument is that a set point change, essentially a step change, would be the most severe disturbance a plant would need to encounter (load disturbances usually being far less abrupt) – thus design for the worst case.

The closed-loop transfer function

Returning to the general case of Figure 4.1, we can write

$$\varepsilon = r - \phi \tag{4.1}$$

$$u = \varepsilon G_c \tag{4.2}$$

$$p = uG_v \tag{4.3}$$

$$y' = pG_p \tag{4.4}$$

$$y'' = dG'_p \tag{4.5}$$

$$y = y' + y'' \tag{4.6}$$

$$\phi = yG_d \tag{4.7}$$

Ultimately, for most analyses we are primarily interested in the response of the controlled variable (y in our generalised system) to changes in plant inputs (r, d in our generalised system). We therefore eliminate unwanted variables in the above by successive substitution to give

$$y = (r - yG_d)G_cG_vG_p + dG'_p$$

which leads to the following:

$$y = \frac{G_cG_vG_p}{(1 + G_cG_vG_pG_d)}r + \frac{G'_p}{(1 + G_cG_vG_pG_d)}d \tag{4.8}$$

Equation (4.8) confirms that the system can be described by two overall transfer functions – one for each input. Note that the denominator of each output–input relationship will be identical and includes the product of all transfer functions around the loop – called the *open-loop transfer function*. In fact, for a plant with only one controlled variable of interest there will be as many overall transfer functions as there are inputs. The numerators of these transfer functions will in general be the product of the transfer functions in the path formed by the input to the output, and all denominators will be identical. Such systems are sometimes called *multiple-input single-output* (MISO) systems.

For the simplest case in which there is only one output and one input (e.g. set point), we have the *single-input single-output* (SISO) system.

The overall transfer function which relates the controlled variable to the set point is called the *closed-loop transfer function* which can be stated as the forward path transfer function divided by one plus the open-loop transfer function.

Many practical systems involve many inputs and outputs – the *multiple-input multiple-output* (MIMO) case – which we will look at in detail in Chapter 6.

The equality of denominator among these overall transfer functions for a given output in fact turns out to be very significant as we shall see a little later. Meanwhile, let us take a look at how we might deal with the more complex block diagram plant model, which describes many practical systems.

Block diagram algebra

The generalised system of Figure 4.1 does not frequently present itself in this convenient form without some initial effort, especially if the starting point for the model is a set of theoretical expressions. However, the block diagram concept has the very helpful property of being easy to manipulate and re-fashion through simple algebraic operations.

Consider for example the situation of Figure E3.4.2 in Chapter 3. This is a complex open-loop MISO system – no controller or control element exists and the loop has not yet been closed with a feedback path. We need to rationalise it in a form similar to the generalised system of Figure 4.1 if we are to make any further progress with it. If we take a closer look at this system, we see that it consists of two interdependent groups of transfer functions – those representing fabric and those representing the heating plant and space. Both subsystems are linked by the room air temperature, fabric surface temperature and heating plant return water temperature and both share one of the disturbance inputs – the external air temperature.

Consider first the fabric subsystem which can be represented as shown in Figure 4.2 in which the transfer functions occupy the same notional positions as in Figure E3.4.2. A simple set of symbols for the transfer functions is used and we will use the general (non-subscripted) form of notation for convenience. Figure 4.2 results in the following:

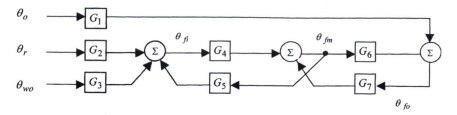

Figure 4.2 Rationalised subsystem of Figure E3.4.2.

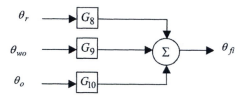

Figure 4.3 Reduced fabric subsystem of Figure 4.2.

$$\theta_{fi} = \theta_r G_2 + \theta_{wo} G_3 + \theta_{fm} G_5 \qquad (4.9)$$

$$\theta_{fm} = \theta_{fi} G_4 + \theta_{fo} G_7 \qquad (4.10)$$

$$\theta_{fo} = \theta_o G_1 + \theta_{fm} G_6 \qquad (4.11)$$

We require θ_{fi} in terms of inputs θ_o, θ_r and θ_{wo} since θ_{fi} forms an input to the plant–space subsystem. We therefore eliminate θ_{fm} and θ_{fo} to give

$$\theta_{fi} = \theta_r G_8 + \theta_{wo} G_9 + \theta_o G_{10} \qquad (4.12)$$

where

$$G_8 = \frac{(G_2 - G_2 G_6 G_7)}{(1 - G_6 G_7 + G_4 G_5)}$$

$$G_9 = \frac{(G_3 - G_3 G_6 G_7)}{(1 - G_6 G_7 + G_4 G_5)}$$

$$G_{10} = \frac{G_1 G_5 G_7}{(1 - G_6 G_7 + G_4 G_5)}$$

The fabric subsystem now reduces to Figure 4.3.

Now consider the plant–space subsystem (Figure 4.4), which again is based on Figure E3.4.2. Figure 4.4 results in the following:

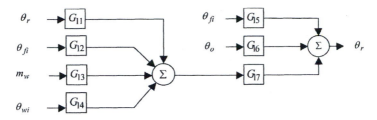

Figure 4.4 System–space subsystem of Figure E3.4.2.

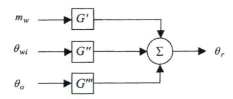

Figure 4.5 Reduced block diagram of Figure E3.4.2.

$$\theta_{wo} = \theta_r G_{11} + \theta_{fi} + m_w G_{13} + \theta_{wi} G_{14} \tag{4.13}$$

$$\theta_r = \theta_{fi} G_{15} + \theta_o G_{16} + \theta_{wo} G_{17} \tag{4.14}$$

Eliminating θ_{fi} and θ_{wo} by combining equations (4.12), (4.13) and (4.14) leads to

$$\theta_r = G' m_w + G'' \theta_{wi} + G''' \theta_o \tag{4.15}$$

where

$$G' = \frac{G_9 G_{13} G_{15} + G_{13} G_{17}}{1 - [G_9 G_{12} + G_8 G_{15} + G_9 G_{11} G_{15} + G_{11} G_{17} + G_8 G_{12} G_{17}]}$$

$$G'' = \frac{G_9 G_{14} G_{15} + G_{14} G_{17}}{1 - [G_9 G_{12} + G_8 G_{15} + G_9 G_{11} G_{15} + G_{11} G_{17} + G_8 G_{12} G_{17}]}$$

$$G''' = \frac{G_{10} G_{15} + G_{16} + G_{10} G_{12} G_{17} + G_9 G_{10} G_{12} G_{15} - G_9 G_{10} G_{12} G_{15} - G_9 G_{12} G_{16}}{1 - [G_9 G_{12} + G_8 G_{15} + G_9 G_{11} G_{15} + G_{11} G_{17} + G_8 G_{12} G_{17}]}$$

Our final fabric–plant–space block diagram model is now Figure 4.5, which represents a considerable rationalisation of the original model. Nevertheless, derivation of G', G'' and G''' from the original transfer functions will

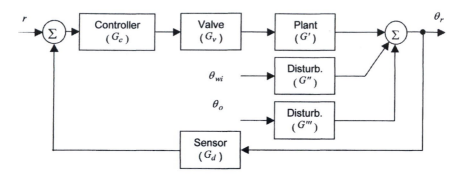

Figure 4.6 Feedback control applied to the rationalised open-loop system of Example E3.4.2.

require substantial algebraic manipulation which for systems of higher order will prove unfeasible – this point will become clearer in a later worked example. Meanwhile, if we complete the rationalised block diagram model of the heating plant example described above by adding a feedback control system which seeks to maintain the room air temperature, θ_r, by varying the heating plant water flow rate, we end up with the system of Figure 4.6. Here θ_{wi} and θ_o form *load disturbances* to the plant (there may of course be other load disturbances had our original model equations included them – such as casual heat gains).

From equation (4.8), an overall closed-loop model for the heating plant and its feedback control system will be

$$\theta_r = \frac{G_c G_v G'}{[1 + G_c G_v G' G_d]} r + \frac{G''}{[1 + G_c G_v G' G_d]} \theta_{wi} + \frac{G'''}{[1 + G_c G_v G' G_d]} \theta_o \qquad (4.16)$$

Now let us take a look at the significance of the overall closed-loop plant model for feedback control. First, what significance does it have as far as the *steady-state* performance of the control system is concerned?

Steady-state control error

Consider the general closed-loop model of equation (4.8) and suppose that we express continuous-time transfer functions in the blocks using Laplace transforms. In the steady state, $s = 0$ and $G_c(s), \ldots, G_d(s) = K_c, \ldots, K_d$ – i.e. the various transfer functions become the values of their steady-state gains. The steady-state closed-loop model will therefore be

$$y_{ss} = \frac{K_{FP}(r)}{1 + K} r_{ss} + \frac{K_{FP}(d)}{1 + K} d_{ss} \qquad (4.17)$$

where y_{ss} = the steady-state value of $\theta_o(s)$, $K_{FP}(r)$ = the forward path gain between set point (r_{ss}) and y_{ss}, K = the open-loop gain and $K_{FP}(d)$ = the forward path gain between load disturbance (d_{ss}) and y_{ss}.

For good control, $r_{ss} - y_{ss} = 0$, but from equation (4.17) we find that

$$r_{ss} - y_{ss} = r_{ss} - \left[\frac{K_{FP}(r)}{1 + K} r_{ss} + \frac{K_{FP}(d)}{1 + K} d_{ss} \right] \tag{4.18}$$

However, we have noted that the gain of a sensor is always expressed as unity in block diagram modelling (section 3.5). Hence $K_{FP}(r) = K$. Equation (4.18) therefore reduces to the following:

$$r_{ss} - y_{ss} = \frac{r_{ss} - K_{FP}(d) \times d_{ss}}{1 + K} \tag{4.19}$$

i.e. a *finite* value for $r_{ss} - y_{ss}$ is evident whilst good control should of course ensure that it is zero at all times. Equation (4.19) implies that unless K is infinity, then in the eventual steady-state conditions following some disturbance or change in set point there will be a difference between what we want, r_{ss}, and what we get, y_{ss}. This amounts to a steady-state error condition by the feedback control system, more commonly referred to as *offset*.

Two points are worth noting. Firstly, since the set point in most practical systems is rarely changed, or certainly not changed frequently, we can assume for most practical purposes that $r_{ss} = 0$ in equation (4.19). Hence for a given K, offset depends mainly on the *load disturbance(s)* that the plant has to compensate for. Thus, if we denote offset by Δy_{ss}, equation (4.19) reduces to

$$\Delta y_{ss} = \frac{K_{FP}(d) \times d_{ss}}{1 + K} \tag{4.20}$$

Secondly, we can eliminate offset completely by ensuring that $K = \infty$. Many early HVAC controllers were purely proportional in which case $G_c = K_c$ and the latter condition could therefore only be achieved by setting $K_c = \infty$. In many practical instances, this could not be done because unstable operation would result as we shall see later. This is in fact why offset was endemic in early (pre-digital) HVAC control systems. However, the general PID controller (depicted by equation (3.45)) introduces an *integrator* ($K_c/i_t s$) in the controller transfer function which affords us the condition $K = \infty$.

Thus for *accuracy* in feedback control, an integral term in the controller or control algorithm will always be necessary unless we have the rare freedom of being able to fix a high controller gain within the limits of stable plant operation.

Example 4.1

Suppose the heating plant of Figure 4.6 has a proportional controller and a controller with a gain of $3.428 \text{kgs}^{-1} (\%)^{-1}$. For controller gains of $10\% \text{K}^{-1}$ and $50\% \text{K}^{-1}$ what will be the steady-state offset of the controlled variable, θ_r, if (a) the heating water inlet temperature changes by 5K, and (b) the external temperature changes by 5K.

Solution

The open-loop gain, K, will be the product of the gains due to the controller, K_c, valve, K_v, and plant, K'. We also note that the sensor gain, K_d, will be unity. Hence

$$K = K_c K_v K'$$

Based on equation (4.15),

$$K' = \frac{K_9 K_{13} K_{15} + K_{13} K_{17}}{1 - [K_9 K_{12} + K_8 K_{15} + K_9 K_{11} K_{15} + K_{11} K_{17} + K_8 K_{12} K_{17}]} \qquad \text{(E4.1.1)}$$

Referring to Figure E3.4.2, we evaluate the various transfer function gains for equation (E4.1.1) by setting $s = 0$. For example,

$$K_8 = \frac{K_2 - K_2 K_6 K_7}{1 - K_6 K_7 + K_4 K_5} = \frac{0.348 - 0.348 \times 0.209 \times 0.171}{1 - 0.209 \times 0.171 + 0.829 \times 0.071} = 0.328$$

Similarly, $K_9 = 0.517$ which leads to $K' = 0.0179$. Therefore

$$K|_{K_c = 10\% K^{-1}} = 10 \times K_v K' = 10 \times 3.428 \times 0.0179 = 0.614$$

and

$$K|_{K_c = 50\% K^{-1}} = 50 \times 3.428 \times 0.0179 = 3.07$$

Now, for the load disturbance path gains, $K_{FP(\theta_{oss})} = K'''$ (based on equation (4.15)), and this can be calculated to give 0.0094. Similarly, $K_{FP(\theta_{wiss})} = K''$ = 0.606.

Based on equation (4.20) we can now calculate the offset,

$$\Delta\theta_{rss}|_{\theta_{wiss}} = \frac{K_{FP}(\theta_{wiss}) \times \theta_{wiss}}{1 + K}$$

$$= \frac{0.606 \times 5}{1 + 0.614} = 1.88 \text{K} \qquad \text{when } K_c = 10\% \text{K}^{-1}$$

and

$$\Delta\theta_{rss}|_{\theta_{wiss}} = \frac{0.606 \times 5}{1 + 3.07} = 0.744 \text{K} \qquad \text{when } K_c = 50\% \text{K}^{-1}$$

Similarly, for a disturbance in the external temperature, θ_o, the respective results are 0.85K and 0.34K.

In linear control system analysis, principles of superposition apply such that the offset arising from simultaneous disturbances in both inputs can be obtained simply by adding the individual results.

We see from these results that the offset is inversely proportional to the controller gain and that the external temperature in this example is less influential than the heating plant inlet flow water temperature for disturbances of the same magnitude. In fact, some of the less resolute temperature sensors are capable of detecting temperature changes no smaller than ±1K and on this basis the offset results at the higher controller gain may well be tolerable.

In many cases however, we see from this example just how unacceptable proportional control on its own can be and that some form of integral action, when correctly tuned, provides us with a remedy. We will come back to the question of controller design in later sections and chapters, but in the meantime, let us take a look at the first priority in control system design – that of achieving stability.

Conditions for plant stability

In a control system, *stability* is the most important design consideration since an unstable system will achieve none of its fundamental objectives and, in some applications, may be unsafe. In general, a stable system is one in which, for any set of inputs, the controlled variable remains within finite bounds. System stability depends on the *system characteristic equation*, which is

$$1 + G = 0 \tag{4.21}$$

where G is the system *open-loop transfer function* ($G_c G_v G_p G_d$ in the generalised system of Figure 4.1).

If we examine the generalised system closed-loop transfer function of equation (4.8), we see that the characteristic equation is the denominator of the closed-loop transfer function equated to zero. We also see from this equation (and for that matter the more complex specific example of equation (4.16)) that the characteristic equation for a system is unique irrespective of the number of inputs.

This is a very important point as far as stability is concerned for it means that the intrinsic stability of a system is independent of the number and nature of load disturbance inputs (provided that the open-loop path transfer functions are themselves stable). Given this precondition, we need not consider any load disturbance path when assessing stability, we need merely consider the basic block diagram model of the SISO case with set point as input.

Roots (i.e. possible solutions) of the characteristic equation (*CE*) can be mapped onto a plane of the Laplace variable, s, since s is a complex variable, i.e. $s = x + j\omega$ (Figure 4.7). Roots of *CE*(s) that lie in the left-hand region of the s-plane are *stable* whilst roots that lie in the right-hand region are *unstable*. A condition of transitionary or *marginal stability* exists when the roots lie on the imaginary axis. Here, a system response to a step change in input will oscillate in a purely sinusoidal manner at some natural undamped frequency ω_n, hence this condition is sometimes referred to as an undamped response. Positive damping exists for roots that lie in the left-hand region and negative damping for roots that lie in the right-hand region. Kuo (1975) gives a good description of these conditions for a hypothetical second-order system.

We will now move on to take a look at two simple methods for identifying roots of the characteristic equation on the s-plane and in so doing establish stable and unstable system performance criteria.

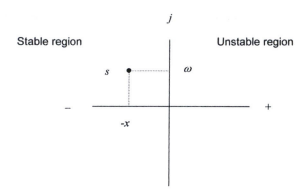

Figure 4.7 The s-plane.

4.2 Stability tests

Root locus

This method was developed by Evans (1948) and is still frequently used today in control system design. It is a graphical method for expressing the roots of the system characteristic equation as the open-loop gain, K, increases. The merit of the method is that degrees of stability can be analysed (i.e. a damping criterion can be assigned and the resulting system gain obtained).

A comprehensive application of the method can be found in a number of sources, see for example Houpis & Lamont (1992). A brief summary of the main rules for generating a root locus plot is given below.

1   *Step 1.* The system open-loop model is expressed in factorised or *pole-zero* form

$$G(s) = \frac{K(s + a)(s + b)\dots}{(s + x)(s + y)\dots} \tag{4.22}$$

From this, *zeros* are values of s which make the open-loop numerator equal to zero ($-a$, $-b$, etc., in the above), and *poles* are assigned values of s which make the open-loop denominator equal to zero (in the above, $-x$, $-y$, etc.). Note that the poles and zeros can have real and imaginary parts.

2   *Step 2.* Poles and zeros are plotted on real and imaginary axes. For the simpler models which yield real factors, all poles and zeros will therefore lie on the real axis. The number of root locus branches is equal to the number of open-loop poles, n_p. Root locus branches *start at open-loop poles and end either at an open-loop zero or infinity*. A locus branch will lie on the real axis when the number of poles plus zeros to the right of a point on the real axis is odd, and will not exist when even. Thus, any locus branches on the real axis can now be drawn in.

3   *Step 3*. Root loci that approach infinity do so along *asymptotes*. There will
    be $n_p - n_z$ asymptotes (where n_z is the number of open-loop zeros)
    which radiate from a centre of gravity, g_{pz}, of poles and zeros given by

$$g_{pz} = \frac{\sum_1^{n_p} v_p - \sum_1^{n_z} v_z}{n_p - n_z} \qquad (4.23)$$

where v_p and v_z are the values of poles and zeros respectively. Asymptotes leaving the imaginary axis make angles of $\pi(2\alpha + 1)/(n_p - n_z)$,
where $\alpha = 0, 1, 2, \ldots, (n_p - n_z - 1)$.

4   *Step 4*. Loci moving towards adjacent zeros which lie on the real axis will
    enter the real axis at a point between these zeros. Conversely, loci
    moving from adjacent poles will leave the real axis at a point between
    these poles. The point of entry or exit from the real axis is at p_e,

$$\sum_1^{n_p} \frac{1}{p_e - v_p} = \sum_1^{n_z} \frac{1}{p_e - v_z} \qquad (4.24)$$

As far as stability is concerned, loci to the left of the imaginary axis
represent damped or critically damped regions of the characteristic
equation. As the open-loop gain is increased, the loci transfer to the
right-hand side of the imaginary axis in which region the system is
unstable. Hence a point on the imaginary axis at which a locus branch
passes represents a critical value of the open-loop gain at which the
system will be at marginal stability. The value of gain at this point can be
estimated from the following:

$$\frac{K|s - a| \times |s - b| \times \ldots}{|s - x| \times |s - y| \times \ldots} = 1 \qquad (4.25)$$

where $|s - a|$, $|s - b|$, etc., are the vectors from zeros to the point of
interest on the locus, and $|s - x|$, $|s - y|$, etc., are the vectors from poles
to the point of interest. These vectors can be easily measured using a
ruler with any scale, provided consistent.

In practice, the construction of root locus plots for high-order system
models (particularly those containing complex poles/zeros) can be very
complicated and a computer program will always tend to be the best resort.
A clear understanding of the above procedures is however helpful both for
rough sketching of the root locus as well as for checking and verifying
computer-generated results.

The Routh stability criterion

This simple method tells us whether a system will be stable or not – it does
not tell us by how much (Routh, 1877). Also, complex system models, such
as those depicted by Example 3.4.2, become unmanageable by manual
calculation of this type. The procedure is summarised as follows.

1   *Step 1.* Express the system characteristic equation, $CE(s)$, as a polynomial in s:

$$a_n s^n + a_{n-1} s^{n-1} + a_{n-2} s^{n-2} + \ldots + a_1 s + a_0 = 0 \qquad (4.26)$$

2   *Step 2.* Form the first two rows of an array. The first row consists of the highest-order coefficient of s (a_n) and alternate coefficients in the series, until a zero is encountered. The second row is formed from the second-highest-order coefficient of s (a_{n-1}) and alternate coefficients until a zero is encountered. Thus

first row	a_n	a_{n-2}	a_{n-4} $\ldots$	0
second row	a_{n-1}	a_{n-3}	a_{n-5} $\ldots$	0

3   *Step 3.* Subsequent rows are formed by a diagonal multiplication operation of 2×2 matrices formed by the two rows immediately preceding (a procedure similar to finding a determinant of the second order):

third row
$$\frac{a_{n-1} a_{n-2} - a_n a_{n-3}}{a_{n-1}} \qquad \frac{a_{n-1} a_{n-4} - a_n a_{n-5}}{a_{n-1}} \qquad \ldots \qquad 0$$
$$(= b) \qquad\qquad\qquad (= c)$$

and, similarly

fourth row
$$\frac{b a_{n-3} - a_{n-1} c}{b} \qquad \frac{b a_{n-5} - a_{n-1} d}{b} \qquad \ldots \qquad 0$$

The array is complete when a full row of zeros is encountered; there will be $n + 1$ rows containing non-zero elements so the array is complete after a few rows for most practical cases.

4   *Step 4.* Interpretation is now possible. Examine the *first column* of the array. If all elements are positive, all roots of the characteristic equation will lie in the left half of the s-plane (i.e. the system will be *stable*). Conversely, if one or more elements in the first column are negative, one or more roots of the characteristic equation will lie in the right half of the s-plane and the system will be unstable. The number of sign changes in the first column is equal to the number of roots with a positive real part which lie in the right half of the s-plane. Note that if a negative coefficient exists as the highest-order or second-highest-order coefficient of s in the characteristic equation (equation (4.26)), then the system will definitely be unstable and there is no need to proceed with the array formation.

A Routh array can be used in two ways:

*   The overall system gain (including the controller gain) is known. The Routh array can then be used to determine whether or not the system is stable.
*   The overall system gain (or the controller gain part of it) is not known. The Routh array can be used to find limits of stable operation for the system – that is, the value of the overall system gain at *marginal stability.*

Example 4.2 illustrates these applications of the Routh array.

Example 4.2

A valve, system and detector can each be represented by first-order lag transfer functions with time constants of 0.5 min, 1.5 min and 0.3 min respectively. The overall gain of these components is 10 and a proportional controller is used with a gain also of 10.

Establish whether the system will be stable according to Routh's criterion.

Solution

The open-loop transfer function is

$$G(s) = \frac{10 \times 10}{(0.5s + 1)(1.5s + 1)(0.3s + 1)}$$

and the characteristic equation is

$$1 + \frac{100}{(0.5s + 1)(1.5s + 1)(0.3s + 1)} = 0$$

which results in the polynomial

$$0.225s^3 + 1.35s^2 + 2.3s + 101 = 0$$

The Routh array will be

first row	0.225	2.3	0
second row	1.35	101	0
third row	$\dfrac{1.35 \times 2.3 - 0.225 \times 101}{1.35}$ $= -14.53$	$\dfrac{1.35 \times 0 - 0.225 \times 0}{1.35}$ $= 0$	
fourth row	$\dfrac{-14.53 \times 0 - 1.35 \times 0}{-14.53}$ $= 101$	$\dfrac{-14.53 \times 0 - 1.35 \times 0}{-14.53}$ $= 0$	
fifth row	$\dfrac{101 \times 0 - (-14.53) \times 0}{101}$ $= 0$		

This system will therefore be unstable, since the first column contains a negative element at row 3 (two sign changes in the first column – from positive to negative and back again – implies that there will be two roots of the characteristic equation in the right half of the s-plane).

Clearly the overall system gain (including the controller) is too high. Our only line of action is of course through the controller gain since we can adjust it. We can now use the Routh array to determine what value the controller gain will be at the limit of stability. Let the controller gain be K_c and the characteristic equation becomes

$$1 + \frac{10K_c}{(0.5s + 1)(1.5s + 1)(0.3s + 1)} = 0$$

This leads to the polynomial

$$0.225s^3 + 1.35s^2 + 2.3s + (1 + 10K_c) = 0$$

and the Routh array will be

first row	0.225	2.3	0
second row	1.35	$(1 + 10K_c)$	0
third row	$\dfrac{1.35 \times 2.3 - 0.225(1 + 10K_c)}{1.35}$ $= (2.13 - 1.67K_c)$	0	
fourth row	$\dfrac{(2.13 - 1.67K_c)(1 + 10K_c) - 1.35 \times 0}{(2.13 - 1.67K_c)}$ $= (1 + 10K_c)$	0	
fifth row	0		

As a generality, K_c itself cannot be negative (though there are some instances where it may be by convention). We can therefore discount row 4 column 1 which leaves only one stability constraint at row 3 column 1. For stable operation,

$$2.13 - 1.67K_c \geq 0 \qquad K_c \leq \frac{2.13}{1.67} = 1.28$$

Marginal stability will occur when the controller gain is at 1.28. The choice of practical controller gain will therefore need to be less than this. One rule of thumb (see section 7.3) is to fix the practical gain setting at 50% of its value at marginal stability – so in this case we might expect satisfactory system operation with a controller gain setting of 0.64.

One further property of the system is its natural or undamped frequency, ω_n, which can be helpful for 'tuning' integral and derivative controller terms. Noting that s is a complex variable which will lie on the imaginary axis at $s = j\omega_n$ we can express our characteristic equation at the point of marginal stability where $K_c = 1.28$, as follows:

$$0.225s^3 + 1.35s^2 + 2.3s + (1 + 10 \times 1.28) = 0.225s^3 + 1.35s^2 + 2.3s + 13.8 = 0$$

$$-0.225j\omega_n^3 - 1.35\omega_n^2 + 2.3j\omega_n + 13.8 = 0 \tag{E4.2.1}$$

We can now determine the critical frequency from either the real part or the imaginary part of equation (E4.2.1). Taking the real part for instance,

$$-1.35\omega_n^2 + 13.8 = 0 \qquad \omega_n = \sqrt{\frac{-13.8}{-1.35}} = 3.197 \, \text{radmin}^{-1}$$

At marginal stability, the system will exhibit a sinusoidal response with a frequency of 3.197 radians per minute, or a period of $2\pi/3.197 = 1.97 \, \text{min}$.

It is now a good time to take a look at some of the more practical applications of these methods in order to explore the influence of both system and controller characteristics on control system stability. We will do this through a series of worked examples.

Example 4.3

The cooling coil of Example 3.5 (Chapter 3) is to be controlled using a three-port diverting valve which maintains the coil leaving air temperature by varying the chilled water flow rate (i.e. case (a) in Example 3.5 applies). The control loop is closed by a resistance temperature detector mounted close to the coil in the supply duct. The control valve can be represented by a first-order lag with a time constant of one minute. If only the proportional term of a three-term controller is to be used, find the limit of controller gain for satisfactory operation of the control loop using both a root locus and a Routh array.

Solution

The plant is configured for control as shown in Figure E4.3.1. We recall the cooling coil transfer function from Example 3.5,

$$\frac{\theta_{ao}(s)}{m_w(s)} = \frac{-8.54}{(13.26s + 1)}$$

the negative gain signifying a *cooling* application in which an increase in chilled water mass flow rate, $m_w(s)$, results in a *decrease* in the coil leaving air temperature, $\theta_{ao}(s)$.

In this example, to avoid some unmanageably large numbers, it will be convenient to use a timebase in minutes and to adopt a control signal range in the interval $0 \rightarrow 1$. Note that the control signal range can be in any units for these problems provided that the units applied to the receiving element (a valve in this case) are consistent. The gain in the above presently has the units $K/(kgs^{-1})$ and therefore requires, along with the time constant, to be

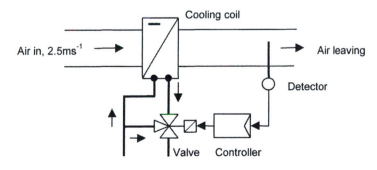

Figure E4.3.1 Control system.

divided by 60 to produce our desired system of units to a timebase in minutes, i.e.

$$\frac{\theta_{ao}(s)}{m_w(s)} = \frac{-0.142}{(0.221s + 1)}$$

Now, assuming linear control over a full range of operation of the valve and coil and noting that the flow range of the valve will be 0–$1.905\mathrm{kgs}^{-1} = 0$–$114.3\mathrm{kg\,min}^{-1}$, we can write down the transfer function for the valve with a 1 min time constant, receiving a control signal, $u(s)$, in the range 0–1,

$$\frac{m_w(s)}{u(s)} = \frac{114.3}{(s + 1)}$$

Hence the transfer function for the plant, consisting of the coil and valve, is

$$G_p(s) = \frac{\theta_{ao}(s)}{u(s)} = \frac{-16.3}{(s + 1)(0.221s + 1)} = \frac{-16.3}{(0.221s^2 + 1.221s + 1)}$$

For the duct-mounted temperature detector, the air velocity in the duct is $2.5\,\mathrm{ms}^{-1}$ according to Example 3.5. Using equations (3.38) and (3.39) for a resistance temperature detector (RTD), the time constant and dead time will be

$$\tau_d = 198.9 - 95.4 \times 2.5 + 23.16 \times 2.5^2 - 2.5 \times 2.5^3 + 0.097 \times 2.5^4$$
$$= 69.9\mathrm{s} = 1.165\mathrm{min}$$

$$t_{dd} = 21.69 - 8.518 \times 2.5 + 1.942 \times 2.5^2 - 0.1974 \times 2.5^3 + 0.0073 \times 2.5^4$$
$$= 9.73\mathrm{s} = 0.162\mathrm{min}$$

which results in the detector transfer function

$$G_d(s) = \frac{\phi(s)}{\theta_{oa}(s)} = \frac{\exp(-0.162s)}{(1.165s + 1)}$$

Noting that the transfer function for a purely proportional controller is $G_c(s) = K_c$ (equation (3.45), where $i_t = \infty$ and $d_t = 0$), we can now draw the SISO block diagram for the plant (Figure E4.3.2).

The characteristic equation can now be written. Note however that we need to express the dead-time term of the detector transfer function in a more convenient form if we are to proceed further. The exponential function will not allow pole–zero formation in a root locus plot, nor will it allow the essential full set of linear coefficients of s needed in a Routh array. We can deal with this by using a Padé approximation for the dead-time term – as described in section 3.5 – and our open-loop transfer function will be

$$G(s) = \frac{16.3K_c(0.081s - 1)}{(s + 1)(0.221s + 1)(1.165s + 1)(0.081s + 1)} \tag{E4.3.1}$$

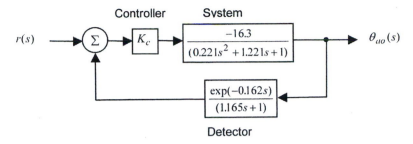

Figure E4.3.2 Block diagram model for cooling coil control.

The characteristic equation $1 + G(s) = 0$ will become

$$(s + 1)(0.221s + 1)(1.165s + 1)(0.081s + 1) + 16.3K_c(0.081s + 1) = 0$$

In polynomial form,

$$0.021s^4 + 0.391s^3 + 1.836s^2 + (2.467 + 1.32K_c)s + (1 - 16.3K_c) = 0 \qquad (E4.3.2)$$

Root locus stability

Expressing the open-loop transfer function of equation (E4.3.1) in zero–pole form,

$$G(s) = \frac{16.3 \times 0.081K_c}{(0.221 \times 1.165 \times 0.081)} \times \frac{(s - 12.346)}{(s + 1)(s + 1/0.221)(s + 1/1.165)(s + 1/0.081)}$$

and putting $K_c = 1$ for the time being, the zero–pole form of the system open-loop transfer function is

$$G(s) = 63.31 \times \frac{(s - 12.346)}{(s + 1)(s + 4.525)(s + 0.858)(s + 12.346)} \qquad (E4.3.3)$$

Hence we have one zero at 12.346, four poles at -1, -4.525, -0.858 and -12.346, and an open-loop gain of 63.31.

For convenience, we can plot this root locus by computer. Here, we will use the MATLAB control toolbox (MATLAB, 1992). The following commands produce the plot – note that MATLAB will require the polynomial form of the open-loop transfer function (equation (E4.3.1)) from which it will deduce the equivalent zero–pole–gain form. In MATLAB,

```
>num=[0,0,0,1.32,-16.3];
>den=[0.021,0.391,1.836,2.467,1];
>rlocus(num,den)
```

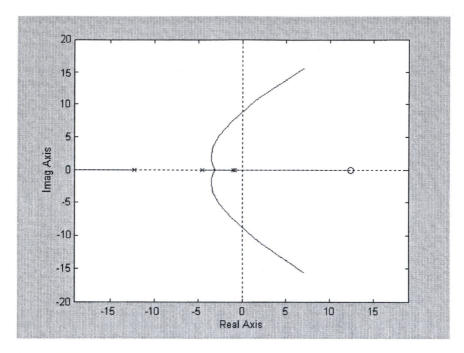

Figure E4.3.3 Root locus plot.

Note that the zero–pole–gain form of the above can be expressed in MATLAB by issuing the transfer function to zero–pole (tf2zp) command >[z,p,k]=tf2zp(num,den). The rlocus command results in the root locus plot of Figure E4.3.3.

Interpretation is now possible. The three poles are all negative and therefore fall on the real axis to the left of the imaginary axis, whilst the positive zero lies on the right-hand side. Two poles to the right are very close together. One of these sends a locus branch which terminates at the zero – noting that where the total number of poles and zeros to the right of a point is even, a locus will not exist on the real axis, whilst it does exist when the number of poles and zeros is odd. Hence the branch from the left-most pole heads to infinity along the real axis to the left of this pole, whilst the middle two poles form branches which meet and then head to infinity via asymptotes, eventually crossing the imaginary axis. Note that the pole which sends a branch to the zero along the real axis is a *dominant pole* since this will very quickly influence stability upon an increase in gain.

We can conveniently find the values of gain at marginal stability if we issue the command

```
>rlocfind(num,den)
```

A crosshair is placed in the plot window which can be used to locate the values of open-loop gains at any point on the loci. The results lead to maximum values for K_c at 0.069 and 92.42. Clearly, the first value is critical and the second value academic. Indeed we can verify this result using equation (4.25) based on the values of poles and zeros themselves since they all lie on the real axis at which point the critical gain lies,

$$\frac{12.346K}{1 \times 4.525 \times 0.858 \times 12.346} = 1 \quad \Rightarrow \quad K = 3.88$$

And, since the overall open-loop gain is 63.31 (equation (E4.3.3)), then the controller gain $K_c = 3.88/63.31 = 0.061$ at this critical condition. The *practical* controller gain must be less than this value (implying a negative value) to ensure stable operation of the loop, but we shall come back to this later.

Routh stability criterion

From the polynomial form of the system characteristic equation (equation (E4.3.2)), we can form the Routh array as follows:

first row	0.021	1.836	$(1 - 16.3K_c)$ 0
second row	0.391	$(2.467 + 1.32K_c)$	0
third row	$(1.704 - 0.0709K_c)$	$(1 - 16.3K_c)$	0
fourth row	$\dfrac{-0.0935K_c^2 + 8.447K_c + 3.813}{1.704 - 0.0709K_c}$	0	
fifth row	$(1 - 16.3K_c)$	0	
sixth row	0		

Stability constraints therefore lie in the first column at rows 3, 4 and 5.

- At first column, row 3: $1.704 - 0.0709K_c \geq 0$ for stability, so $K_c \leq 24$.
- At first column, row 4: two roots are evident from this quadratic element. Solving the quadratic leads to two further criteria at $K_c \geq 90.79$ and $K_c \geq -0.45$. Owing to the constraint at row 3, we can clearly discard the higher of these two values.
- At first column, row 5: $1 - 16.3K_c \geq 0$, hence $K_c \leq 0.061$, which clearly discards the condition given at row 3.

The Routh array has led us to the additional criterion for stable operation at $K_c \geq -0.45$. In fact, according to the Routh criterion, we must specify $-0.45 \leq K_c \leq 0.061$, whilst according to the root locus we must merely specify $K_c \leq 0.069$. We have already noted that the controller gain is not in general negative but may be by convention. However, this is a special case in which, upon an increase in controlled variable value, we require to *increase* the control signal (i.e. the flow rate of chilled water to the coil). Similarly, a decrease in controlled variable value needs to be met with reduced control effect; the bias of control action is opposite that of the conventional case and a negative controller gain becomes appropriate.

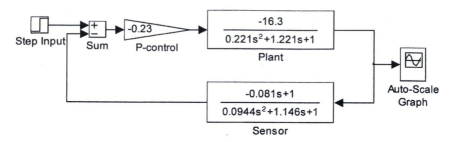

Figure E4.3.4 `Simulink` model for closed-loop coil control.

Hence the negative limit suggested by the Routh array would seem to be the correct way ahead. The best way to test this is through a simulation of the closed-loop system.

Choice of practical controller gain

We now know what the limit of controller gain needs to be for our coil control loop, but what value of practical controller gain setting do we need to specify? It must be some value within the limit. There are various criteria that can be used – such as the use of a *damping factor* or, for root locus plots, a *damping angle*. This is the angle made by a line radiating from the origin of the root locus plot which will intersect with a root locus branch at a point at which a gain value giving suitable damping can be found from equation (4.25). A typical practical value is 30°.

If computer simulation can be used to verify the appropriate value of gain, then trial-and-error search for a suitable value can be done. For proportional control, we fix the practical gain at one-half of the critical value (this and other control 'tuning' criteria will be discussed in detail in Chapter 7). On this basis, using the results from the Routh array, we might try a practical controller gain setting of $0.5 \times (-0.45) = -0.23\mathrm{K}^{-1}$, giving the required negative controller gain for this application.

Using `Simulink` (Simulink, 1996), a simple block diagram model for the system employing this controller setting is shown in Figure E4.3.4. Results of a unit step response in set point are shown in Figure E4.3.5 which suggests that the 50% criterion for the controller gain seems reasonable in that there is sufficient damping to assure eventual stability. There is however unacceptable offset, which arises from the use of purely proportional control in a plant which has low innate damping. This underlines the need for integral control action with this type of plant.

Example 4.3 demonstrated the use of both root locus and Routh array techniques for searching for marginally stable values of open-loop system gain. We were able to use this knowledge to establish a satisfactory value of proportional controller gain which we tested using computer simulation.

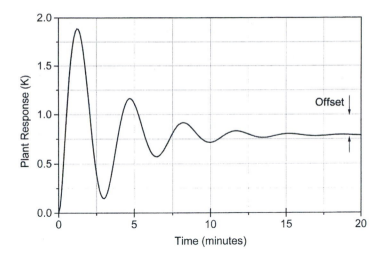

Figure E4.3.5 Step response for closed-loop coil control.

We also found that the Routh array led us conveniently to all possible values of critical gain for this unusual case of cooling coil control, whilst the root locus did not. In fact, in many cases where the algebra does not get to be too complicated (i.e. up to fourth- or fifth-order systems), the Routh array will often prove the most convenient tool to use for establishing limits for the open-loop system (and, hence, controller) gain.

Moving on a little, Example 4.3 represented a local HVAC control loop such as might occur as part of an air handling sequence controller. Though very common in practice, many HVAC control applications also incorporate a space at some point. We already know from Example 3.4 that the space can include fabric elements with at least one third-order transfer function in addition to the plant itself. It should be immediately clear that, even for the relatively simple techniques of SISO system stability analysis, the algebra involved in room-coupled situations will become unmanageable, certainly for manual calculations. Example 4.4 looks at a useful short-cut for such cases.

Example 4.4

Suppose the cooling coil control loop of Example 4.3 provides air conditioning for the space of Example 3.4. Neglecting the effect of the fabric in the space, assess the range of expected stable operation for the control system in terms of controller gain, using Routh's criterion. By means of computer simulation of the closed-loop system, assess how valid this simplified model is in relation to a model that includes the effect of the fabric. Assume that the temperature sensor is mounted in the return air duct from the space, at the same conditions of local air velocity as in Example 4.3.

Solution

Note firstly that this simplified room modelling case represents the *air capacity room model* described in section 3.1.

The appropriate room model for this case is 'case 1' as identified in section 3.4, since the plant is decoupled from the space. Hence equation (3.28) applies and, since the effect of the fabric is to be neglected, $\bar{\theta}_f(s) = 0$ and the fabric term in equation (3.28) disappears. Furthermore, we know that load disturbances do not affect inherent system stability (section 4.1) which allows us to fix $m_a(s) = 0$ and $\theta_o(s) = 0$ and these corresponding terms also disappear in equation (3.28).

Our 'room' transfer function is, simply,

$$G_r(s) = \frac{\theta_r(s)}{\theta_{ao}(s)} = \frac{K_{\theta_{ao}}}{(\tau_r s + 1)}$$

For the simplified 'room time constant', and based on the data in Example 3.4,

$$\tau_r = \frac{V_r \rho_a c_{pa}}{\left[m_{ass} c_{pa} + \sum(AU_i) + n_v V_r / 3 \right]}$$

$$= \frac{12 \times 7 \times 3 \times 1.2 \times 1.025}{\left[2 \times 1.2 \times 1.025 + 0.13 + 0.5 \times 12 \times 7 \times 3 \times 10^{-3} / 3 \right]} = 117.7s$$

Keeping to our timebase of minutes, the 'room time constant' therefore is 1.96 min. The simplified room gain will be

$$K_{\theta_{ao}} = \frac{m_{ass} c_{pa}}{\left[m_{ass} c_{pa} + \sum(AU_i) + n_v V_r / 3 \right]}$$

$$= \frac{2 \times 1.2 \times 1.025}{\left[2 \times 1.2 \times 1.025 + 0.13 + 0.5 \times 12 \times 7 \times 3 \times 10^{-3} / 3 \right]} = 0.935$$

This leads to the room transfer function

$$G_r(s) = \frac{0.935}{(1.96s + 1)}$$

Therefore, the open-loop model for this case will be $G(s) = G_r(s) G_p(s)$ and $G_p(s)$ is as in Example 4.3 (note that the temperature sensor transfer function in the latter is unchanged also, since it experiences the same in-duct air velocity as before). Therefore

$$G(s) = \frac{16.3 K_c (0.081s - 1)}{(s + 1)(0.221s + 1)(1.165s + 1)(0.081s + 1)} \times \frac{0.935}{(1.96s + 1)}$$

which leads to the characteristic equation, expressed as a fifth-order polynomial,

$$0.041s^5 + 0.788s^4 + 4s^3 + 6.679s^2 + (4.43 + 1.235 K_c)s + (1 - 15.24 K_c) = 0$$

Forming the Routh array:

first row	0.041	4	$(4.43 + 1.235K_c)$	0
second row	0.788	6.679	$(1 - 15.24K_c)$	0
third row	3.653	$(4.378 + 2.028K_c)$	0	
fourth row	$(5.735 - 0.438K_c)$	$(1 - 15.24K_c)$	0	

fifth row $\dfrac{-0.888K_c^2 + 65.39K_c}{5.735 - 0.438K_c}$

$+ \dfrac{21.455}{5.735 - 0.438K_c}$ 0

sixth row $(1 - 15.24K_c)$ 0

seventh row 0

The Routh array identifies stability constraints at column 1, rows 4, 5 and 6.

- At row 4: $5.735 - 0.438K_c \geq 0$, so $K_c \leq 13.09\text{K}^{-1}$.
- At row 5: $(-0.888K_c^2 + 65.39K_c + 21.455)/(5.735 - 0.438K_c) \geq 0$, so $K_c \geq 73.99\text{K}^{-1}$ or $K_c \geq -0.31\text{K}^{-1}$.
- At row 6: $1 - 15.24K_c \geq 0$, so $K_c \leq 0.066\text{K}^{-1}$.

By the same reasoning as before, the criterion for stable operation will therefore be $-0.31 \leq K_c \leq 0.066$ (i.e. for practical purposes $-0.31 \leq K_c \leq 0$). Thus adopting the previous 'one-half gain rule' we would adopt a controller gain value of -0.15K^{-1} (once again with the minus sign to account for the cooling case).

Now how acceptable will this value be and how adequate is our assumption that fabric elements in the room transfer function be neglected? Taking the various blocks from Figure E3.4.1 (which is the convective case of Example 3.4) and adding the controller, control valve and closing the control loop with the temperature detector gives us the Simulink block diagram model for a unit step change in set point shown in Figure E4.4.1. Note that the timebase in this model has been converted to *hours*, for convenience. Figure E4.4.2 shows the case with the simple room air capacity model, neglecting fabric effect (the timebase has been kept as *minutes* for this case).

Both room and plant control system models adopt the value of controller gain calculated above. A comparative response to a unit step change (increase) in set point from the two model forms is given in Figure E4.4.3.

Figure E4.4.3 shows that the responses from both models are very similar – essentially the fabric response is itself so highly damped that it makes virtually no contribution to the room air temperature response for the short period of interest. Indeed, for most control system stability studies of this type, in which the time constants of plant components are a few minutes at the most, stability is determined entirely by the response behaviour over a very short period of time. The use of a simple room air capacity model is perfectly adequate therefore. Only in cases where control system

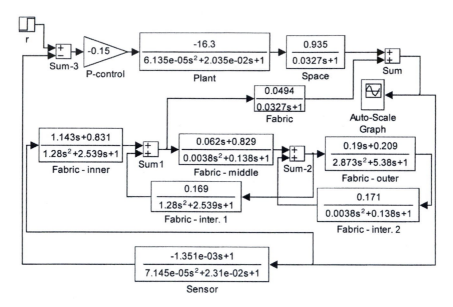

Figure E4.4.1 Simulink comprehensive room and plant model.

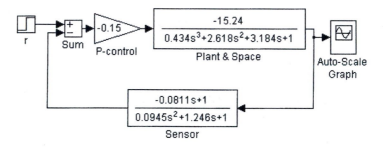

Figure E4.4.2 Simulink simple room and plant model.

performance over a long period of time (several hours) or where the response of plant with high thermal capacity (e.g. underfloor heating) is of interest, need the detailed room model be expressed. Note once again that the offset in this example is unacceptably high – we will look at remedies for this in later chapters.

In Example 4.4, we have confirmed that, for control stability and short-term response analysis, a simple air capacity model will suffice and the fabric elements can be neglected leading to a dramatically simplified characteristic equation from which the analysis can be conducted.

Now let us take a look at one final example involving stability analysis – the case of space heating in which the room and plant are intrinsically coupled.

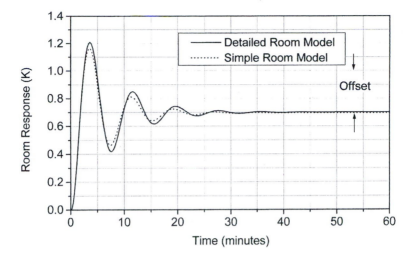

Figure E4.4.3 Comparative response for detailed and simple room models.

Example 4.5

For the convector heating case of Example 3.4, add a control valve with a time constant of one minute, a proportional controller and a room temperature sensor of the resistance wire type; thereby forming a feedback control loop. Assume that the mean air velocity passing the temperature sensor is $0.1 ms^{-1}$ and use a simplified air capacity transfer function description for the space. Determine the value of controller gain at which a reasonable response to a step change in control set point might be expected.

Solution

To determine the controller gain through a stability analysis, we can neglect the boundary conditions $(\theta_{wi}(s), \theta_o(s)$ – see Figure E3.4.1). Hence based on stage (c) of Example 3.4, we can write down transfer functions for the heating system (note that a timebase of minutes has been adopted in this case):

$$G_h(s) \quad \Rightarrow \quad \frac{\theta_{wo}(s)}{m_w(s)} = \frac{1.38}{(6.84s + 1)} \tag{E4.5.1}$$

$$G_h'(s) \quad \Rightarrow \quad \frac{\theta_{wo}(s)}{\theta_r(s)} = \frac{0.2}{(6.84s + 1)} \tag{E4.5.2}$$

(We see from these equations that the coupling of the heating system and space introduces an additional term in the heating system transfer function.)

The room air capacity transfer function will be, based on stage (b) of Example 3.4 with the time constant in minutes for this example,

$$G_r(s) \quad \Rightarrow \quad \frac{\theta_r(s)}{\theta_{wo}(s)} = \frac{0.367}{(19s + 1)} \tag{E4.5.3}$$

For a design heating water flow rate of $5.713\,\mathrm{kg\,min^{-1}}$, with a linear response across full range, and a control signal in the interval $0 \rightarrow 1$, the valve transfer function will be

$$G_v(s) \quad \Rightarrow \quad \frac{m_w(s)}{u(s)} = \frac{5.713}{(s + 1)} \tag{E4.5.4}$$

where $u(s)$ is the control signal and, hence, valve position.

Finally the sensor. For a mean air velocity of $0.1\,\mathrm{ms^{-1}}$ passing the sensor, using equations (3.38) and (3.39) the sensor time constant and dead time are respectively $3.16\,\mathrm{min}$ and $0.35\,\mathrm{min}$. Thus the temperature sensor transfer function will be

$$G_d(s) \quad \Rightarrow \quad \frac{\phi(s)}{\theta_r(s)} = \frac{\exp(-0.35s)}{(3.16s + 1)} \tag{E.4.5.5}$$

Now the block diagram for this system can be drawn up (Figure E4.5.1).

To determine the required controller gain at the limit of loop stability, we require the characteristic equation (e.g. for a Routh array). This cannot be written down as trivially as usual in this case because of the heating–space interaction; we need to derive it from scratch. Writing down the block diagram equations for this case, we have

$$\varepsilon(s) = r(s) - \phi(s)$$

$$u(s) = \varepsilon(s)G_c(s)$$

$$m_w(s) = u(s)G_v(s) \qquad\qquad\qquad\qquad\qquad\qquad \text{(see over)}$$

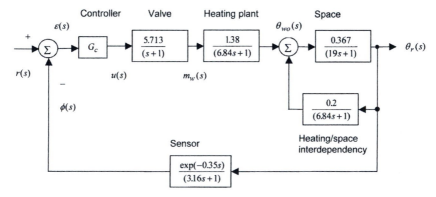

Figure E4.5.1 Block diagram model.

$\theta_{wo}(s) = m_w(s)G_h(s) + \theta_r(s)G_h'(s)$ (from previous page)

$\theta_r(s) = \theta_{wo}(s)G_r(s)$

$\phi(s) = \theta_r(s)G_d(s)$

Eliminating unwanted variables ($(\varepsilon(s),\ \phi(s),\ u(s),\ m_w(s)$ and $\theta_{wo}(s))$ by substitution leads to the overall closed-loop transfer function

$$\frac{\theta_r(s)}{r(s)} = \frac{G_c(s)G_v(s)G_h(s)G_r(s)}{[1 - G_h'(s)G_r(s) + G_c(s)G_v(s)G_h(s)G_r(s)G_d(s)]}$$

from which the system characteristic equation is

$$1 - G_h'(s)G_r(s) + G_c(s)G_v(s)G_h(s)G_r(s)G_d(s) = 0 \qquad \text{(E4.5.6)}$$

Substituting the actual transfer functions from the above, using the usual Padé approximation for the dead-time component of the temperature sensor, and noting that, for proportional control, $G_c(s) = K_c$, we obtain

$$1 - \frac{0.0734}{[(6.84s + 1)(19s + 1)]} + \frac{2.893K_c(1 - 0.175s)}{[(s + 1)(6.84s + 1)(19s + 1)(3.16s + 1)(0.175s + 1)]}$$
$$= 0$$

which, expressed as a polynomial in s, is

$$9367s^7 + 69564s^6 + 100079s^5 + 49717s^4 + (10913 - 65.9K_c)s^3$$
$$+ (1139 + 363K_c)s^2 + (53.9 + 74.3K_c)s + (0.93 + 2.89K_c) = 0 \qquad \text{(E4.5.7)}$$

This seventh-order system is going to be very cumbersome to deal with manually by any of the methods of stability analysis discussed earlier. In fact, these methods are fine for systems of up to about fourth or fifth order – systems of higher order where there is a degree of freedom (in this case the degree of freedom is the controller gain) become difficult to manage. In a Routh array, the above would lead to quadratic and cubic functions of K_c which are not convenient to solve by manual means.

One way round this is to substitute values for K_c which then make a Routh array fairly easy to manage. For example, if we take K_c to be $1.0\,\text{K}^{-1}$, the characteristic equation will be

$$9367s^7 + 69564s^6 + 100079s^5 + 49717s^4 + 10847s^3 + 776s^2 + 128.2s + 3.82 = 0$$

leading to the Routh array

first row	9367	100079	10847	128.2	0
second row	69564	49717	1502	3.82	0
third row	93385	10645	127.7	0	
fourth row	41787	1407	3.82	0	
fifth row	7501	119.2	0		
sixth row	743	3.82	0		
seventh row	80.6	0			
eighth row	3.82	0			
ninth row	0				

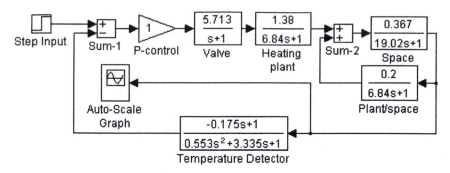

Figure E4.5.2 Simulink model of the heating system control.

Thus the plant will be stable at a controller gain of 1.0. We note from the array that the first column descends to some relatively low numbers which in fact suggests that marginal stability will be realised at a controller gain not much higher than 1.0. If we repeat the exercise with $K_c = 2.0$ we again find a positive set of elements in the first column, but at $K_c = 3.0$ we obtain the following:

$$9367s^7 + 69564s^6 + 100079s^5 + 49717s^4 + 10715s^3 + 2228s^2 + 277s + 9.6 = 0$$

and the Routh array

first row	9367	100079	10715	277	0
second row	69564	49717	2228	9.6	0
third row	93385	10415	276	0	
fourth row	41959	2022	9.6	0	
fifth row	5915	255	0		
sixth row	213	9.6	0		
seventh row	−11.6	0			
eighth row	9.6	0			

Thus the limit of stability under proportional control for this plant will lie in the interval 2→3.

In general however, the higher-order systems will be difficult to resolve using these techniques and computer modelling will be the best resort (see for example Stoecker *et al.*, 1978).

Representing the block diagram system of Figure E4.5.1 in Simulink (Figure E4.5.2), we can readily simulate the closed-loop step response to a change in set point which confirms the marginal stability criterion we found in the above. It also tells us that a controller gain of $1.0\,\mathrm{K}^{-1}$ (i.e. about one-half of the marginally stable range for K_c) will lead to a reasonably damped response (Figure E4.5.3), though once again exhibiting proportional offset.

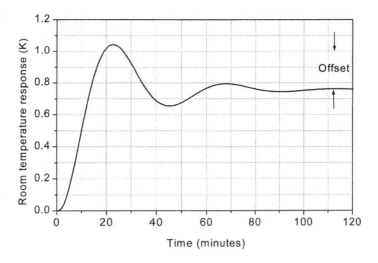

Figure E4.5.3 Room temperature unit step response.

Notice that the controller gain for this highly damped situation of heating system control is numerically higher than in the case of air conditioning (Example 4.4) which confirms that heating control in general tends to be less prone to instability than conventional air conditioning. In the latter, the plant is physically decoupled from the space and tends to be much more responsive than the much less responsive space heating plant.

4.3 Feedforward control

It is important to recognise that the primary role of any control system, whether it be an air conditioning plant, an auto-pilot, a guided missile or the Treasury's management of the economy, is to accommodate load disturbances or to *reject disturbances*. In this disturbance-rejection role, it must seem logical to track the disturbance itself and take some preemptive action, instead of waiting for it to upset the controlled condition (the essential mechanism in the feedback case). This can be achieved using *feedforward compensation*. Several examples exist in HVAC control. Here are a few:

- Weather-compensated heating (the external temperature is fed forward and used to reset the set point of the heating flow water temperature; the heating flow water temperature is the controlled variable).
- Supply temperature scheduling in variable-air-volume air conditioning systems (the supply air temperature is 'scheduled' from some other condition, usually external temperature).
- Cascade control in air conditioning systems (in which the supply air temperature is the controlled variable but its set point is fixed according to the prevailing value of room temperature).

There are other practical examples – see for example Peat (1990) in which heating flow and return temperature is used to control domestic heating. The flexibility of a building energy management system (BEMS) and digital control makes the application of feedforward compensation virtually limitless. In principle, feedforward action can improve control by arresting the influence that disturbances have on a system as they happen. Nevertheless, there are some vital conditions for successful implementation.

- Each disturbance variable must be accounted for with a feedforward compensator. Consider for example the over-heating problems associated with compensated heating control when solar and internal casual heat gains are ignored.
- It must be possible to measure the disturbance. This may not be as trivial as it seems; imagine the difficulty in compensating for the heat gains produced by people in a densely populated space such as an auditorium.
- The relationship between the ultimate controlled variable and each disturbance variable must be known. This informs the design of the feedforward compensator.

Because of these factors, it is generally advisable to combine feedforward compensation with feedback control. The feedforward compensator in theory handles most of the control action, leaving the feedback loop to provide 'trim' (in any case, a feedback loop will be needed if set point adjustments require periodically to be made).

Design of feedforward compensators

Clearly, the feedforward compensator must enter the control system at the only place it can influence events, the input to the control element. The generalised block diagram system model can therefore be modified as shown in Figure 4.8.

Design of the feedforward compensator simply requires that its transfer function, G'_c, be found. The objective of the feedforward compensator is to ensure that the controlled variable, y, is held constant whilst disturbance, d, varies and, ideally, all control objectives but set point are satisfied by the action of this path. Hence, $\varepsilon = 0$, $u' = 0$, so $u = u''$, from which

$$y = y' + y'' = \phi' G'_c G_p + d G'_p$$

But $y = 0$ (i.e. constant), and $d = \phi' / G'_d$. Hence,

$$\phi' G'_c G_p + \phi' \frac{G'_p}{G'_d} = 0$$

from which

$$G'_c = -\frac{G'_p}{G'_d G_p} \tag{4.27}$$

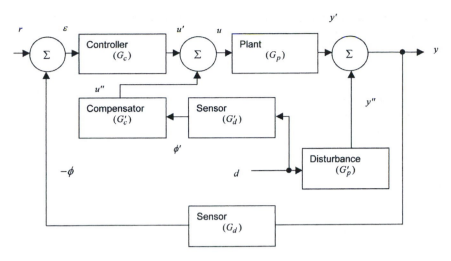

Figure 4.8 Block diagram system with feedforward compensation.

Thus compensation can be achieved if the system, control element, disturbance and disturbance sensor transfer functions are known. In many instances, the disturbance will be measurable by a sensor of rapid response or a digital state (e.g. the switching of a fixed load such as lighting or other heat-producing equipment) in which case G'_d in equation (4.27) either plays no part or can be neglected. In such cases, the feedforward compensator will simply be the disturbance transfer function multiplied by the inverse of the plant transfer function (including the control element).

A feedforward compensator derived in this way will have *dynamic characteristics*. In some cases, the transfer functions of one or more of the participating components may be difficult to arrive at or may simply be too complex for a practical implementation of feedforward compensation. In these cases, some improvement might be achieved through the use of a simple gain compensator restricted to steady-state characteristics. As an alternative, a first-order lag transfer function might arbitrarily be chosen with a 'tunable' time constant and gain in order to achieve some improvement. We will take a look at the design of feedforward compensation and the crucial issue of its implementation, in the next example.

Example 4.6

In Example 3.5, two transfer functions of a chilled water cooling coil were derived respectively for chilled water flow rate disturbances and chilled water inlet temperature disturbances to the coil. Suppose that the coil receives its chilled water from a thermostatically controlled chiller which results in very rapid temperature deviations in the chilled water temperature due to the switching of the chiller stages within dead-band limits. The coil itself is under control to maintain the local in-duct air temperature by varying the chilled water flow rate, through the use of a valve which can be described by a first-order lag with a time constant of one minute.

Neglecting the feedforward temperature sensor, design a feedforward compensator and use simulation to evaluate its performance when compared with conventional proportional feedback control.

Solution

The plant transfer function here is that which relates the coil outlet air temperature to control signal (i.e. the coil transfer function as derived in Example 3.5 multiplied by the valve transfer function). Expressing with a timebase in minutes,

$$G_p(s) = -\frac{16.3}{(s+1)(0.22s+1)}$$

The disturbance transfer function can also be taken from Example 3.5 which, with a Padé representation of the dead-time term, will be

$$G_p'(s) = \frac{3.25(1 - 0.10s)}{(0.24s + 1)(0.10s + 1)}$$

A feedforward compensator will therefore have the following form:

$$G_c'(s) = -\frac{G_p'(s)}{G_p(s)} = \frac{(s+1)(0.22s+1)}{16.3} \times \frac{3.25(1 - 0.10s)}{(0.24s + 1)(0.10s + 1)}$$

$$G_c'(s) = \frac{-0.005s^3 + 0.019s^2 + 0.22s + 0.2}{0.025s^2 + 0.34s + 1} \qquad (E4.6.1)$$

The arrangement is shown in Figure E4.6.1.

Now, what about implementation? We have two choices: we can attempt an *inverse Laplace transform* for an analytical solution, or we can discretise equation (E4.6.1) to enable a numerical solution.

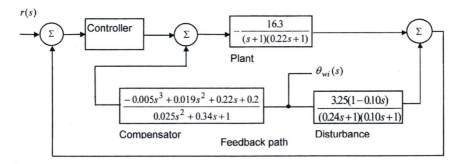

Figure E4.6.1 Coil model with disturbance.

Inevitably, the more complex higher-order functions do not readily lend themselves to inverse transformation and we may recall that, to carry out an inverse transformation, we require the degree of the numerator of the transfer function of interest to be less than that of the denominator. We see from equation (E4.6.1) that this is not the case (indeed since most compensators of this type will include a lead element this essential condition frequently will not apply). Whilst it is often possible to express the transfer function in the required form this can become complex and, in most practical cases, discretisation will be the simplest option.

A more detailed treatment of discretisation is given in section 5.2. For this example, we will find that the simple *backward discretisation* will suffice. In general for the first-order terms,

$$\frac{dy}{dt} = y(s)s \cong \frac{1}{T}[y(n) - y(n-1)] \tag{E4.6.2}$$

Hence for second- and third-order terms,

$$\frac{d^2y}{dt^2} = y(s)s^2 \cong \frac{1}{T^2}[y(n-2) - 2y(n-1) + y(n)] \tag{E4.6.3}$$

$$\frac{d^3y}{dt^3} = y(s)s^3 \cong \frac{1}{T^3}[-y(n-3) + 3y(n-2) - 3y(n-1) + y(n)] \tag{E4.6.4}$$

where n is the time instant and T the sampling time interval.

From the above, we can calculate the discrete-time output of our compensator at successive instants in time ($n = 0, 1, 2, \ldots, \infty$) for chosen and equidistant intervals in time, T. The only complication is the choice of T. Too large a value will result in a 'coarse' and inaccurate construction of the compensator response needed. Too small a value will be accurate enough but will be expensive in computational effort; a compromise is needed. Though we shall consider the choice of T in detail later when we look at discrete-time control, one helpful rule of thumb is to fix T at one-tenth of the dominant time constant present in the model. The dominant time constant can be thought of as the one which locates a pole nearest the origin in a root locus plot – in the case of the system in this example, this would be the valve time constant at 1 min; so $T = 0.1$min might be appropriate.

Substituting this value and equations (E4.6.2)–(E4.6.4) into equation (E4.6.1) produces the following discrete-time expression for the compensator if we let $G'_c(s) = u(s)/\theta_{wi}(s)$:

$$u(n) = 1.217u(n-1) - 0.362u(n-2) + 0.725\theta_{wi}(n-3)$$
$$-1.9\theta_{wi}(n-2) + 1.304\theta_{wi}(n-1) - 0.102\theta_{wi}(n) \tag{E4.6.5}$$

We now simulate the performance of the compensated control system using the Simulink model shown in Figure E4.6.2, where the function block (f(u)) is in the form of equation (E4.6.5). Initial simulation results revealed that the control system would give a reasonable response under

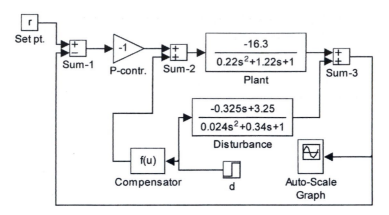

Figure E4.6.2 Simulink model of the compensated loop.

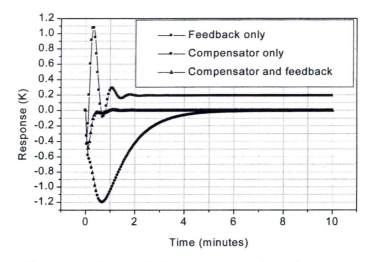

Figure E4.6.3 Response of feedback, feedforward and combined control.

feedback proportional control using a (cooling application) controller gain of $-1K^{-1}$ (though with a small offset). In Figure E4.6.3, a comparison of the response in controlled variable to a unit step disturbance in chilled water flow temperature is given for proportional feedback control, feedforward compensation only and feedforward compensation supplemented with proportional feedback control.

Figure E4.6.3 confirms that feedforward compensation acting alone will give the desired corrective action and is offset-free; but it responds undesirably slowly. However, a combination of feedforward compensation and feedback control gives an excellent offset-free response in which the initial oscillation present in the feedback loop is largely eliminated.

As a general conclusion, feedforward compensation will rarely work perfectly on its own but when combined with feedback control will give substantial improvement to the performance of a feedback control system. Since feedback will generally be required in any case for periodic set point adjustments, this combination is very convenient.

4.4 Model reduction

In a number of places in this chapter, we have seen how typical HVAC system models are often likely to be of high order, severely hampering the analysis of these systems using rational techniques other than computer simulation. In particular, system models higher than third or fourth order generally become quite unmanageable. In this section, we look at some of the methods for representing high-order models with low-order approximants, a technique called *model reduction*.

Model reduction using Routh approximation

Hutton & Friedland (1975) give an account of model reduction using the *Routh approximation method*. The method makes use of two arrays including a Routh array of the denominator polynomial of the model to be reduced. Hutton and Friedland's procedure is summarised as follows.

1 *Step 1*. Express the numerator and denominator of the model to be reduced, $G(s)$, as polynomials in s. Assuming for convenience the order of the numerator to be one less than the denominator,

$$G(s) = \frac{a_{n-1}s^{n-1} + \ldots + a_0}{b_n s^n + \ldots + b_0} \qquad (4.28)$$

2 *Step 2*. Express the *reciprocal transformation*, which, in effect, reverses the order of the polynomial coefficients,

$$\hat{G}(s) = \frac{a_0 s^{n-1} + \ldots + a_{n-1}}{b_0 s^n + \ldots + b_n} \qquad (4.29)$$

3 *Step 3*. Derive a Routh array using the denominator of the transformed model (see section 4.2 for this procedure), together with a supplementary array based on the numerator coefficients. For the latter, the first two rows are formed as in the Routh case. Subsequent rows however are calculated using the Routh cross-multiplication procedure but with the *immediately previous row substituted by the corresponding row from the complementary Routh (denominator) array.*

4   *Step 4.* Now express the following coefficients. For the *i*th row of *m* rows of non-zero first column elements,

$$\alpha_i = \frac{x_{i,1}}{x_{i+1,1}} \tag{4.30}$$

$$\beta_i = \frac{y_{i,1}}{y_{i+1,1}} \qquad (i = 1, 2, \ldots, m-1) \tag{4.31}$$

(where x and y are the elements of the denominator-based and numerator-based arrays respectively).

5   *Step 5.* Routh convergents, which result in approximations for reduced-order polynomial coefficients, can now be arranged into the following form for first-, second- and third-order (etc.) approximations:

- first order

$$G'(s) = \frac{\beta_1}{s + \alpha_1} \tag{4.32}$$

- second order

$$G'(s) = \frac{\beta_2 s + \alpha_2 \beta_1}{s^2 + \alpha_2 s + \alpha_1 \alpha_2} \tag{4.33}$$

- third order

$$G'(s) = \frac{(\beta_1 + \beta_2)s^2 + \alpha_3 \beta_2 s + \alpha_2 \alpha_3 \beta_1}{s^3 + (\alpha_1 + \alpha_2)s^2 + \alpha_2 \alpha_3 s + \alpha_1 \alpha_2 \alpha_3} \tag{4.34}$$

Example 4.7

The forward path system model of Example 4.5 contains a control valve, heat emitter and space with emitter/space interdependence. This group of components can be expressed as a single fourth-order model as shown in the following.

The existing four-block arrangement is as shown in Figure E4.7.1(a). This is now to be reduced to the equivalent single block as shown in Figure E4.7.1(b). The block diagram equations representing Figure E4.7.1(a) are, from Figure E4.5.1,

$$m_w(s) = u(s)G_v(s)$$

$$\theta_{wo}(s) = m_w(s)G_h(s) + \theta_r(s)G'_h(s)$$

$$\theta_r(s) = \theta_{wo}(s)G_r(s)$$

which, by substitution, reduce to

$$G(s) = \frac{\theta_r(s)}{u(s)} = \frac{G_v(s)G_h(s)G_r(s)}{1 - G'_h(s)G_r(s)} \tag{E4.7.1}$$

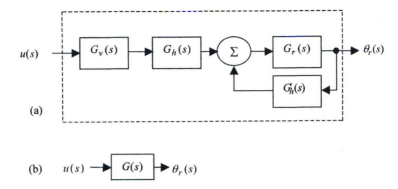

(a)

(b) $u(s) \rightarrow \boxed{G(s)} \rightarrow \theta_r(s)$

Figure E4.7.1 (a) High-order model. (b) Form of reduced model.

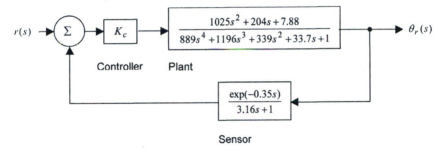

Figure E4.7.2 Rationalised block diagram model of the high-order plant.

which represents the single block of Figure E4.7.1(b). Inserting the appropriate transfer functions from Figure E4.5.1 leads to the rationalised transfer function for the heating system, valve and space of Example 4.5:

$$G(s) = \frac{1025s^2 + 204s + 7.88}{889s^4 + 1196s^3 + 339s^2 + 33.7s + 1} \tag{E4.7.2}$$

Figure E4.5.1 can alternatively be redrawn as the rationalised block diagram model for the control system (Figure E4.7.2).

Tasks

(a) Derive and compare the open-loop signal step responses of first-, second- and third-order approximants of the rationalised fourth-ordr system model using the Routh approximation method. (b) Reassess stability using the reduced-order models.

Table E4.7.1 Model reduction arrays

Routh (denominator) array				Supplementary (numerator) array		
1	339	889	0	7.88	1025	0
33.7	1196	0		204	0	
303.5	889	0		745.3 (*)	0	
1097.3	0	0		−597.5 (**)	0	
889	0			0		
0						

Note the following operations in the supplementary array:

$$(*) \quad 1025 - \frac{7.88 \times 1196}{33.7}$$

and

$$(**) \quad 0 - \frac{204 \times 889}{303.5}$$

Solution

First, we consider the reduced-order approximants. The reciprocal transformation of equation (E4.7.2) will be

$$\hat{G}(s) = \frac{7.88s^2 + 204s + 1025}{s^4 + 33.7s^3 + 339s^2 + 1196s + 889}$$

The denominator (Routh) array and the supplementary numerator array can now be expressed as in Table E4.7.1.

Now the α and β coefficients can be calculated (equations (4.30) and (4.31)):

$$\alpha_1 = \frac{1}{33.7} = 0.0297, \qquad \alpha_2 = \frac{33.7}{303.5} = 0.1110$$

$$\alpha_3 = \frac{303.5}{1097.3} = 0.2766, \qquad \alpha_4 = \frac{1097.3}{889} = 1.234$$

$$\beta_1 = \frac{7.88}{33.7} = 0.2338, \qquad \beta_2 = \frac{204}{303.5} = 0.6722$$

$$\beta_3 = \frac{745.3}{1097.3} = 0.6792, \qquad \beta_4 = \frac{-597.5}{889} = -0.6721$$

Using equations (4.32), (4.33) and (4.34), approximations to the fourth-order system model of equation (E4.7.2) can now be expressed,

- first order

$$G'(s)_{R-1} = \frac{7.88}{33.7s + 1} \qquad\qquad\qquad (E4.7.3)$$

- second order

$$G'(s)_{R-2} = \frac{204s + 7.88}{303s^2 + 33.7s + 1} \qquad\qquad\qquad (E4.7.4)$$

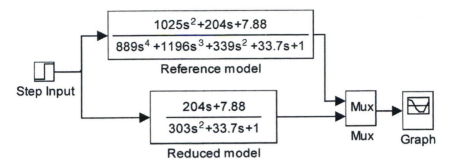

Figure E4.7.3 Testing the reduced model using Simulink.

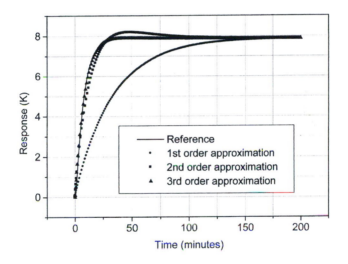

Figure E4.7.4 Comparison of model approximations using Routh's method.

- third order

$$G'(s)_{R-3} = \frac{1001s^2 + 206s + 7.88}{1097s^3 + 336s^2 + 33.7s + 1} \qquad (E4.7.5)$$

Clearly, first- and second-order approximants represent a considerable simplification of the original fourth-order system model but we must now move on to take a look at just how good our approximations are. The fourth-order model (reference model) has been used to generate the response to an open-loop unit step change in signal alongside the various reduced-order approximants, using the Simulink block diagram of Figure E4.7.3 (shown here is the second-order approximant). The results have been plotted on Figure E4.7.4.

The results show that the third-order approximant gives a response which is practically identical to the original fourth-order system model. However, the second-order model gives a very good approximation and is far simpler. Clearly, the first-order model is completely inadequate. All reduced models agree precisely with the reference model as far as ultimate steady-state results are concerned (this can be easily verified from the models themselves by putting $s = 0$). From the viewpoint of maximum reduction for least error, it is clear that the second-order approximant would be the best choice; certainly as far as system response investigations might be concerned.

Now consider how adequate the reduced model might be when assessing system stability using the methods discussed in this chapter.

Using the second-order approximant (equation (E4.7.4)) and adding the sensor transfer function and the controller gain thus forming an open-loop model of the entire control system, a manageable fourth-order model results if a Padé approximation is used for the sensor dead-time term. Using the Routh stability criterion, a controller gain at marginal stability of $5.88\,\mathrm{K}^{-1}$ is obtained (readers should have no difficulty in verifying this for themselves). The conclusion using the unwieldy reference model in Example 4.5 suggested a value of between 2 and $3\,\mathrm{K}^{-1}$. Evidently then, conventional stability analysis using a reduced-order system model may imply greater inherent stability than actually exists and caution should be exercised.

Model reduction using Padé approximations

Shamash (1975) describes an alternative and simpler method of model reduction due to Padé approximations. He confirms that the polynomial form of high-order transfer function (equation (4.28)) can be expanded into a power series of system time moments taking the form

$$G(s) = c_0 + c_1 s + c_2 s^2 + \dots \tag{4.35}$$

in which

$$c_0 = \frac{a_0}{b_0} \tag{4.36}$$

$$c_i = \frac{1}{b_0}\left[a_i - \sum_{j=1}^{i} b_j c_{i-j} \right] \qquad (i > 0) \tag{4.37}$$

Assuming that the reduced model can be expressed as

$$G'(s) = \frac{d_{k-1}s^{k-1} + \dots + d_1 s + d_0}{e_k s^k + \dots + e_1 s + e_0} \tag{4.38}$$

then the following Padé approximations can be expressed:

$$d_0 = e_0 c_0$$
$$d_1 = e_0 c_1 + e_1 c_0$$
$$\vdots$$
$$d_{k-1} = e_0 c_{k-1} + e_1 c_{k-2} + \ldots + e_{k-1} c_0 \tag{4.39}$$

which can be solved simultaneously for the reduced model coefficients.

Example 4.8

Derive reduced-order models for the fourth-order system model considered in Example 4.7 using the Padé method.

Solution

First we find the coefficients of the expanded expression (i.e. equations (4.36) and (4.37)),

$$c_0 = \frac{a_0}{b_0} = \frac{7.88}{1} = 7.88$$

$$c_1 = \frac{1}{b_0}\left[a_1 - \sum_{j=1}^{1} b_j c_{i-j}\right] = \frac{1}{1}[204 - (33.7 \times 7.88)] = -61.6$$

(Similarly, $c_2 = 429.6$, $c_3 = -3020$, $c_4 = 22794$, $c_5 = -203417$.)

For the first-order Padé approximant, $d_1 = 0$ and $e_1 = 1$ and equations (4.39) reduce to

$$d_0 = 7.88 e_0$$
$$0 = -61.6 e_0 + 7.88$$

Therefore

$$d_0 = 1.009 \quad \text{and} \quad e_0 = 0.128$$

and

$$G'(s)_{P-1} = \frac{7.88}{7.81s + 1} \tag{E4.8.1}$$

For the second-order Padé approximant, $d_2 = 0$ and $e_2 = 1$ and equations (4.39) reduce to

$$d_0 = 7.88 e_0$$
$$d_1 = -61.6 e_0 + 7.88 e_1$$
$$0 = 429.6 e_0 - 61.6 e_1 + 7.88$$
$$0 = -3020 e_0 + 429.6 e_1 - 61.6$$

which are solved to give

$$d_0 = -1.86, \quad d_1 = 0.338, \quad e_0 = -0.236, \quad e_1 = -1.802$$

resulting in the second-order approximant

$$G'(s)_{P-2} = \frac{1.43s - 7.88}{4.24s^2 - 7.64s - 1} \qquad \text{(E4.8.2)}$$

Note that the coefficient of s in the denominator of the above is negative which implies that the second-order Padé approximant will be *unstable* even though the original fourth-order model is stable.

Finally, for the third-order approximant, $d_3 = 0$ and $e_3 = 1$ and equations (4.39) will be for this case

$$d_0 = 7.88e_0$$
$$d_1 = -61.6e_0 + 7.88e_1$$
$$d_2 = 429.6e_0 - 61.6e_1 + 7.88e_2$$
$$0 = -3020e_0 + 429.6e_1 - 61.6e_2 + 7.88$$
$$0 = 22794e_0 - 3020e_1 + 429.6e_2 - 61.6$$
$$0 = -203417e_0 + 22794e_1 - 3020e_2 + 429.6$$

and give the third-order Padé approximant

$$G'(s)_{P-3} = \frac{18.92s^2 + 158.4s + 7.88}{160s^3 + 166.1s^2 + 27.92s + 1} \qquad \text{(E4.8.3)}$$

Figure E4.8.1 shows the comparative performance of the first- and third-order Padé approximants alongside the reference model (the unstable and therefore unsuitable second-order approximant is not shown).

Surprisingly, the first-order approximant gives a very accurate representation of the fourth-order reference system model and certainly considerably better than the first-order Routh approximant. The Padé method, though a little easier than the Routh method to apply, can produce unsta-

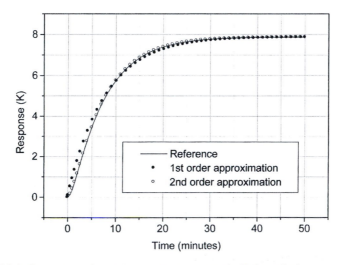

Figure E4.8.1 Comparison of model approximations using the Padé method.

ble low-order models as we found with the second-order example here. The Routh method, though plainly not as accurate with low-order approximants, on the other hand will give stable results provided that the original high-order model is stable. This point is confirmed and discussed further by Shamash (1980), though it is also challenged by Ashoor & Singh (1982) who report that there are instances where Routh approximants give more accurate results than Padé.

Though the Routh method is based on the order of the numerator being one less than the order of the denominator, it is possible to enable the order of the numerator to be fixed independently by combining both methods. In this case, the denominator polynomial coefficients for the reduced model are obtained using a Routh array whilst the numerator polynomial coefficients are obtained using the Padé method. This is a simplification of the Routh method whilst still retaining essential stability conditions (Shamash, 1975).

The issue of accuracy has been addressed by Warwick (1985) who introduces an error polynomial as a basis for reduced-order modelling. A cost function is defined as a weighted sum of squares of the error polynomial and the reduced model is obtained by minimising the cost function though the method is computationally more demanding than the methods described earlier.

Note that whilst we have restricted our consideration of model reduction to continuous-time systems, the techniques discussed in this section can be extended to discrete-time systems (which are the subject of the next chapter) – see Shamash (1974).

References

Ashoor, N., Singh, V. (1982) A note on low order modelling. *IEEE Transactions on Automatic Control*, **AC-27** (5), 1124–1126.

Evans, W.R. (1948) Graphical analysis of control systems. *AIEE Transactions*, **67**, 547–551.

Houpis, C.H., Lamont, G.B. (1992) *Digital Control Systems – Theory, Hardware, Software*. McGraw-Hill, New York.

Hutton, M.F., Friedland, B. (1975) Routh approximations for reducing order of linear, time-invariant systems. *IEEE Transactions on Automatic Control*, **AC-20** (3), 329–337.

Kuo, B.C. (1975) *Automatic Control Systems*. Prentice-Hall, Englewood Cliffs, NJ.

MATLAB (1992) *Control System Toolbox User's Guide*. The Mathworks Inc., Natick, MA.

Peat, B.J. (1990) Novel room temperature control. *Building Services Engineering Research and Technology*, **11** (1), 9–12.

Routh, E.J. (1877) *Dynamics of a System of Rigid Bodies*. Macmillan, London.

Shamash, Y. (1974) Continued fraction methods for the reduction of discrete-time dynamic systems. *International Journal of Control*, **20** (2), 267–275.

Shamash, Y. (1975) Model reduction using the Routh stability criterion and the Padé approximation technique. *International Journal of Control*, **23** (3), 475–484.

Shamash, Y. (1980) Failure of the Routh–Hurwitz method of reduction. *IEEE Transactions on Automatic Control*, **AC-25** (2), 313–314.

Simulink (1996) *SIMULINK 2 Dynamic System Simulation for MATLAB*. The Mathworks Inc., Natick, MA.

Stoecker, W.F., Rosario, L.A., Heidenreich, M.E., Phelan, T.R. (1978) Stability of an air-temperature control loop. *ASHRAE Transactions*, **84** (1), 35–53.

Warwick, K. (1985) Weighting factor functions for reduced order models. *Proceedings of the IEE International Conference: Control '85*, vol. 2, pp. 679–684.

5 Discrete-time systems

5.1 The z-transform

So far, we have considered systems which behave as a continuous function of time, t. The main vehicle for the analysis of these systems has been the Laplace transform. In continuous-time systems, the feedback signal is converted to a control signal using an analogue controller whose output is a smooth continuous function of time. Today, few electromechanical analogue control systems exist in HVAC control – these having mostly given way to stand-alone application-specific or universal digital controllers (see section 1.3).

Digital control systems are constrained by the finite time taken for feedback data to be converted from analogue measurement to digital number and then processed to form a control signal. Essentially, the digital control system functions at discrete intervals in time and a further variable characterises the digital loop beyond those variables considered for continuous control systems – that of the sampling interval, T.

Discrete control systems can be analysed using methods similar to those developed previously but a new operator is introduced – that of the *z-transform*. For a practical discussion see Rohrer & Stoecker (1986).

In a discrete-time system, the substitution

$$z = \exp(Ts) \tag{5.1}$$

converts the Laplace-domain mathematical model into a z-domain model appropriate to digital control. In effect, z can be visualised as a unit time lead, and z^{-1} a unit time delay.

5.2 Representing discrete-time systems

Figure 5.1 shows the component blocks of the generalised discrete-time feedback block diagram model. The hold is sometimes referred to as a *filter* and its function is to delay the received control signal in a specified way. The most commonly used hold is that in which the value of the last sample of data is held until the next sample is taken which is called a *zero-order hold*.

The transfer function of a zero-order hold can be shown to be

$$D(s) = \frac{1 - \exp(-Ts)}{s} = \frac{1}{s}[1 - \exp(-Ts)] \tag{5.2}$$

and the z-transform of $1 - \exp(-Ts)$ is $1 - z^{-1} = (z - 1)/z$. The z-transform of the *combined plant and hold* can therefore be expressed as

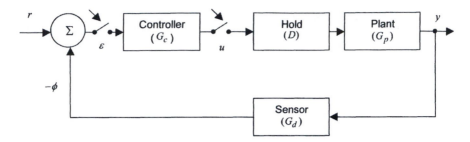

Figure 5.1 Generalised discrete-time control system.

$$G(z) = \frac{z-1}{z} G'(z) \tag{5.3}$$

in which $G'(z)$ is the z-transform of $G(s)/s$ and $G(s) = G_p(s)\,G_d(s)$.

Modelling a plant, $G(s)$, from first principles (e.g. using the methods in Chapter 3) will result in a Laplace-domain representation. Therefore, in most cases, discrete-time system analysis and design involve converting the plant transfer function to a z-domain representation, then deriving an appropriate model for the controller. As will be seen later, z-domain expressions can be readily expressed as finite-difference equations for direct implementation in, for instance, programmable control systems as well as in simulation models. We will therefore now consider the process of converting s-domain transfer functions to equivalent or approximate discrete-time expressions.

Standard transform tables

Tables of exact s–z transform conversions are widely available. See for example Houpis & Lamont (1992) or Coughanowr (1991). Table 5.1 gives a summary of some leading transform pairs, together with their time-domain equivalents.

Modified transforms

General transforms will inform plant response and stability at times consistent with integer multiples of the sampling interval, T. However, there are two cases where this will not be adequate:

- where information is required on plant behaviour *between* sampling instants, and
- where the plant model contains a dead-time term (quite common in HVAC systems).

In such circumstances, a modified z-transform may be used (see for example Coughanowr, 1991). There are broadly three interpretations of these.

Table 5.1 Some common transform pairs

s-Transform	z-Transform	Time function
Integrator		
$\dfrac{a}{s}$	$\dfrac{z}{z-1}$	at
First-order lag		
$\dfrac{1}{s+a}$	$\dfrac{z}{z-\exp(-aT)}$	$\exp(-at)$
First-order lag + integrator		
$\dfrac{a}{s(s+a)}$	$\dfrac{z[1-\exp(-aT)]}{(z-1)[z-\exp(-aT)]}$	$a[1-\exp(-at)]$
Second order		
$\dfrac{b-a}{(s+a)(s+b)}$	$\dfrac{z[\exp(-aT)-\exp(-bT)]}{[z-\exp(-aT)][z-\exp(-bT)]}$	$\exp(-at)-\exp(-bt)$
Second order + integrator		
$\dfrac{ab}{s(s+a)(s+b)}$	$\dfrac{z}{z-1}+\dfrac{bz}{(a-b)[z-\exp(-aT)]}$ $-\dfrac{az}{(a-b)[z-\exp(-bT)]}$	$\dfrac{b}{(a-b)}\exp(-aT)$ $-\dfrac{a}{(a-b)}\exp(-bT)$

1 In circumstances where a dead-time term exists in the plant model and the dead time, t_d, is *less* than the sampling interval, T, such that $t_d = \lambda T$ (and $\lambda < 1$),

$$Z\big[G(s)|_{(t-\lambda T)}\big] = G(z,md) \tag{5.4}$$

where $G(z,md)$ is the *modified* z-transform, in which $t_d = \lambda T$. This case is also useful in situations where the analysis requires to explore system behaviour *in between* sampling instants.

2 In circumstances where a dead-time term exists where the dead time is an integer number (i) of sampling intervals T, such that $t_d = iT$,

$$Z\big[G(s)|_{(t-iT)}\big] = z^{-i}G(z) \tag{5.5}$$

3 Where a dead-time term exists in which the dead time is a non-integer number of sampling intervals, such that $t_d = (i+\lambda) \times T$,

$$Z\big[G(s)|_{(t-(i+\lambda)T)}\big] = z^{-i}G(z,md) \tag{5.6}$$

Some common modified z-transforms are given in Table 5.2, in which $m = 1 - \lambda$.

Discretisation

Some of the more complex component transfer functions may prove intractable to convert from s-domain to z-domain using standard values.

Table 5.2 Common modified z-transforms

s-Transform	Modified z-transform	Time function
Dead time		
$\exp(-asT)$	z^{-a}	$f(t) = 0$ for $0 \le t \le aT$
Dead time + simple lag		
$\dfrac{\exp(-asT)}{(s + b)}$	$\dfrac{z^{-1}\exp(-amT)}{1 - z^{-1}\exp(-aT)}$	$\exp[-a(t - bT)]$
Dead time + simple lag + integrator		
$\dfrac{\exp(-asT)}{s(s + b)}$	$\dfrac{z^{-1}}{a}\left(\dfrac{1}{1 - z^{-1}} - \dfrac{\exp(-amT)}{1 - z^{-1}\exp(-aT)}\right)$	$1 - \exp[-a(t - bT)]$

Approximations can made by simply discretising the original *s*-domain function. Essentially, this takes the form of substituting *s* in the original expression with a numerical approximation expressed in terms of the operator *z*.

We will take a look at a simple method here. The simplest form of discretisation can be made by fixing a forward- or backward-difference equation substitution for the derivative. Take, for example, the forward-difference equation

$$\frac{\mathrm{d}y}{\mathrm{d}t} = sy(s) \cong \frac{y(t + T) - y(t)}{T}$$

which is, in effect, the basis of initial-value Euler integration. Introducing the *z* operator,

$$sy(z) \cong \frac{zy(z) - y(z)}{T}$$

$$s \cong \frac{z - 1}{T} \tag{5.7}$$

The simple substitution of equation (5.7) approximates the *s*-domain expression to a *z*-domain expression.

Differencing

For a controller (or any other object) designed in discrete time (i.e. $G_c(z)$) it will be necessary to translate the *z*-domain expression into a time-domain expression for on-line implementation. The same applies if we want to simulate the performance of a plant in the *z*-domain.

One approach is to invert the *z*-domain expression using the standard table of transforms (Table 5.1). Very often however, particularly when we are seeking to invert a higher-order closed-loop system model, this will prove intractable. The *z*-domain expression can instead be translated to a *difference equation* – essentially a numerical approximation to the true time-domain expression.

A *backward-in-time* difference equation from the z-domain model can be accomplished with the following substitution. The general z-function for an output, $y(z)$, and an input, $u(z)$, is expressed as

$$z^0 k_1 y(z) + z^{-a} k_2 y(z) + z^{-b} k_3 y(z) + \ldots = z^{-a} l_1 u(z) + z^{-b} l_2 u(z) + z^{-c} l_3 u(z) + \ldots \quad (5.8)$$

and this can be represented by the following time sequence,

$$k_1 y(n) + k_2 y(n - a) + k_3 y(n - b) + \ldots = l_1 u(n - a) + l_2 u(n - b) + l_3 u(n - c) + \ldots$$
$$(5.9)$$

where, $n =$ sampling instant, $k_1,\ k_2,\ k_3, \ldots =$ coefficients of the output variable, y, and $l_1,\ l_2,\ l_3, \ldots =$ coefficients of the input variable, u.

Note that the above is *physically realisable*. That is, it depends on *previous* values of the input and output variables in order to generate a current value of output, $y(n)$.

Example 5.1

A plant has a time constant of 10 min, dead time 2 min and an open-loop gain of unity. Using a sampling interval of one-quarter of the dead time, compare the open-loop unit step response of the system using a forward-difference discretisation with that of the continuous-time function.

Solution

For the continuous-time function with a unit step input excitation,

$$G_p(s) = \frac{\exp(-2s)}{s(10s + 1)}$$

which gives the inverse transformation

$$y(t) = 1 - \exp\left(\frac{2 - t}{10}\right) \qquad \text{for } t \geq t_d \qquad (E5.1.1)$$

For forward-difference discretisation we make the substitution given by equation (5.7) (the unit step operator, $1/s$, which appears in the above is not included in the discretised expression):

$$s = \frac{z - 1}{T} = \frac{z - 1}{2/4} = 2(z - 1)$$

and therefore

$$G_p(z) = \frac{y(z)}{u(z)} = \frac{z^{-4}}{10 \times 2(z - 1) + 1} = \frac{z^{-5}}{20 - 19z^{-1}}$$

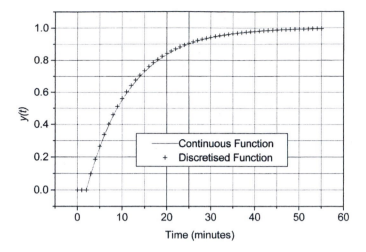

Figure E5.1.1 Comparison of discretised and continuous-time responses.

(Note that the dead-time term, $\exp(-2s)$, is a time delay of magnitude $4 \times T$, i.e. z^{-4}.)

We now express the following difference equation which can be solved recursively,

$$20\,y(z) - 19\,y(z)z^{-1} = u(z)z^{-5}$$
$$y(z) = 0.95\,y(z)z^{-1} + 0.05u(z)z^{-5}$$

i.e. at the nth time instant,

$$y(n) = 0.95\,y(n-1) + 0.05u(n-5) \tag{E5.1.2}$$

Solving equations (E5.1.1) and (E5.1.2) leads to the comparison between the analytical (exact) method and the difference equation result which is shown in Figure E5.1.1, confirming that the latter gives an excellent approximation in this case.

In practice however, many systems will not easily lend themselves to a simple backward-in-time or forward-in-time difference equation solution. Results are crucially dependent on the choice of sampling interval and in some of the more complex cases (including closed-loop transient response) a forward-in-time difference approximation may not be good enough, requiring more sophisticated methods to be sought. For a further discussion on this, see Leigh (1992).

5.3 Stability tests in discrete-time systems

The w-transformation

Established stability tests can in principle be used with z-domain system models in a manner similar to that of continuous-time systems. However, we recall that the expression $z = \exp(sT)$ relates the z-domain to the s-domain, so to use methods developed for the latter, the z-plane needs to be mapped onto the equivalent s-plane which can be achieved through another transformation, the *bilinear* or *w-transformation*,

$$w = \frac{z+1}{z-1} \tag{5.10}$$

This transformation maps the roots of the z-domain characteristic equation with reference to a circle of unit radius. Briefly, the justification for this is as follows (for a fuller discussion, see for example Coughanowr, 1991).

1 We recall from Figure 4.7 that a typical stable root of the s-plane lies at $s = -x + j\omega$.
2 We have $z = \exp(sT) = \exp(-xT)\exp(j\omega T)$, in the stable region.
3 We have $x \neq 0$, i.e. $\exp(-xT) < 1$, thus provided that the real roots of the discrete-time characteristic equation, $1 + G(z) = 0$, are ≤ 1, or lie within a circle of unit radius, the discrete-time system will be stable.

For a special (and significant) case of stability analysis for a discrete-time system, it is valuable to look at the familiar case of a plant modelled by a first-order lag with dead time, i.e.

$$G(s) = \frac{K_p \exp(-t_d s)}{(\tau s + 1)} = \frac{K_p \exp[-(i + \lambda) \times Ts]}{(\tau s + 1)} \tag{5.11}$$

where $t_d = (i + \lambda) \times T$.
For the case where the dead time is an integer number, i, of the sampling interval, T, the shifted transform of equation (5.5) is appropriate. Selecting the relevant s–z transform conversion from Table 5.1, adding a controller term (a gain, K_c) and a zero-order hold for the discrete-time equivalent system, we have $i \neq 0$, $\lambda = 0$, $m = 1 - \lambda = 1$, and

$$G(z)\big|_{\lambda=0,i\neq0} = \frac{(1 - z^{-1})}{z^i} K_c K_p Z\left[\frac{1}{s}G(s)\right]$$

i.e.

$$G(z)\big|_{\lambda=0,i\neq0} = \frac{(1 - z^{-1})}{z^i} K_c K_p Z\left[\frac{1/\tau}{s(s + 1/\tau)}\right]$$

which reduces to

$$G(z)\big|_{\lambda=0,i\neq0} = \frac{K_c K_p}{z^{i+1}}\left[\frac{1 - \exp(-A)}{1 - z^{-1}\exp(-A)}\right] \tag{5.12}$$

where $A = T/\tau$.

For the case where the dead time is less than the sampling interval, we have $i = 0, 0 < \lambda < 1$, and $m = 1 - \lambda$. The modified z-transform becomes appropriate (Table 5.2),

$$G(z,md)|_{\lambda \neq 0, i=0} = \left(1 - z^{-1}\right) K_c K_p Z\left[\frac{1/\tau}{s(s + 1/\tau)}\right]$$

which reduces to

$$G(z,md)|_{\lambda \neq 0, i=0} = K_c K_p \left[\frac{[\exp(-B) - \exp(-A)]z^{-2} + [1 - \exp(-B)]z^{-1}}{1 - z^{-1}\exp(-A)}\right] \qquad (5.13)$$

where, $B = (1 - \lambda)T/\tau$.

The final situation is where the dead time is a non-integer number of sampling intervals, T, for which $i \neq 0, 0 < \lambda < 1$, $m = 1 - \lambda$, and the modified z-transform is appropriate,

$$G(z,md)|_{\lambda \neq 0, i \neq 0} = \frac{\left(1 - z^{-1}\right)}{z^i} K_c K_p \left[\frac{1/\tau}{s(s + 1/\tau)}\right]$$

which reduces to

$$G(z,md)|_{\lambda \neq 0, i \neq 0} = \frac{(K_c K_p)}{z^i}\left[\frac{[\exp(-B) - \exp(-A)]z^{-2} + [1 - \exp(-B)]z^{-1}}{1 - z^{-1}\exp(-A)}\right] \qquad (5.14)$$

Example 5.2

An open-loop step response test has resulted in the following plant model, in which the incoming control signal is a percentage scale and the output variable is temperature:

$$G(s) = \frac{0.146\exp(-37.5s)}{(217.5s + 1)}$$

Find the maximum proportional gain at critical roots of the system characteristic equation for (a) the continuous-time plant model, (b), the discrete-time plant model with $T = t_d$, (c) $T = 0.6t_d$ and (d), $T = 2t_d$.

Solution

This is a special case because it contains a significant dead time in relation to the time constant. Significant dead time is quite common in HVAC control applications due to the mix of highly responsive and slow-acting components found in many systems.

(a)   For continuous-time operation, the system open-loop model, using a Padé approximation for the dead-time term, and adding a controller gain term, will be

$$G(s) = \frac{0.146K_c(1 - 18.75s)}{(217.5s + 1)(18.75s + 1)}$$

which from the characteristic equation $1 + G(s) = 0$ leads to

$$4078.0s^2 + 236.3s - 2.74K_c s + 1 + 0.146K_c = 0$$
$$4078.0s^2 + (236.3 - 2.74K_c)s + (0.146K_c + 1) = 0$$

For this control system, we are seeking a positive value for K_c and we can see that if a Routh array were to be formed, only the s^1 coefficient can possibly influence stability. That is,

$$236.3 - 2.74K_c \leq 0 \qquad K_c \leq 86.24\%\mathrm{K}^{-1}$$

(b)   For discrete-time operation with $T = t_d$, we have $\lambda = 0$, $i = 1$, and equation (5.12) applies:

$$G(z) = \frac{K_c K_p}{z^{i+1}}\left[\frac{1 - \exp(-A)}{1 - z^{-1}\exp(-A)}\right]$$

where

$$A = \frac{T}{\tau} = \frac{37.5}{217.5} = 0.172$$

and so

$$G(z) = \frac{0.146K_c}{z^2}\left[\frac{1 - \exp(-0.172)}{1 - z^{-1}\exp(-0.172)}\right] = \frac{0.023K_c}{(z^2 - 0.842z)}$$

This results in the discrete-time characteristic equation $1 + G(z) = 0$ and thus

$$z^2 - 0.842z + 0.023K_c = 0$$

We now make a w-transformation using the substitution $z = (w + 1)/(w - 1)$ (equation (5.10)), and expressing the resulting characteristic equation in the form of a polynomial in w,

$$(0.158 + 0.023K_c)w^2 + (2 - 0.046K_c)w + (1.842 + 0.023K_c) = 0$$

Again, as in the case of the continuous-time situation, only the w term can possibly force instability unless

$$2 - 0.046K_c \leq 0 \qquad K_c \leq 43.5\%\mathrm{K}^{-1}$$

(c)   For discrete-time operation with $T = 0.6t_d$, we have $t_d = 1.67T$, $\lambda = 0.67$, $m = 1 - \lambda = 0.33$, $i = 1$, and equation (5.14) applies:

$$G(z,md) = \frac{K_c K_p}{z^i}\left[\frac{[\exp(-B) - \exp(-A)]z^{-2} + [1 - \exp(-B)]z^{-1}}{1 - z^{-1}\exp(-A)}\right]$$

where

$$A = \frac{T}{\tau} = \frac{0.6 \times 37.5}{217.5} = 0.103$$

and

$$B = (1 - \lambda)\frac{T}{\tau} = (1 - 0.67)\frac{0.6 \times 37.5}{217.5} = 0.034$$

which gives the modified discrete-time model

$$G(z,md) = \frac{0.0094K_c z^{-2} + 0.0049K_c z^{-1}}{z - 0.902}$$

and leads to the characteristic equation

$$z^3 - 0.902z^2 + 0.0049K_c z + 0.0094K_c = 0$$

This, in turn, leads to the w-transformed characteristic equation

$$(0.098 + 0.0143K_c)w^3 + (2.098 - 0.0143K_c)w^2 + (2.098 + 0.0233K_c)w$$
$$+ (1.902 - 0.0045K_c) = 0$$

Critical roots for this expression are far less straightforward than in the previous second-order case. The Routh array needs to be set up and a stability-critical quadratic function of K_c can be identified at

$$(2.098 - 0.0143K_c)^2 - (0.098 + 0.0143K_c)(1.902 - 0.0045K_c) \le 0$$

the solution of which leads to the gain criterion at

$$K_c \le 59.3\%\mathrm{K}^{-1}.$$

(d) Finally, for the discrete-time case when $T = 2t_d$, we have $T = 75.0$s, $\lambda = 0.5$, $n = 0$, $m = 1 - \lambda = 0.5$, and equation (5.13) applies:

$$G(z,md) = K_c K_p \left[\frac{[\exp(-B) - \exp(-A)]z^{-2} + [1 - \exp(-B)]z^{-1}}{1 - z^{-1}\exp(-A)} \right]$$

where

$$A = \frac{T}{\tau} = \frac{75.0}{217.5} = 0.345$$

and

$$B = (1 - \lambda)\frac{T}{\tau} = (1 - 0.5)\frac{75.0}{217.5} = 0.172$$

giving the modified discrete-time model

$$G(z,md) = \frac{0.134K_c z^{-2} + 0.158K_c z^{-1}}{1 - 0.708z^{-1}}$$

which leads to the characteristic equation

$$z^2 + (0.158K_c - 0.708)z + 0.134K_c = 0$$

In turn, this leads to the w-transformed system characteristic equation

$$(0.292 + 0.292K_c)w^2 + (2.0 - 0.268K_c)w + (1.708 - 0.024K_c) = 0$$

This gives a critical stability criterion for the controller gain based on the coefficient of w^1,

$$2.0 - 0.268K_c \leq 0 \qquad K_c \leq 7.46\%\text{K}^{-1}$$

We see from these results that a further consideration affecting system stability is introduced with discrete-time systems which is not present in continuous-time systems – that of the sampling interval T.

If one applies a Ziegler–Nichols tuning rule (see section 7.3), applicable (for the sake of argument) to purely proportional control, $K_c = 0.5K'_c$ (where K'_c takes on the above values at marginal stability), the notional controller settings for the various cases would be:

- Continuous-time system, $K_c|_{contin.} = 43.1\%\text{K}^{-1}$
- Discrete-time system, $T = t_d$, $K_c|_{T=t_d} = 21.8\%\text{K}^{-1}$
- Discrete-time system, $T = 0.6t_d$, $K_c|_{T=0.6t_d} = 29.7\%\text{K}^{-1}$
- Discrete-time system, $T = 2t_d$, $K_c|_{T=2t_d} = 3.73\%\text{K}^{-1}$

It is clear from this that the stability of the equivalent discrete-time system when compared with continuous-time performance is inherently lower (i.e. requires the imposition of a lower controller gain) and, in this example, the higher the sampling interval, the lower the required gain. This is more or less in line with what we are already aware of in that the sampling interval is, in effect, a form of system dead time and we know from earlier examples that systems exhibiting high dead time tend to be less stable.

The Jury–Blanchard test

The Jury–Blanchard test (often referred to simply as Jury's test) offers a much simpler method of establishing system stability for systems of low order than the Routh test, since it cuts out the need to make a w-transformation of the discrete-time characteristic equation (Jury & Blanchard, 1961). Indeed, Jury's test is to discrete-time systems as Routh's test is to continuous-time systems.

The method is summarised as follows. Further discussions on the application of the method can be found in a number of texts on control – see for example Houpis & Lamont (1992) or Leigh (1992).

1 The discrete-time characteristic equation in polynomial form is, in general,

$$a_n z^n + a_{n-1} z^{n-1} + \ldots + a_1 z + a_0 = 0 \qquad\qquad (5.15)$$

where $a_0 \neq 0$ and $a_n > 0$.

2 For *second-order systems*, $n = 2$ and provided that the constraints

$CE(z)|_{z=1} > 0$ (constraint 1)
$CE(z)|_{z=-1} > 0$ (constraint 2)
$|a_0| < a_n$ (constraint 3)

are satisfied, no roots of the characteristic equation will lie on or outside the unit circle and the system will be stable. The system will not be stable if any one of these conditions does not apply.

3 For higher-order systems where $n > 2$, further constraints must be imposed.

  a Firstly, constraints 1, 2 and 3 are tested for, but constraint 2 now becomes

$$CE(z)|_{z=-1} > 0 \text{ (for } n \text{ even) or} < 0 \text{ (for } n \text{ odd)} \qquad \text{(constraint 3}')$$

     ($CE(z)$ refers to the discrete-time characteristic equation).

  b Next, determine the remaining constraints, j_{max}, since

$$j_{max} = n - 2 \qquad\qquad (5.16)$$

  c The Jury–Blanchard array can now be set up. There will be $n + 1$ maximum columns and a maximum of $2j_{max} + 1$ rows.

  d The array consists of row pairs – the second row of each pair always taking the form of the first row in reverse order. The first row of the array consists of the coefficients of the characteristic equation in ascending order,

first row	a_0	a_1	$\ldots$	a_{n-1} $\quad a_n$
second row	a_n	a_{n-1}	$\ldots$	a_1 $\quad a_0$

  e Now form subsequent row pairs until a pair of rows each with three elements is encountered to complete the array,

third row	b_0	b_1	$\ldots$	b_{n-1}
fourth row	b_{n-1}	b_{n-2}	$\ldots$	b_0

$\vdots$

$2j_{max}$th row	x_0	x_1	x_2
final row	x_2	x_1	x_0

The first row of subsequent row pairs is formed from the determinants of 2×2 matrices whose elements are taken from the two preceding rows. For the third row for example, this results in the following operations:

$$b_0 = a^2{}_0 - a^2{}_n$$
$$b_1 = a_0 a_1 - a_n a_{n-1}$$
$$b_2 = a_0 a_2 - a_n a_{n-2}$$
etc. (5.17)

The fourth row then consists of the third row in reverse. Row pairs c, $d, \ldots, x$ are then formed in a similar manner.

f   The remaining constraints to be tested for are based on comparing the absolute values of the first and final elements in the first row of each row pair:

$$|b_0| > |b_{n-1}| \qquad \text{(constraint 4)}$$

$$\vdots$$

$$|x_0| > |x_2| \qquad \text{(constraint } n+1\text{)}$$

That is, if any of the absolute values of the final element in each of the first row of each row pair is greater than the corresponding final element, the system will be unstable.

Clearly, the method is very simple to carry out for second-order systems but becomes progressively more cumbersome to apply as the system order increases.

Example 5.3

Re-assess stability in Example 5.2 using Jury's test.

Solution

(a)   For the case $T = t_d$, we recall the characteristic equation in polynomial form,

$$z^2 - 0.842z + 0.023K_c = 0$$

Here, $n = 2$, hence

- first constraint:

$$|a_0| < a_2 \quad \Rightarrow \quad |0.023K_c| < 1 \qquad \text{i.e. } |K_c| < 43.5\%\text{K}^{-1}$$

- second constraint:

$$CE(z)|_{z=1} = 1^2 - 0.842 \times 1 + 0.023K_c > 0 \qquad \text{i.e. } K_c > -6.87\%\text{K}^{-1}$$

- third constraint:

$$CE(z)|_{z=-1} = -1^2 - 0.842 \times (-1) + 0.023K_c > 0 \qquad \text{i.e. } K_c > -80.1\%\text{K}^{-1}$$

For stability, we require that $-6.87 < K_c < 43.5\%\text{K}^{-1}$ (which is in agreement with the findings of Example 5.2).

(b)   For the case $T = 0.6t_d$, we recall the higher-order system model,

$$z^3 - 0.902z^2 + 0.0049zK_c + 0.0094K_c = 0$$

Here, $n = 3$, hence

- first constraint:

$$|a_0| < a_3 \quad \Rightarrow \quad |0.0094K_c| < 1 \quad \text{i.e. } |K_c| < 106.4\%\text{K}^{-1}$$

- second constraint:

$$CE(z)|_{z=1} = 1^3 - 0.902 \times 1^2 + 0.0049 \times 1 \times K_c + 0.0094K_c > 0$$
$$\text{i.e. } K_c > -6.85\%\text{K}^{-1}$$

Now, since n is odd in this case, constraint 3' becomes applicable. Hence

- third constraint:

$$CE(z)|_{z=-1} = -1^3 - 0.902 \times \left(-1^2\right) + 0.0049 \times (-1) \times K_c + 0.0094K_c < 0$$
$$\text{i.e. } K_c < 422.7\%\text{K}^{-1}$$

So far then, we require that $-6.85 < K_c < 106.4\%\text{K}^{-1}$ for this case, but since $n = 3$ we must apply one more constraint which will evolve from the Jury–Blanchard array.

For the array, $j_{max} = n - 2 = 1$, hence the array will have, $2j_{max} + 1 = 3$ rows. The first two rows can be written down immediately,

first row	$0.0094K_c$	$0.0049K_c$	-0.902	1
second row	1	-0.902	$0.0049K_c$	$0.0094K_c$

The third row (b elements) is obtained from determinants based on the first two rows according to equations (5.17),

$$b_0 = 0.0094^2 K_c^2 - 1^2 = 8.836 \times 10^{-5} K_c^2 - 1$$
$$b_1 = 0.0094K_c \times 0.0049K_c - (-0.902) \times 1 = 4.606 \times 10^{-5} K_c^2 + 0.902$$
$$b_2 = 0.0094K_c \times (-0.902) - 0.0049K_c \times 1 = -1.338 \times 10^{-2} K_c$$

The third and fourth rows are now apparent,

third row		
$8.836 \times 10^{-5}K_c^2 - 1$	$4.606 \times 10^{-5}K_c^2 + 0.902$	$-1.338 \times 10^{-2}K_c$
fourth row		
$-1.338 \times 10^{-2}K_c$	$4.606 \times 10^{-5}K_c^2 + 0.902$	$8.836 \times 10^{-5}K_c^2 - 1$

In fact, we do not need to concern ourselves with the fourth row, since the final constraint of interest lies in the third row.

- fourth constraint:

$$|b_0| > |b_2| \quad \Rightarrow \quad \left|8.836 \times 10^{-5}K_c^2 - 1\right| > \left|-1.338 \times 10^{-2}K_c\right|$$

A trial-and-error solution of the quadratic leads to an approximate value for K_c which is a maximum at $K_c \cong 55\%\mathrm{K}^{-1}$. Hence for this case, $-6.85 < K_c < 55\%\mathrm{K}^{-1}$. The maximum value here is close to the result predicted by the Routh test in Example 5.2 $(59.3\%\mathrm{K}^{-1})$.

(c)   The third case in which $T = 2t_d$, in common with the first case, is a second-order system and can therefore be tackled in much the same way as (a) above. This results in the minimum condition $-1 < K_c < 7.46\%\mathrm{K}^{-1}$, which is, again, in agreement with the finding of Example 5.2.

Using a root locus in discrete-time systems

If a discrete-time model has formed the starting point for a root locus construction, then we must interpret stable and unstable regions in relation to a unit circle (whose origin lies at the origin of the real and imaginary axes), for the reason given in the previous section. The stable region lies within the unit circle. Therefore, a critically stable response can be expected at a point where the locus leaves the unit circle. The open-loop gain can then be found in the same way as dealt with for the continuous-time case.

Once again, we will re-assess Example 5.2 using the root locus method by adopting the MATLAB control toolbox (MATLAB, 1992) in the following example.

Example 5.4

The MATLAB control toolbox functions rlocus and rlocfind can be used to generate a root locus plot for a given open-loop system model as we found in Chapter 4.

Consider first the discrete-time case with $T = 0.6t_d$ in Example 5.2. Setting K_c equal to 1 as the basis for constructing the root locus plot will lead to a value of plant gain at critical stability where branches lie at the boundary of the unit circle:

$$G(z)\big|_{T=0.6t_d} = \frac{0.0049z + 0.0094}{z^3 - 0.902z^2} = \frac{0.0049(z + 1.9184)}{z^2(z - 0.902)}$$

(note the factored form). Setting the numerator and denominator coefficients in MATLAB requires two lines,

```
>num=[0,0,0.0049,0.0094];
```

```
>den=[1,-0.902,0,0];
```

Now construct the basic root locus plot,

```
>rlocus(num,den)
```

The plot will appear in the graphics window. We can conveniently obtain a value of critical gain by using the rlocfind function which places a cross-wire in the graphics window, but we need to know where the boundary of

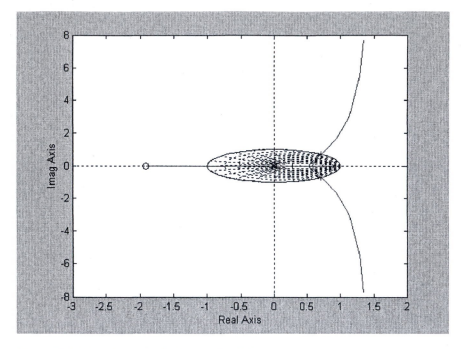

Figure E5.4.1 Root locus plot for the case $T = 0.6t_d$.

the unit circle lies in relation to the loci. This we can do by overlaying a map of damping and natural frequency contours onto our plot using the zgrid function,

```
>zgrid
```

```
>rlocfind(num,den)
```

In the resulting graphics window, locate a point at which one of the symmetrical root loci crosses the boundary of the damping map using the cross-wire and a value of critical gain (which is in essence the critical K_c) is returned to the MATLAB command window (Figure E5.4.1). The value obtained is $K_c = 55.5$ (units are %K^{-1}) which compares with our earlier result using a Routh array with a w-transformed model of 59.3%K^{-1}.

In a similar manner, results for the cases $T = t_d$ and $T = 2t_d$ are obtained as 43.4%K^{-1} and 7.5%K^{-1}, respectively, giving good agreement with the previous results.

We are now in a position to take a look at what sort of time-series response these alternative system gain and sampling times will give. Using MATLAB/Simulink, a block diagram representation of the system can be constructed and simulated for any defined input disturbance (Simulink, 1996). Figure E5.4.2 shows the Simulink block diagram for this for the case in which $T = 0.6t_d$. Notice that two separate block diagram systems

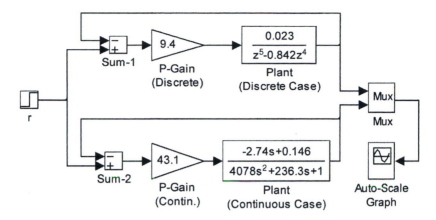

Figure E5.4.2 Block diagram Simulink model.

have been constructed side by side for a convenient comparison – the continuous system case and the discrete system case. Results from the simulation of Figure E5.4.2 and the remaining cases for which $T = t_d$ and $T = 2t_d$ are compared in Figures E5.4.3, E5.4.4 and E5.4.5.

Notice the significant 'staircasing' in the discrete-time responses. This is the result of the combined effect of the hold and the system dead time and will be pronounced in situations where the system dead time in relation to time constant is high and the sampling interval is also high (we note that as the sampling interval reduces for this system with high dead time, so does the staircasing characteristic of the response).

Hence a lower sampling interval will tend to force a system response closer to that characterised by a continuous-time system and the main problem with the coarser sampling interval is that a highly delayed response becomes predominant. In fact, the choice of sampling interval is not at all straightforward. We have seen from the above that not only will the sampling interval initiate response delay when sampling at low frequency, but that the innate stability of the system is also affected. It is therefore a good place to take a look at some of the considerations behind the choice of T.

5.4 The sampling interval

Leigh (1992) identifies four degrading effects resulting from increasing or decreasing the sampling interval from some optimum:

- *Dead-time effect.* Increasing T effectively increases the dead time of the system (earlier examples have illustrated the destabilising effect of this).
- *Input disturbance frequency.* The sampling frequency must be higher than the highest anticipated frequency of the input.

- *Discretisation.* A control algorithm which has been derived by numerically discretising an equivalent continuous-time function can lead to significant error when a high T is used.
- *Information loss effect.* This relates to very low choices of T, where sampling takes place at a higher frequency than the dynamics of the feedback system. For example, a sensor may take 3s to construct a signal accurately whilst the controller is sampling at 2s intervals; hence information is lost between sampling instants.

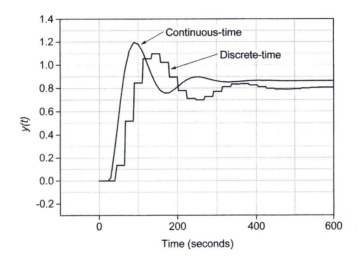

Figure E5.4.3 Comparison of plant response at $T = 0.6t_d$.

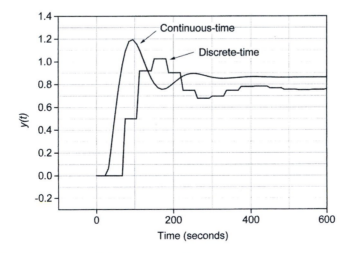

Figure E5.4.4 Comparison of plant response at $T = t_d$.

Aliasing and Shannon sampling

In Figure 5.2, the influence of sampling rates on a periodic signal is shown (Strang & Nguyen, 1996). The situation on the left shows that by sampling at an interval, π, the full signal is reconstructed, and the sampling period will be $T = \pi/\omega$ (where ω is the frequency of the signal).

Now consider the sequence on the right in which sampling takes place at some lower rate and it is possible to reconstruct both the original signal and a slower one (shown dotted). Thus the slower signal is an *alias* of the original signal. Where there are several frequency components present in the signal being measured (arising from several independent disturbances) the highest frequency signal is often referred to as the *Nyquist frequency*. To reconstruct all of the data, the sampling rate must therefore be based on

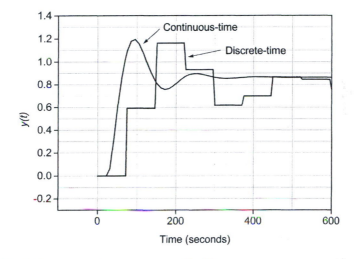

Figure E5.4.5 Comparison of plant response at $T = 2t_d$.

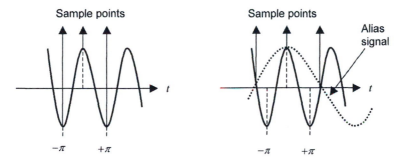

Figure 5.2 Sampling rates and aliasing.

the Nyquist frequency, ω_{max}. This can be stated formally as *Shannon's sampling theorem* (Oliver *et al.*, 1948): 'a function of time, $e(t)$, which contains no frequency components greater than ω_{max} can be reconstructed by the values of $e(t)$ at any set of sampling points that are spaced apart by $T < \pi/\omega_{max}$'.

Thus a sampling frequency criterion can be fixed at $\omega_s > 2\omega_{max}$ (in which $T = 2\pi/\omega_s$ and ω_s is the sampling frequency).

Practical choice of T

In many practical cases, particularly energy system control, a lower bound on T is generally unnecessary since the practical value used will always be high enough for any likely inter-sample fluctuations in set point or load disturbance to be properly tracked. There are however stability limitations to this argument as will be seen later and where measurement noise is expected, some minimum limit on T may be desirable after all.

Practical suggestions for T have evolved mainly through experience. Bennett (1988) for example gives the following summary in cases where characteristics of plant model are known.

1   If a *dominant* time constant, τ_{dom}, is known, then

$$T < \frac{\tau \, dom}{10} \tag{5.18}$$

2   If a model, in the form of a first-order lag plus dead time can be fitted to the plant or system, then

$$0.05 < \frac{T}{t_d} < 0.3 \tag{5.19}$$

3   Alternatively, if the closed-loop response of the equivalent continuous-time system is known and if the response has a finite settling time, t_s, then

$$T < \frac{t_s}{10} \tag{5.20}$$

4   Finally, where the natural frequency, ω_n, is known

$$T > \frac{0.2\pi}{\omega_n} \tag{5.21}$$

Example 5.5

For Example 5.2, establish and compare the sampling interval using the various criteria given above.

Solution

Using the dominant time-constant criterion, we note that the dominant time constant will usually be the highest present in the model (i.e. the lowest *pole* on the *s*-plane), provided that there is a zero on the real axis in the right half-plane (note that a first-order Padé approximation to a dead-time term ensures this). Hence, in Example 5.2, $\tau_{dom} = 217.5$s and

$$T < \frac{\tau\,dom}{10} = \frac{217.5}{10} = 21.75\text{s}$$

Using the dead-time criterion, $0.05 t_d < T < 0.3 t_d$ and $t_d = 37.5$s, and therefore

$1.88 < T < 11.25$s

Using the settling-time criterion, from Figure E5.4.3 for the equivalent continuous-time system we see that the settling time is approximately 400s, i.e.

$$T < \frac{t_s}{10} = \frac{400}{10} = 40.0\text{s}$$

We can obtain the natural undamped frequency, ω_n, if we substitute the gain at marginal stability for the continuous-time system into its characteristic equation. Setting $s = j\omega$ determines imaginary roots of the characteristic equation at the zero damping condition – i.e. those symmetrical roots that lie on the imaginary ordinate. The associated ω value at this point is ω_n, the system natural frequency. We recall that the gain at marginal stability was found to be $86.24\% \, \text{K}^{-1}$ for the continuous-time case and

$$4078 j^2 \omega_n^2 + (236.3 - 2.74 \times 86.24) j \omega_n + (1 + 0.146 \times 86.24) = 0$$
$$-4078 \omega_n^2 + 13.59 = 0$$

which gives

$\omega_n = 0.058 \text{rads}^{-1}$

Hence,

$$T > \frac{0.2\pi}{\omega_n} = \frac{0.2\pi}{0.058} = 10.8\text{s}$$

Four results for T have been obtained in the above with the lower (and possibly more cautious) values suggesting a sampling interval of no greater than about 10 or 11 seconds. In fact if we fix the sampling interval at 9.375s, the dead time is a convenient integer multiple of the sampling interval ($n = 4$).

Extending the analysis carried out in Examples 5.2, 5.3 and 5.4 but with $T = 9.375$s, we first obtain a new plant model,

$$G(z) = \frac{0.023 K_c}{z^5 - 0.842 z^4}$$

This leads to a proportional controller gain of $9.4\% \text{K}^{-1}$ according to the methods used in either Example 5.2 or 5.4. Now, repeating the Simulink

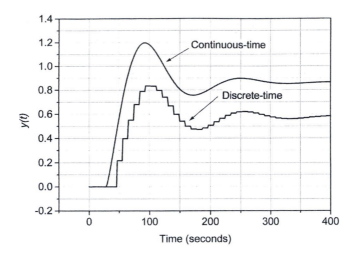

Figure E5.5.1 Comparative responses with $T = 0.25t_d$.

modelling but with the modified system transfer function and new sampling interval we arrive at the responses shown in Figure E5.5.1.

Figure E5.5.1 confirms that as the sampling interval is reduced, the discrete-time system response begins to resemble that of a continuous-time system. However, there is one major difference in that the required proportional gain needed to achieve stability in the discrete-time system is low, and we therefore end up with quite unacceptable offset. Evidently, a discrete-time system with a low sampling interval tends to behave like the equivalent continuous-time system with increased dead time and, correspondingly, reduced stability. It is clear then that the 'controller' in a discrete-time system requires special consideration if we are to remedy this.

In the above analysis, we have adopted a simple proportional controller simply to illustrate the significance that the sampling interval can have in discrete-time system stability yet one of the major merits of discrete-time control lies in flexibility to choose the precise characteristics of the control algorithm for the best response obtainable. This is addressed in the next section.

5.5 Digital control algorithms

Discretised PID

The most common form of digital control algorithm in HVAC control is the discretised form of the familiar PID controller, the continuous-time form of which appears as equation (3.45) in Chapter 3, i.e.

$$\frac{u(s)}{\varepsilon(s)} = G_c(s) = K_c\left(1 + \frac{1}{i_t s} + d_t s\right)$$

Forward-difference discretisation (section 5.2) is inappropriate because it leads us to the need to know *future* values of the control error. Instead, a backward-difference approximation can be used successfully,

$$s \cong \frac{\left(1 - z^{-1}\right)}{T} \tag{5.22}$$

Substituting for s in the continuous-time version of the PID transfer function leads to the following discrete-time version:

$$G_c(z) = \frac{az^2 - bz + c}{z^2 - z} \tag{5.23}$$

where

$$a = \frac{K_c}{Ti_t} \times \left(T^2 + Ti_t + i_t d_t\right) \tag{5.24}$$

$$b = \frac{K_c}{T} \times (T + 2d_t) \tag{5.25}$$

$$c = \frac{K_c}{T} \times d_t \tag{5.26}$$

Expressing this in finite-difference form,

$$u(n) = a\varepsilon(n) - b\varepsilon(n - 1) + c\varepsilon(n - 2) + u(n - 1) \tag{5.27}$$

In this form, this algorithm can be readily implemented in digital control, is widely used, well understood and reasonably successful when correctly tuned. Tuning however is not easy – there are four degrees of freedom in the algorithm; K_c, T, i_t, d_t. We have extensively explored ways of considering three of these parameters but the previous section has highlighted the special considerations needed for T.

Another consideration here is that the discretised PID algorithm represents the only practical choice for control when the plant dynamics cannot be identified (i.e. modelled or model-fitted). However, where an accurate system model can be identified there are a variety of simple methods for designing a control algorithm, based on specifying favourable response features as we shall see in the next few sections. For a further discussion on the treatment of discretised PID control, see Dexter (1988).

The dead-beat algorithm

The dead-beat (or minimal response) algorithm is so called because it is designed to satisfy the following (somewhat ambitious) criteria:

- the settling time, t_s, must be finite;
- the rise time, t_r, must be a minimum; and
- the response must be offset-free.

Interpreting these criteria for a unit step change in control error leads to a response which will have zero error at all sampling instants after the first. That is,

$$y(z) = z^{-1}r(z)$$

Neglecting the feedback element for this illustration (i.e. for unity negative feedback), since the closed-loop representation of the generalised discrete-time control scheme of Figure 5.1 can be expressed as

$$\frac{y(z)}{r(z)} = \frac{G_c(z)G(z)}{1 + G_c(z)G(z)}$$

and, from the above

$$\frac{y(z)}{r(z)} = z^{-1}$$

hence

$$G_c(z) = \frac{1}{G(z)} \times \frac{1}{(z-1)} \tag{5.28}$$

Equation (5.28) represents the required transfer function of the dead-beat controller, from which a time-domain algorithm can be derived by differencing, provided that $G(z)$ is known. This algorithm will tend to maintain the desired performance at the sampling instants *though not necessarily in between*.

The Dahlin algorithm

In highly damped systems, dead-beat response is rarely feasible unless the plant capacity in relation to control demand is high, since few systems will enable the response to match the change in error in the space of one sampling instant.

As a practical alternative, Dahlin (1968) suggested that the closed-loop system should behave like a continuous first-order lag with dead time, undergoing a step response. We recall the various transformations for discrete-time first-order lag with dead-time responses (Table 5.2). For a practical interpretation of the Dahlin algorithm, it is convenient to consider the case where the dead time, t_d, is an integer number of sampling intervals, T. Indeed, as we will see later, this and the dead-beat algorithm are not generally feasible for cases where $T \le t_d$. Again, for the case with unity negative feedback, the closed-loop model will be

$$\frac{y(z)}{r(z)} = \frac{G_c(z)G(z)}{1 + G_c(z)G(z)}$$

from which, for a unit step change in $r(z)$,

$$G_c(z) = \frac{1}{G(z)} \times \frac{y(z)}{[1 - y(z)]} \tag{5.29}$$

We recall equation (5.12) which gives us the discrete-time transfer function for a first-order lag plus dead time. For a unit response,

$$y(z) = \frac{1}{z^i} Z\left[\frac{1/\tau}{s(s + 1/\tau)}\right]$$

which gives the required Dahlin response,

$$y(z) = z^{-i-1} + \frac{(1 - E)}{\left(1 - z^{-1}E\right)} \tag{5.30}$$

Combining equations (5.29) and (5.30) leads to a general interpretation of the Dahlin control algorithm,

$$G_c(z) = \frac{1}{G(z)} \times \frac{(1 - E)z^{-i-1}}{\left[1 - Ez^{-1} - (1 - E)z^{-i-1}\right]} \tag{5.31}$$

In many practical applications it will be feasible to set $i = 0$ and 'tune' the algorithm through τ. It will often also be desirable to introduce an additional gain tuning parameter for flexibility, K_d. Hence,

$$G_c(z) = \frac{K_d}{G(z)} \times \frac{(1 - E)z^{-1}}{\left(1 - z^{-1}\right)} \tag{5.32}$$

Note that $E = \exp(-T/\tau)$. This algorithm leads to a stable response which is generally highly damped and will work well therefore in situations where stability is the overriding consideration and speed of response of secondary interest.

The Kalman algorithm

This is a more practically realisable form of the dead-beat response (Kalman, 1954) and seeks to satisfy the following criteria:

- The system must respond in minimum *possible* time.
- There must be no overshoot.
- There must be no offset.

It seeks to alleviate the inherent weaknesses in the dead-beat algorithm by achieving these objectives in two or more sampling intervals and, in a departure from previous methods which have tended to consider response quality only, the Kalman method considers both the response, $y(z)$, and the controller output (or plant input), $u(z)$. By considering these variables together it is possible in principle to remove oscillation from the system response.

The reasoning behind this algorithm is as follows.

1   In the first time interval after a disturbance, the controller output is changed immediately by a large amount, Δu_0, and the system tends to change rapidly towards its final value, changing by Δy_0 at the end of the interval.

2   During the second interval after the disturbance, the controller output is reduced to a lower value, Δu_1, so as to prevent overshoot and the resulting change in response is Δy_1.

3   The plant input and corresponding response continue to change until the following criteria are satisfied:

$$\Delta u_0 + \Delta u_1 + \Delta u_2 + \ldots + \Delta u_j = 1$$

and correspondingly,

$$\Delta y_0 + \Delta y_1 + \Delta y_2 + \ldots + \Delta y_{j+1} = 1$$

Plant and controller responses can therefore be expressed as

$$\frac{y(z)}{r(z)} = \Delta y_0 z^{-1} + \Delta y_1 z^{-2} + \Delta y_2 z^{-3} + \ldots + \Delta y_{i+1} z^{-(j+2)} = Y$$

and

$$\frac{u(z)}{r(z)} = \Delta u_0 + \Delta u_1 z^{-1} + \Delta u_2 z^{-2} + \Delta u_3 z^{-3} + \ldots + \Delta u_j z^{-j} = X$$

But

$$G(z) = \frac{y(z)}{u(z)} = \frac{Y}{X}$$

and since (with unity negative feedback for this illustration)

$$G_c(z) = \frac{1}{G(z)} \times \frac{y(z)/r(z)}{1 - y(z)/r(z)}$$

it follows that

$$G_c(z) = \frac{X}{1 - Y} \tag{5.33}$$

The coefficients of the plant transfer function, $G(z)$, can be directly related to the Kalman control algorithm as given in equation (5.33). Note that the following conditions must apply for this to occur:

$$\Delta u_0 + \Delta u_1 + \Delta u_2 + \ldots + \Delta u_j = 1$$

and

$$\Delta y_0 + \Delta y_1 + \Delta y_2 + \ldots + \Delta y_{j+1} = 1$$

This is achieved by normalising the system transfer function. The normalised coefficients are arrived at by setting the system transfer function gain to unity and dividing the numerator and denominator by the sum of the coefficients of the numerator, as will be seen in the example which follows.

Example 5.6

Mixing dampers represent a special problem in HVAC control. When controlled from air supply temperature, they tend to have a wide operating range (the same can be said for heating coils in full fresh air plant) because the mixed air condition which the dampers are seeking to maintain is strongly influenced by fresh air temperature which varies substantially throughout the year. Thus a damper control system commissioned in mid-season conditions may have difficulty in maintaining stable control in winter due to the increase in damper gain with respect to the temperature control function. Formally stated, the control of mixing dampers tends to be non-linear due to the variation in fresh air intake temperature and its influence on plant gain.

Consider then the case of a mixing damper set with a time constant of 20s and dead time 3s. Assume that the dampers have been carefully selected and matched to system requirements such that a linear response over a full range of operation by the dampers can be expected at a given inlet air temperature.

Assuming unity negative feedback, compare the various control algorithms when designed for operation on a mild day when the mixed temperature range is 12°C (fresh air damper fully open) to 20°C (fresh air damper fully closed) contrasting sampling intervals of $t_d/2$ and $2t_d$.

Solution

(a)   Firstly, the system model. In continuous-time control,

$$K_p = \frac{\text{output variable range}}{\text{positioning range}} = \frac{20 - 12(K)}{100\%} = 0.08 K(\%)^{-1}$$

and therefore

$$G_p(s) = G(s) = \frac{0.08\exp(-3s)}{(20s + 1)}$$

When $T = 0.5t_d$, we can use the transform of equation (5.12) to obtain a discrete-time equivalent model for $G(s)$ incorporating a zero-order hold. Noting that i in equation (5.12) is in this case 2 (i.e. the dead time is equivalent to twice the sampling interval), then

$$G(z) = \frac{0.08}{z^3}\left[\frac{1 - \exp(-1.5/20)}{1 - z^{-1}\exp(-1.5/20)}\right] = \frac{0.00578}{z^3 - 0.928z^2}$$

When $T = 2t_d$, we obtain the discrete-time transformation from equation (5.13) which is a modified transform since t_d is a non-integer multiple of T. Thus, for this case,

$$G(z) = \frac{0.0111z + 0.0096}{z^2 - 0.714z}$$

(b)   Now, for discrete-time PID control, the required controller transfer function will be (equation (5.23)),

$$G_c(z) = \frac{az^2 - bz + c}{z^2 - z}$$

But how do we determine the coefficients? We can adopt the established tuning rules for an equivalent continuous-time system as a starting point, though we will need to adopt a 'trial-and-error' approach from there. Using the Ziegler–Nichols tuning rules (discussed extensively in Chapter 7) as might be applied to the equivalent continuous-time system, we can arrive at the following notional controller parameters:

$K_c = 107.5(\%)\mathrm{K}^{-1}$ $i_t = 4.4\mathrm{s}$ and $d_t = 1.1\mathrm{s}$

Representing the system and its discrete-time controller transfer function in $\mathtt{Simulink}$, we obtain a reasonable discrete-time response to a unit step change in error, with zero offset for $T = 1.5\mathrm{s}$ (Figure E5.6.1), at the following controller parameters:

$K_c = 53.8(\%)\mathrm{K}^{-1}$ $i_t = 4.4\mathrm{s}$ and $d_t = 1.1\mathrm{s}$

and therefore

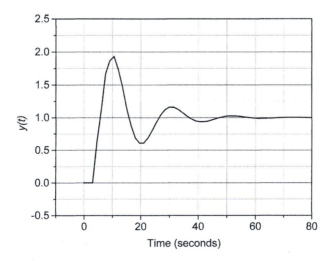

Figure E5.6.1 Tuned PID response at $T = 1.5\mathrm{s}$.

$$a = 111.5(\%)\mathrm{K}^{-1} \qquad b = 132.6(\%)\mathrm{K}^{-1} \qquad \text{and} \qquad c = 39.4(\%)\mathrm{K}^{-1}$$

Similarly, when $T = 6\mathrm{s}$ (Figure E5.6.2),

$$K_c = 26.9(\%)\mathrm{K}^{-1} \qquad i_t = 4.4\mathrm{s} \qquad \text{and} \qquad d_t = 1.1\mathrm{s}$$

and therefore

$$a = 68.5(\%)\mathrm{K}^{-1} \qquad b = 36.8(\%)\mathrm{K}^{-1} \qquad \text{and} \qquad c = 4.9(\%)\mathrm{K}^{-1}$$

Thus discrete-time PID controller transfer functions for the damper set are

$$G_c(z)\Big|_{T=1.5}^{\mathrm{PID}} = \frac{111.5z^2 - 132.6z + 39.4}{z^2 - z}$$

and

$$G_c(z)\Big|_{T=6.0}^{\mathrm{PID}} = \frac{68.5z^2 - 36.8z + 4.9}{z^2 - z}$$

Differencing leads to the time-domain controller algorithms as a function of sampling instant, n,

$$u(z)\Big|_{T=1.5}^{\mathrm{PID}} = 111.5\varepsilon(n) - 132.6\varepsilon(n-1) + 39.4\varepsilon(n-2) + u(n-1) \qquad \text{(E5.6.1)}$$

$$u(z)\Big|_{T=6.0}^{\mathrm{PID}} = 68.5\varepsilon(n) - 36.8\varepsilon(n-1) + 4.9\varepsilon(n-2) + u(n-1) \qquad \text{(E5.6.2)}$$

For PID control then, we see that the controller gain in this case needs to be much lower than that of the equivalent continuous-time system and, indeed, will require to vary in inverse proportion to the sampling interval

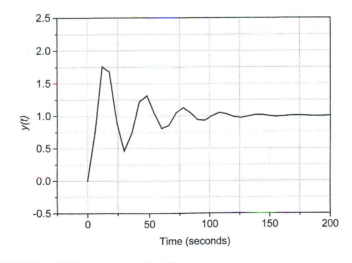

Figure E5.6.2 Tuned PID response at $T = 6.0$s.

for a stable outcome. Thus the larger the sampling interval, the more damped the discrete-time control response will be.

(c)   For *dead-beat control*, we can obtain the required controller transfer function from equation (5.28) and, because this control method is fully model-dependent, it requires no tuning. Hence,

$$G_c(z) = \frac{1}{G(z)} \times \frac{1}{(z-1)}$$

and, at a sampling interval of 1.5s,

$$G_c(z)\big|_{T=1.5}^{\text{dead-beat}} = \frac{z^3 - 0.928z^2}{0.00578} \times \frac{1}{(z-1)} = \frac{z^2 - 0.928z}{0.00578(1 - z^{-1})}$$

which produces the time-domain control algorithm

$$u(n)\big|_{T=1.5}^{\text{dead-beat}} = 173.0\varepsilon(n + 2) - 160.6\varepsilon(n + 1) + u(n - 1) \qquad \text{(E5.6.3)}$$

We see that this algorithm depends on *future values* of control error, ε. Clearly therefore, it will not be possible to implement this (unless reliable *predictions* of control error can be made). In general, where $G_c(z)$ leads to a numerator order which is higher than the denominator order, the resulting control algorithm will be *unfeasible*.
    However, when $T = 6$s,

$$G_c(z)\big|_{T=6.0}^{\text{dead-beat}} = \frac{z^2 - 0.741z}{0.0111z + 0.0096} \times \frac{1}{(z-1)} = \frac{z^2 - 0.741z}{0.0111z^2 - 0.0015z - 0.0096}$$

and

$$u(n)\big|_{T=6.0}^{\text{dead-beat}} = 90.1\varepsilon(n) - 66.8\varepsilon(n - 1) + 0.14u(n - 1) + 0.87u(n - 2) \qquad \text{(E5.6.4)}$$

Equation (E5.6.4) is feasible. As a rule, dead-beat control will only be feasible for systems exhibiting dead time when the sampling interval is greater than the dead time.

(d)   The Dahlin algorithm, in its simplest form, can be obtained from equation (5.32),

$$G_c(z) = \frac{K_d}{G(z)} \times \frac{(1 - E)}{(z - 1)}$$

where $E = \exp(-T/\tau)$. Note that when $E = 0$ and $K_d = 1$, the Dahlin algorithm expressed in this simple form reduces to dead-beat control. As a consequence, when T is less than the dead time of the system model, the Dahlin algorithm expressed here suffers from the same limitation as the dead-beat algorithm in that the control algorithm ends up depending on future values of control error which is unfeasible. (This could be remedied by selecting the form of Dahlin algorithm given by equation (5.31) and tuning the algorithm with a high value of i, but this tends to lead to an ill-conditioned response.)

However, at $T = 6s$ for Dahlin control, a feasible algorithm emerges,

$$G_c(z)|_{T=6.0}^{\text{Dahlin}} = K_d \times \frac{0.0111z + 0.0096}{z^2 - 0.741z} \times \frac{(1 - E)}{(z - 1)}$$

$$= K_d \times \frac{0.0111z(1 - E) + 0.0096(1 - E)}{z^3 - 1.741z^2 + 0.741z} \quad \text{(E5.6.5)}$$

Like the PID algorithm, this needs to be tuned; in this case through K_d and τ. Setting up a Simulink model and by trial and error, we obtain a stable offset-free response at $K_d = 7$ and $\tau = 0.6$ (Figure E5.6.3). However, the response has much greater damping than in either of the PID cases; this may be an advantage where stability and freedom from offset are crucial, but a distinct disadvantage when a rapid response is needed. Thus, for Dahlin control,

$$G_c(z)|_{T=6.0}^{\text{Dahlin}} = \frac{0.078z + 0.067}{z^3 - 1.741z^2 + 0.741z}$$

giving

$$u(n)|_{T=6.0}^{\text{Dahlin}} = 0.078\varepsilon(n - 2) + 0.067\varepsilon(n - 3) + 1.741u(n - 1) - 0.741u(n - 2) \quad \text{(E5.6.6)}$$

(e)   For Kalman control, we use equation (5.33) which is based on a normalised system transfer function, obtained by dividing the numerator and denominator of the system transfer function by the sum of the numerator coefficients when the gain is unity.

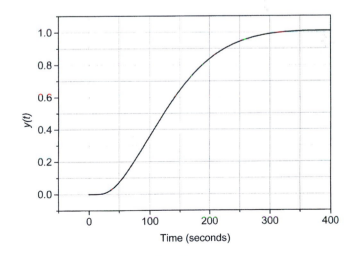

Figure E5.6.3 Tuned Dahlin response at $T = 6.0$s.

For the case when $T = 1.5\text{s}$,

$$\frac{G(z)}{0.08} = \frac{0.0723}{z^3 - 0.928z^2}$$

Normalising gives

$$\frac{1}{13.83z^3 - 12.83z^2} = \frac{Y}{X}$$

and therefore

$$G_c(z)\big|_{T=1.5}^{\text{Kalman}} = \frac{X}{1 - Y} = \frac{1}{0.08} \times \frac{13.83z^3 - 12.83z^2}{z^3 - 1}$$

Note that the result is scaled to the system gain – the control algorithm according to this method will therefore be

$$u(n)\big|_{T=1.5}^{\text{Kalman}} = 172.9\varepsilon(n) - 160.4\varepsilon(n - 1) + 12.5u(n - 3) \qquad \text{(E5.6.7)}$$

Similarly, when $T = 6\text{s}$,

$$G_c(z)\big|_{T=6.0}^{\text{Kalman}} = \frac{1}{0.08} \times \frac{3.86z^2 - 2.86z}{z^2 - 0.54z - 0.46}$$

resulting in the control algorithm

$$u(n)\big|_{T=6.0}^{\text{Kalman}} = 48.3\varepsilon(n) - 35.8\varepsilon(n - 1) + 6.7u(n - 1) + 5.8u(n - 2) \qquad \text{(E5.6.8)}$$

(f) All algorithms are compared in Figure E5.6.4 for a unit step disturbance in control variable (caused for example by a sudden change in

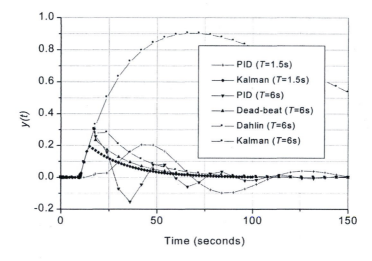

Figure E5.6.4 Comparison of controller responses at low system gain.

return air temperature). Clearly, with the exception of the Dahlin algorithm, all methods achieve control objectives more or less in a short period of time.

(g)   Now, what happens in winter when the external (fresh air) temperature is at design conditions (say $-3°C$)? The new damper gain will be $[20-(-3)]/100\% = 0.23\,K(\%)^{-1}$. If we scale $G(z)$ by $0.23/0.08 = 2.875$, we obtain new responses based on the various control algorithms set up for the mid-season conditions (Figures E5.6.5, E5.6.6 and E5.6.7). Clearly, control

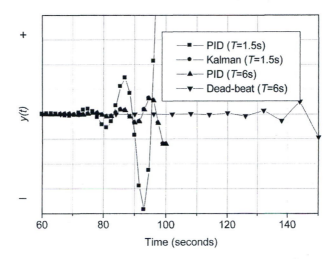

Figure E5.6.5 Comparison of controller responses at high system gain.

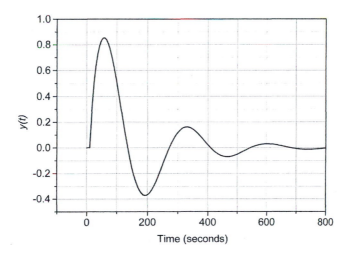

Figure E5.6.6 Dahlin controller response at high system gain and $T = 6s$.

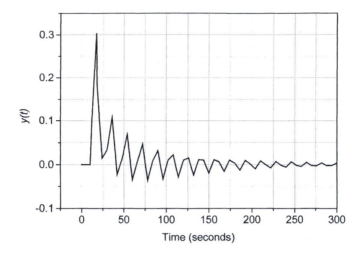

Figure E5.6.7 Kalman controller response at high system gain and $T = 6s$.

is now completely unsatisfactory (Figure E5.6.5) for all but Dahlin (Figure E5.6.6) and Kalman at $T = 6s$ (Figure E5.6.7) and even in these cases control is not entirely satisfactory.

This example represents one of the main cases for the use of *adaptive control* which we will look at later in Chapter 8. Nevertheless, Dahlin and Kalman control in this example have confirmed that they can achieve considerable *robustness*.

In summary, for most problems in discrete-time control, the starting point will be a continuous-time model of the system from which an equivalent z-domain model is generated and many of the stability tests applicable for continuous-time analysis can be used in the discrete domain with appropriate adjustments. Discrete-time control presents us with an additional degree of freedom over that of continuous-time control – that of the sampling interval, T, the choice of which can have a major bearing on the stability and quality of response of a control system. We will come back to this a little later in Chapter 7 when we look at control system identification and tuning.

References

Bennett, S. (1988) *Real Time Control: An Introduction.* Prentice-Hall, Englewood Cliffs, NJ.

Coughanowr, D.R. (1991) *Process Systems Analysis and Control.* McGraw-Hill, New York.

Dahlin, E.B. (1968) Designing and tuning digital controllers. *Instrumentation and Control Systems,* **41** (6), 77–83.

Dexter, A.L. (1988) Control system simulation – computer control. *Energy and Buildings,* **10**, 203–211.

Houpis, C.H., Lamont, G.B. (1992) *Digital Control Systems – Theory, Hardware, Software*. McGraw-Hill, New York.

Jury, E.I., Blanchard, J. (1961) A stability test for linear discrete-time systems in table form (Correspondence). *Proceedings of the Institute of Radio Engineers*, **49** (12), 1947–1948.

Kalman, R.E. (1954) In a discussion of Bergen, A.R., Ragazini, J.R., Sampled data processing techniques for feedback control systems. *AIEE Transactions, Part 2 – Applications & Industry*, **73**, 236–247.

Leigh, J.R. (1992) *Applied Digital Control – Theory, Design and Implementation*. Prentice-Hall, Englewood Cliffs, NJ.

MATLAB (1992) *Control System Toolbox User's Guide*. The Mathworks Inc., Natick, MA.

Oliver, B.M., Pierce, J.R., Shannon, C.E. (1948) The philosophy of pulse code modulation. *Proceedings of the Institute of Radio Engineers*, **36** (11), 1324–1331.

Rohrer, C.E., Stoecker, W. (1986) Z-transforms as an aid to DDC system analysis. *ASHRAE Transactions*, **92** (1B), 185–202.

Simulink (1996) *SIMULINK 2 Dynamic System Simulation for MATLAB*. The Mathworks Inc., Natick, MA.

Strang, G., Nguyen, T. (1996) *Wavelets and Filter Banks*. Wellesley-Cambridge, Wellesley, MA.

6 Multivariable control

6.1 State-space representation of systems

In Chapter 4 we considered the *single-input single-output* (SISO) system, where the input is a set point, and the *multiple-input single-output* (MISO) case in which there are several inputs (a set point and one or more disturbances). Both of these cases have one thing in common; they are characterised by a solitary controlled variable.

In a small number of HVAC applications, we are faced with two or more interacting controlled variables in the same physical domain. In HVAC control, the classic example of this lies in the simultaneous control of both temperature and humidity which is required in close-control air conditioning applications. This is an example of *multivariable* or *multiple-input multiple-output* (MIMO) control.

Particular problems arise in MIMO systems and we will devote much of what is considered in this chapter to these problems and their solutions. We will start by considering methods for handling the representation of these rather more complex interacting systems.

Consider the block diagram representation of the SISO temperature control process of a space modelled using continuous-time transfer functions and involving a controller, control element (i.e. valve), coil, space and a sensor forming the feedback path. Suppose that we have a proportional plus integral (PI) controller and that all other components can be represented by first-order lag transfer functions.

Figure 6.1 represents this general case (symbols have their usual meaning). For the controller, we can write

$$u(s)i_t s = \varepsilon(s)K_c i_t s + \varepsilon(s)K_c = K_c i_t [r(s)s - \phi(s)s] + K_c[r(s) - \phi(s)]$$

This can be expressed as the following differential equation:

$$\frac{du}{dt} - K_c \frac{dr}{dt} + K_c \frac{d\phi}{dt} = \frac{K_c}{i_t}(r - \phi)$$

or, with the more convenient notation,

$$\dot{u} - K_c \dot{r} + K_c \dot{\phi} = \frac{K_c}{i_t} r - \frac{K_c}{i_t} \phi$$

Putting $\dot{u} - K_c \dot{r} + K_c \dot{\phi} = \dot{x}_1$, $-K_c/i_t = a_{15}$, $K_c/i_t = b_{11}$ and $\phi = x_5$ we can now write

$$\dot{x}_1 = a_{15} x_5 + b_{11} r \tag{6.1}$$

We can express the other component blocks in Figure 6.1 in a similar form thus:

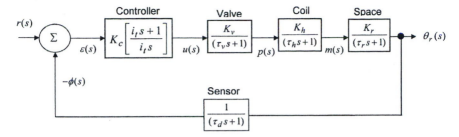

Figure 6.1 Block diagram representation for SISO room temperature control.

$$\dot{x}_2 = a_{21}x_1 + a_{22}x_2$$

where $x_2 = p$, $a_{21} = K_v/\tau_v$ and $a_{22} = -1/\tau_v$,   (6.2)

$$\dot{x}_3 = a_{32}x_2 + a_{33}x_3$$

where $x_3 = m$, $a_{32} = K_h/\tau_h$ and $a_{33} = -1/\tau_h$,   (6.3)

$$\dot{x}_4 = a_{43}x_3 + a_{44}x_4$$

where $x_4 = \theta_r$, $a_{43} = K_r/\tau_r$ and $a_{44} = -1/\tau_r$,   (6.4)

$$\dot{x}_5 = a_{54}x_4 + a_{55}x_5$$

where $a_{54} = 1/\tau_d$ and $a_{55} = -1/\tau_d$.   (6.5)

Equations (6.1) through (6.5) can now be written in the compact matrix–vector form

$$\dot{\mathbf{x}} = \mathbf{A}\mathbf{x} + \mathbf{b}r \tag{6.6}$$

where

$$\dot{\mathbf{x}} = \begin{bmatrix} \dot{x}_1 \\ \dot{x}_2 \\ \dot{x}_3 \\ \dot{x}_4 \\ \dot{x}_5 \end{bmatrix}, \quad \mathbf{A} = \begin{bmatrix} a_{11} & a_{12} & a_{13} & a_{14} & a_{15} \\ a_{21} & a_{22} & a_{23} & a_{24} & a_{25} \\ a_{31} & a_{32} & a_{33} & a_{34} & a_{35} \\ a_{41} & a_{42} & a_{43} & a_{44} & a_{45} \\ a_{51} & a_{52} & a_{53} & a_{54} & a_{55} \end{bmatrix}, \quad \mathbf{x} = \begin{bmatrix} x_1 \\ x_2 \\ x_3 \\ x_4 \\ x_5 \end{bmatrix}, \quad \mathbf{b} = \begin{bmatrix} b_{11} \\ b_{21} \\ b_{31} \\ b_{41} \\ b_{51} \end{bmatrix}$$

and r = a scalar. Here, $x_1, \ldots, x_5$ are called *state variables*. In general, $\mathbf{A}$ will be an $n \times n$ matrix of coefficients (where n is the number of equations forming the model) for the linear time-invariant case. For the non-linear system model, the elements of $\mathbf{A}$ will themselves be variables.

For the SISO case as depicted in this illustration, $\mathbf{b}$ is a column vector and r the single input (i.e. a scalar). For the more general MIMO case, $\mathbf{b}$ will be replaced by $\mathbf{B}$, an $n \times m$ matrix (where m is the number of inputs to the system) and r will be replaced by $\mathbf{r}$, a column vector of input variables.

As to the above illustration, substituting the specific variables involved, the system of Figure 6.1 can be expressed in the following *state-space* or *matrix differential equation* form:

$$
\dot{\mathbf{x}} = \begin{bmatrix} (\ddot{u} + K_c\dot{r} + K_c\dot{\phi}) \\ \dot{p} \\ \dot{m} \\ \dot{\theta}_r \\ \dot{\phi} \end{bmatrix}, \qquad
\mathbf{A} = \begin{bmatrix} 0 & 0 & 0 & 0 & -K_c/i_t \\ K_v/\tau_v & -1/\tau_v & 0 & 0 & 0 \\ 0 & K_h/\tau_h & -1/\tau_h & 0 & 0 \\ 0 & 0 & K_r/\tau_r & -1/\tau_r & 0 \\ 0 & 0 & 0 & 1/\tau_d & -1/\tau_d \end{bmatrix}
$$

$$
\mathbf{x} = \begin{bmatrix} (u + K_c r + K_c \phi) \\ p \\ m \\ \theta_r \\ \phi \end{bmatrix}, \qquad
\mathbf{b} = \begin{bmatrix} K_c/i_t \\ 0 \\ 0 \\ 0 \\ 0 \end{bmatrix}
$$

Zaheer-uddin (1992) applies these methods to a more complex HVAC plant based on linearised differential equations.

Example 6.1

Express the model for the convective heating of a room which was developed in Example 3.4 (Chapter 3) in state-space form.

Solution

This is a MISO system with three inputs (external temperature, heating system flow rate and heating system inlet water temperature) and five state variables (room air temperature, inner layer, outer layer and middle layer fabric temperatures, and the heating system outlet water temperature). The room air temperature forms the single output variable of interest.

First, we reconstruct the individual transfer functions as a set of differential equations. From equation (E3.4.4) we obtain the following for the space air temperature:

$$
\dot{\theta}_r = -3.155\theta_r + 1.158\theta_{wo} + 1.508\theta_{fi} + 0.489\theta_o \tag{E6.1.1}
$$

From equation (E3.4.5) we obtain the following for the heating system outlet water temperature:

$$
\dot{\theta}_{wo} = -8.772\theta_{wo} + 1.754\theta_r + 0.202m_w + 7.018\theta_{wi} \tag{E6.1.2}
$$

From equations (E3.4.1), (E3.4.2) and (E3.4.3) we obtain, after substitution for $\dot{\theta}_r$ from equation (E6.1.1),

$$
\ddot{\theta}_{fi} = -1.984\dot{\theta}_{fi} + 0.566\theta_{fi} - 2.168\theta_r + 1.034\theta_{wo} + 0.437\theta_o + 0.132\theta_{fm} \tag{E6.1.3}
$$

$$
\ddot{\theta}_{fm} = -36.316\dot{\theta}_{fm} - 263.158\theta_{fm} + 16.316\dot{\theta}_{fi} + 218.158\theta_{fi} + 45\theta_{fo} \tag{E6.1.4}
$$

$$
\ddot{\theta}_{fo} = -1.873\dot{\theta}_{fo} - 0.348\theta_{fo} + 0.066\dot{\theta}_{fm} + 0.073\theta_{fm} + 0.275\theta_o \tag{E6.1.5}
$$

We note that fabric equations (E6.1.3) through (E6.1.5) contain second-order terms and, for a state-space representation, these must be reduced to an equivalent first-order expression. This is accomplished as follows. Put

$$\theta'_{fi} = \dot{\theta}_{fi} \tag{E6.1.6}$$

then

$$\dot{\theta}'_{fi} = -1.984\theta'_{fi} + 0.566\theta_{fi} - 2.168\theta_r + 1.034\theta_{wo} + 0.437\theta_o + 0.132\theta_{fm} \tag{E6.1.7}$$

(By substitution, one finds that equations (E6.1.6) and (E6.1.7) are in fact the same as equation (E6.1.3).) In a similar manner,

$$\theta'_{fm} = \dot{\theta}_{fm} \tag{E6.1.8}$$

$$\dot{\theta}'_{fm} = -36.316\theta'_{fm} - 263.158\theta_{fm} + 16.316\theta'_{fi} + 218.158\theta_{fi} + 45\theta_{fo} \tag{E6.1.9}$$

$$\theta'_{fo} = \dot{\theta}_{fo} \tag{E6.1.10}$$

$$\dot{\theta}'_{fo} = -1.873\theta'_{fo} - 0.348\theta_{fo} + 0.066\theta'_{fm} + 0.073\theta_{fm} + 0.275\theta_o \tag{E6.1.11}$$

The matrix differential equation can now be written for this system,

$$\dot{\theta} = A\theta + Br \tag{E6.1.12}$$

(where r is a vector of inputs). Hence,

$$
\begin{bmatrix} \dot{\theta}_r \\ \dot{\theta}_{wo} \\ \dot{\theta}_{fi} \\ \dot{\theta}'_{fi} \\ \dot{\theta}_{fm} \\ \dot{\theta}'_{fm} \\ \dot{\theta}_{fo} \\ \dot{\theta}'_{fo} \end{bmatrix}
=
\begin{bmatrix}
-3.155 & 1.158 & 1.508 & 0 & 0 & 0 & 0 & 0 \\
1.754 & -8.772 & 0 & 0 & 0 & 0 & 0 & 0 \\
0 & 0 & 0 & 1 & 0 & 0 & 0 & 0 \\
-2.168 & 1.0340 & 0.566 & -1.984 & 0.132 & 0 & 0 & 0 \\
0 & 0 & 0 & 0 & 0 & 1 & 0 & 0 \\
0 & 0 & 218.158 & 16.316 & -263.158 & -36.316 & 45 & 0 \\
0 & 0 & 0 & 0 & 0 & 0 & 0 & 1 \\
0 & 0 & 0 & 0 & 0.073 & 0.066 & -0.348 & -1.873
\end{bmatrix}
$$

$$
\times
\begin{bmatrix} \theta_r \\ \theta_{wo} \\ \theta_{fi} \\ \theta'_{fi} \\ \theta_{fm} \\ \theta'_{fm} \\ \theta_{fo} \\ \theta'_{fo} \end{bmatrix}
+
\begin{bmatrix}
0 & 0 & 0.489 \\
0.202 & 7.018 & 0 \\
0 & 0 & 0 \\
0 & 0 & 0.437 \\
0 & 0 & 0 \\
0 & 0 & 0 \\
0 & 0 & 0 \\
0 & 0 & 0.275
\end{bmatrix}
\times
\begin{bmatrix} m_w \\ \theta_{wi} \\ \theta_o \end{bmatrix}
\tag{E6.1.13}
$$

which is the state-space representation of the system model of Example 3.4 and is evidently a far more convenient representation of this system than that described by the block diagram format of Figure E3.4.1. Systems described by such matrix differential equations can be processed using matrix algebra but in most cases these manipulations can be exceedingly tedious and computational tools such as MATLAB will provide useful help (MATLAB, 1996).

6.2 The transfer function matrix

When dealing with stability and other design issues surrounding MISO and MIMO systems, it is often convenient to express the *transfer function matrix* for the entire system. This can be easily formed from the state-space representation of the system.

Taking Laplace transforms of the matrix differential equation (equation (6.6)),

$$L[\dot{x}] = L[Ax + Br] \quad \Rightarrow \quad sx(s) = Ax(s) + Br(s)$$

and therefore

$$x(s)(sI - A) = Br(s)$$

(where I is a unit matrix) or

$$x(s) = G(s)r(s) \tag{6.7}$$

where

$$G(s) = (sI - A)^{-1}B \tag{6.8}$$

In the above, $G(s)$ is the *transfer function matrix* which expresses the relationship between a set of system state variables and their corresponding inputs.

Example 6.2

Neglecting fabric effects in the convective heating case of Example 6.1, express the transfer function matrix.

Solution

If we neglect fabric effects, the deviation in all fabric variables will be zero. Hence rows 3–8 in the matrix differential equation (E6.1.13) all become zero leading to the following state-space representation of this simple system:

$$\begin{bmatrix} \dot{\theta}_r \\ \dot{\theta}_{wo} \end{bmatrix} = \begin{bmatrix} -3.155 & 1.158 \\ 1.754 & -8.772 \end{bmatrix} \times \begin{bmatrix} \theta_r \\ \theta_{wo} \end{bmatrix} + \begin{bmatrix} 0 & 0 & 0.489 \\ 0.202 & 7.018 & 0 \end{bmatrix} \times \begin{bmatrix} m_w \\ \theta_{wi} \\ \theta_o \end{bmatrix} \tag{E6.2.1}$$

We obtain the transfer function matrix from equation (6.8) in which

$$(sI - A) = \begin{bmatrix} s + 3.155 & 1.158 \\ 1.754 & s + 8.772 \end{bmatrix}$$

The determinant of $(sI - A)$ will be

$$|(sI - A)| = (s + 8.772)(s + 3.155) - (-1.158)(-1.754)$$

which reduces to

$$|(sI - A)| = (s + 2.815)(s + 9.112)$$

Now the cofactor of $(s\mathbf{I} - \mathbf{A})$ is

$$\begin{bmatrix} s + 8.772 & -1.754 \\ -1.158 & s + 3.155 \end{bmatrix}$$

and the adjoint of $(s\mathbf{I} - \mathbf{A})$ is the transpose of the cofactor matrix,

$$\mathrm{adj}(s\mathbf{I} - \mathbf{A}) = \begin{bmatrix} s + 8.772 & -1.158 \\ -1.754 & s + 3.155 \end{bmatrix}$$

Hence the inverse of $(s\mathbf{I} - \mathbf{A})$ can now be expressed as

$$(s\mathbf{I} - \mathbf{A})^{-1} = \frac{\mathrm{adj}(s\mathbf{I} - \mathbf{A})}{|(s\mathbf{I} - \mathbf{A})|} = \begin{bmatrix} \dfrac{s + 8.772}{(s + 2.815)(s + 9.112)} & \dfrac{-1.158}{(s + 2.815)(s + 9.112)} \\[3mm] \dfrac{-1.754}{(s + 2.815)(s + 9.112)} & \dfrac{s + 3.155}{(s + 2.815)(s + 9.112)} \end{bmatrix}$$

Finally, the transfer function matrix $\mathbf{G}(s)$ will be given by

$$\mathbf{G}(s) = (s\mathbf{I} - \mathbf{A})^{-1}\mathbf{B} \qquad \text{and} \qquad \mathbf{B} = \begin{bmatrix} 0 & 0 & 0.489 \\ 0.202 & 7.018 & 0 \end{bmatrix}$$

and therefore

$$\mathbf{G}(s) = \begin{bmatrix} \dfrac{-0.233}{(s + 2.815)(s + 9.112)} & \dfrac{-8.127}{(s + 2.815)(s + 9.112)} & \dfrac{0.489(s + 8.772)}{(s + 2.815)(s + 9.112)} \\[3mm] \dfrac{0.202(s + 3.155)}{(s + 2.815)(s + 9.112)} & \dfrac{7.018(s + 3.155)}{(s + 2.815)(s + 9.112)} & \dfrac{-0.858}{(s + 2.815)(s + 9.112)} \end{bmatrix}$$

$$(\text{E6.2.2})$$

We will make use of the transfer function matrix (which applies equally to both s-domain and z-domain systems) in later sections dealing with multivariable control, but first, let us consider where we are likely to come across the multivariable case in HVAC control.

6.3 The multivariable control problem

Some HVAC control processes involve the control of two or more variables in the same physical domain. The classic case in HVAC control is the simultaneous control of temperature and relative humidity in air conditioning systems. We will therefore dedicate our treatment of multivariable control to this special case.

Consider the sensible heating process of Figure 6.2, expressed as a psychrometric process, as air is heated from point 'A' to point 'B' due, for example, to a casual heat gain source in a space. Here θ_r is dry-bulb temperature, S_r the percentage saturation (taken to be the relative humidity), and g_r moisture content. A change (increase) in room temperature, $\delta\theta_r$, also causes a change (decrease) in relative humidity, δS_r. Provision for both temperature and humidity control results in the following sequence of events:

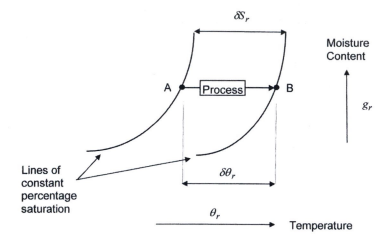

Figure 6.2 Sensible heating process in a room.

- The temperature controller effects a return in air temperature to point 'A'.
- The humidity controller simultaneously takes action to restore the relative humidity prevalent at point 'A'.

The controlled variables are therefore said to be *coupled*, in that a disturbance and subsequent control action in the temperature loop causes a disturbance and independent control action in the humidity loop. The situation is quite insidious, for the usual course of action for humidity control will be to adjust preheat input (i.e. if humidification is achieved through adiabatic spray) or steam injection (practical steam humidifiers result in small changes in temperature besides moisture content). So the control action itself ensures ongoing disturbance in the companion loop.

Some practical approaches to the special case of temperature and humidity control

Control based on moisture content

Figure 6.2 confirms that, for this process at least, if no action were to take place in the humidity loop, the action by the temperature loop in restoring temperature to point 'A' also restores the humidity to its correct location. Clearly, for simultaneous control of temperature and relative humidity, action by the humidity loop is only necessary when the moisture content changes (signifying a latent load disturbance) or in the less frequent event of a change in either set point.

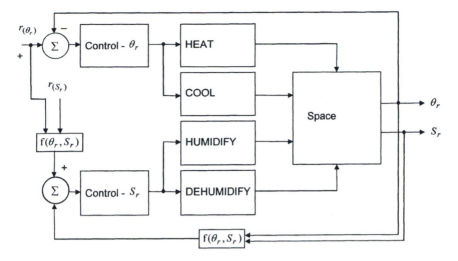

Figure 6.3 Humidity control based on moisture content.

One such strategy for achieving this is proposed by Howell (1988) using the system described by the block diagram of Figure 6.3 for temperature and humidity control in a space (note that the blocks here refer to control functions rather than components, though the practical significance is much the same). Here, the system hardware is the same as that for a conventional temperature and humidity control scheme. Outwardly, control in the humidity loop is with respect to relative humidity but in reality it is with respect to moisture content. Moisture content is 'measured' intelligently by calculating on-line its value based on temperature and relative humidity feedback from the space. Equation (2.4) (Chapter 2) is an example of an equation that could be used for this calculation.

Provision is also made to regulate the relative humidity to its set point, rather than the less convenient moisture content value.

With this scheme, winter-mode heat and humidify functions are fully decoupled. However in summer mode, two specific control functions remain coupled and this arises from the dual function of the cooling system which has to control both temperature and *dehumidification*. The coupled functions are:

- *Cool–humidify*. The action of the cooling system when called on to provide sensible cooling will cause a depression in relative humidity which, in turn, will prompt a call on the humidifier.
- *Dehumidify–heat*. The action of the cooling system when called on to dehumidify will induce a temperature disturbance in the space which, in turn, will prompt a call on the heating system. The situation is further complicated where a coil bypass control strategy is used (Howell, 1986).

These, and other forms of coupled or multivariable control problems, can be improved with the use of *process decouplers* which we shall take a look at in section 6.4.

Constant dew-point–reheat strategy

A *constant dew-point–reheat* control strategy for temperature and humidity entirely decouples both loops. Here, the cooling system has just one function which is to ensure a near-saturated primary supply temperature at design 'apparatus dew-point' conditions, all year round. Thus at all times other than design summer conditions, the supply air to the space(s) is always at a lower dry-bulb temperature and moisture content than the space(s) requires. This then reduces year-round space control functions to those of heat addition and humidification which, when using Howell's proposal (Howell, 1988), results in unconditionally fully decoupled control. Figure 6.4 illustrates the physical plant and control needed for this.

The 'dew-point temperature sensor', θ_{dp}, is in reality just a conventional dry-bulb temperature sensor. However, since the air leaving the cooling coil will always be near-saturated during those occasions when the entering air is at a higher moisture content than that associated with the dew-point value, then the significance of 'dew-point' control becomes clear.

The disadvantage with this approach is that cooling is cancelled with heating for much of the year with consequential implications for energy efficiency. However, the year-round robustness of the method means that it is likely to be the best option where close control over temperature and humidity is of greater priority than energy use, such as is the case in many process air conditioning applications.

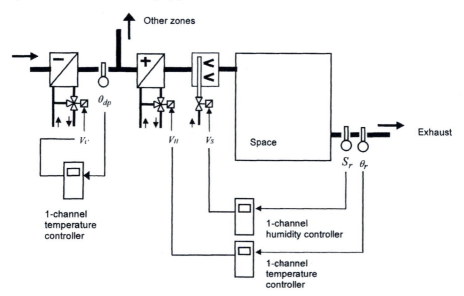

Figure 6.4 Line plant for constant dew-point–reheat control.

Stability in multivariable systems

A multivariable system of two loops (such as temperature and relative humidity control) can be generalised by the block diagram of Figure 6.5. The system output in compact matrix–vector notation will be

$$\mathbf{y} = \mathbf{G_p G_c}[\mathbf{r} - \mathbf{y G_d}] \tag{6.9}$$

where

$$\mathbf{y} = \begin{bmatrix} y_1 \\ y_2 \end{bmatrix}, \quad \mathbf{G_p} = \begin{bmatrix} G_{11} & G_{12} \\ G_{21} & G_{22} \end{bmatrix}, \quad \mathbf{G_c} = \begin{bmatrix} G_{c1} & 0 \\ 0 & G_{c2} \end{bmatrix}, \quad \mathbf{r} = \begin{bmatrix} r_1 \\ r_2 \end{bmatrix}$$

Hence,

$$\mathbf{y}[\mathbf{I} + \mathbf{G_p G_c G_d}] = \mathbf{G_p G_c} \times \mathbf{r}$$

which gives,

$$\mathbf{y} = [\mathbf{I} + \mathbf{G_p G_c G_d}]^{-1} \mathbf{G_p G_c} \times \mathbf{r} \tag{6.10}$$

That is, the two-loop multivariable system results in a similar form of closed-loop model as for the SISO case (equation (4.8)) but in matrix form.

Expressing equation (6.10) as the matrix closed-loop model

$$\frac{\mathbf{y}}{\mathbf{r}} = \frac{\mathrm{adj}[\mathbf{I} + \mathbf{G}]}{|[\mathbf{I} + \mathbf{G}]|} \times \mathbf{G_p G_c} \tag{6.11}$$

(where $\mathbf{G} = \mathbf{G_p G_c G_d}$ is the matrix open-loop transfer function), implies that the denominator of the matrix closed-loop model for a MIMO system, $|[\mathbf{I} + \mathbf{G}]|$, specifies the closed-loop roots for which the multivariable system will be stable or unstable.

Note that whilst we have considered the dual interacting loop situation here, it is possible to show that this condition applies irrespective of the number of interactions present.

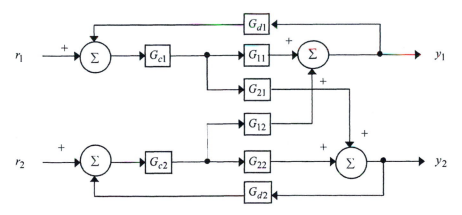

Figure 6.5 Generalised representation of a two-loop multivariable system.

Thus stability in multivariable systems depends on the determinant of a unit matrix plus the open-loop matrix transfer function of the system.

Modelling and stability analysis for the multivariable temperature–humidity case

Based on Figure 6.5, the multivariable system for the (two-loop) temperature–humidity control system can be represented as Figure 6.6. Here $G_{c\theta_r}$ and G_{cg_r} represent temperature and humidity loop controllers, respectively; and $G_{d\theta_r}$ and G_{dg_r} represent temperature and humidity sensors, respectively.

The cool–humidify case

As was discussed earlier, for systems under moisture-content-disturbance-based humidity control, the actions of the cooling system in providing both temperature and humidity control in summer cause two-loop interactions: the need to humidify after sensible cooling (cool–humidify) and the need to heat after dehumidifying (dehumidify–heat).

Let us first consider the development of a transfer function matrix which describes the cool–humidify case. $G_{\theta_r}(s)$ will consist of the transfer function connecting space temperature with temperature loop control. If we assume an all-air system with chilled water cooling coil, then based on equations (3.1) and (3.34) a simple system–space transfer function can be written if we choose to neglect the long-term dynamics of the space. Suppose also that the coil control valve can be described with a first-order expression. Then,

$$G_{\theta_r}(s) = \frac{K_{\theta_r}}{(\tau_{\theta_r}s + 1)(\tau_{cc}s + 1)(\tau_{cv}s + 1)} \tag{6.12}$$

where   (see over)

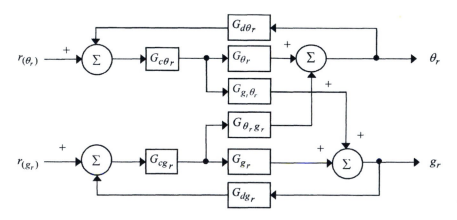

Figure 6.6 The multivariable temperature–humidity control system.

$$K_{\theta_r} = \frac{m_a c_{pa} K_{cc} K_{cv}}{(m_a c_{pa} + \Sigma(AU_i) + n_v V_r/3)}$$

and τ_{θ_r}, τ_{cc} and τ_{cv} are the time constants of the space air volume (see equation (3.6)), cooling coil and coil control valve respectively. K_{cc} and K_{cv} are the cooling coil temperature gain and valve gain, respectively.

$G_{g_r}(s)$ will consist of the transfer function connecting space moisture content with the humidity control loop. There will be a transfer function for the space and one for the humidifier. A room moisture balance can be expressed for the former. In fact moisture transfer in a space is a highly complex phenomenon which is complicated by the long-term moisture absorption/desorption in the fabric of the space (see for example Yik et al., 1995). However, in common with the neglect of detailed fabric effects when dealing with sensible heat transfer, these moisture absorption/desorption dynamics can be neglected since they are dynamically too damped to be of any interest in the special case where short-term control system response is of interest.

For our purposes, the following simple moisture balance will suffice, and for a detailed treatment of moisture transients in spaces, the reader is directed to a number of sources which consider these phenomena in detail (Letherman, 1988; Zaheer-uddin, 1993; Loveday et al., 1993; Townsend et al., 1986):

$$V_r \rho_a \frac{dg_r}{dt} = m_a(g_p - g_r) - \frac{n_v V_r \rho_a}{3600}(g_o - g_r) + m_{source} \tag{6.13}$$

where g_p is the moisture content of air leaving the plant and entering the space. Discarding 'steady-state' terms (involving the external air moisture content, g_o, and moisture source term, m_{source}), and taking Laplace transfoms, we can write

$$g_r(s) = \frac{K_{g_r}}{(\tau_{g_r} s + 1)} \times g_p(s) \tag{6.14}$$

where

$$K_{g_r} = \frac{m_a}{(m_a + n_v V_r \rho_a/3600)} \quad \text{and} \quad \tau_{g_r} = \frac{V_r \rho_a}{(m_a + n_v V_r \rho_a/3600)}$$

The humidifier in common use today is the direct steam-injection type. The steam-injection process into an air stream, though almost isothermal, causes a small temperature increase such that, strictly, there is a humidifier–temperature loop interaction. The temperature gain, $\Delta\theta_{r(g)}$, across a steam-injection process to air can be obtained from a heat balance such that, starting with the equation for air enthalpy (equation (2.6)), it is possible to show that

$$\Delta\theta_{r(g)} = \frac{h_{ai}m_a + m_{st}h_{st} - 2501(g_{ai}m_a + m_{st})}{m_a(1.805 g_{ai} + 1) + 1.805 m_{st}} - \theta_{ai} \tag{6.15}$$

where h represents enthalpy and the subscripts ai and st refer to inlet air and steam, respectively (for a detailed discussion of steam-injection processes to air see Jones (1995)).

The principal dynamic influence over the steam-injection process will be due to the modulating steam valve and its positioner, for which we might adopt a first-order description in common with the cooling coil valve considered earlier. Combining equation (6.14) with a first-order steam valve gives us

$$G_{g_r}(s) = \frac{K_{g_r}}{(\tau_{g_r} s + 1)(\tau_{sv} s + 1)} \tag{6.16}$$

where

$$K_{g_r} = \frac{m_a K_{sv}}{(m_a + n_v V_r \rho_a / 3600)}$$

in which K_{sv} and τ_{sv} are the gain and time constant of the steam valve.

For the temperature–humidity loop interaction resulting from unwanted interference in humidity when the cooling coil is called upon to provide sensible cooling, $G_{g,\theta_r}(s)$ will consist of the transfer function connecting space moisture content (i.e. equation (6.14)) with the *latent* component of the cooling coil and its control valve.

For the latent component of the cooling coil, it is reasonable to suppose that the coil time constant will be the same as for the sensible component, τ_{cc}. Hence,

$$G_{g,\theta_r}(s) = \frac{K_{g,\theta_r}}{(\tau_{g_r} s + 1)(\tau_{cc} s + 1)(\tau_{cv} s + 1)} \tag{6.17}$$

where

$$K_{g,\theta_r} = \frac{m_a K'_{cc} K_{cv}}{(m_a + n_v V_r P_a / 3600)}$$

and K'_{cc} is the cooling coil moisture content gain, the change in air moisture content across the cooling coil per unit mass flow rate of cooling medium.

Finally for the cool–humidify case there is the interaction resulting from the influence in the temperature loop of the steam-injection humidifier. This is included here for completeness though in most cases it will be sufficiently minor to be ignored. The resulting transfer function, $G_{\theta,g_r}(s)$, will be the space temperature transfer function and steam humidifier transfer function combined in which the gain of the latter, K'_{sv}, will be the temperature gain of the humidifier (i.e. the increase in air temperature described by equation (6.15) per unit of steam control valve signal). That is

$$G_{\theta,g_r}(s) = \frac{K_{\theta,g_r}}{(\tau_{\theta_r} + 1)(\tau_{sv} s + 1)} \tag{6.18}$$

where

$$K_{\theta,g_r} = \frac{m_a c_{pa} K'_{sv}}{(m_a c_{pa} + \Sigma(AU_i) + n_v V_r/3)}$$

The matrix transfer function describing the temperature–humidity control system with cool–humidify interaction can now be expressed as

$$\mathbf{G}_{(c-h)}(s) = \begin{bmatrix} \dfrac{K_{\theta_r}}{(\tau_{\theta_r}s + 1)(\tau_{cc}s + 1)(\tau_{cv}s + 1)} & \dfrac{K_{\theta,g_r}}{(\tau_{\theta_r}s + 1)(\tau_{cv}s + 1)} \\ \dfrac{K_{g,\theta_r}}{(\tau_{g_r}s + 1)(\tau_{cc}s + 1)(\tau_{cv}s + 1)} & \dfrac{K_{g_r}}{(\tau_{g_r}s + 1)(\tau_{sv}s + 1)} \end{bmatrix} \times \begin{bmatrix} G_{c\theta_r} & 0 \\ 0 & G_{cg_r} \end{bmatrix}$$

$$\times \begin{bmatrix} G_{d\theta_r} & 0 \\ 0 & G_{dg_r} \end{bmatrix} \tag{6.19}$$

The dehumidify–heat case

For this case, $G_{\theta_r}(s)$ is the heating mode form of the previous case,

$$G_{\theta_r}(s) = \frac{K'_{\theta_r}}{(\tau_{\theta_r}s + 1)(\tau_{hc}s + 1)(\tau_{hv}s + 1)} \tag{6.20}$$

where

$$K'_{\theta_r} = \frac{m_a c_{pa} K_{hc} K_{hv}}{(m_a c_{pa} + \Sigma(AU_i) + n_v V_r/3)}$$

K_{hc} and K_{hv} are the temperature gain of the heating coil and heating coil control valve gain respectively, and τ_{hc} and τ_{hv} are the coil and control valve time constants. For the dehumidify–heat case, $G_{g,\theta_r}(s)$ will be zero since the heating coil does not influence the space moisture content.

$G_{g_r}(s)$ is the transfer function of the latent component of the cooling coil (and control valve) together with the moisture balance transfer function of the space, i.e. equation (6.17) of the cool–humidify case. $G_{\theta,g_r}(s)$ for this case is the same as $G_{\theta_r}(s)$ of the previous case, i.e. equation (6.12).

Hence the overall matrix transfer function for this system and its interactions is as follows:

$$\mathbf{G}_{(d-ht)}(s) = \begin{bmatrix} \dfrac{K'_{\theta_r}}{(\tau_{\theta_r}s + 1)(\tau_{hc}s + 1)(\tau_{hv}s + 1)} & 0 \\ \dfrac{K'_{\theta,g_r}}{(\tau_{\theta_r}s + 1)(\tau_{cc}s + 1)(\tau_{cv}s + 1)} & \dfrac{K'_{g_r}}{(\tau_{g_r}s + 1)(\tau_{cc}s + 1)(\tau_{cv}s + 1)} \end{bmatrix}$$

$$\times \begin{bmatrix} G_{c\theta_r} & 0 \\ 0 & G_{cg_r} \end{bmatrix} \times \begin{bmatrix} G_{d\theta_r} & 0 \\ 0 & G_{dg_r} \end{bmatrix} \tag{6.21}$$

where $K'_{g_r} = K_{g,\theta_r}$ (equation (6.17)) and $K'_{\theta,g_r} = K_{\theta_r}$ (equation (6.12)).

Example 6.3

Assess stability for the cool–humidify and dehumidify–heat interactions for the following fully air conditioned application, neglecting the dynamics of the heating and cooling coils, and using simple proportional controllers for temperature and humidity.

Data

Space volume (V_r): $1000\,\text{m}^3$
Infiltration rate (n_v): $3.0\,\text{h}^{-1}$
$\Sigma(AU_i)$: $2.7\,\text{kWK}^{-1}$
System air mass flow rate (m_a): $2.0\,\text{kgs}^{-1}$
Valve time constants (all valves, $\tau_{cv}, \tau_{hv}, \tau_{sv}$): $1.0\,\text{min}$
Humidity and temperature sensor time constants $(\tau_{d\theta_r}, \tau_{dg_r})$: 0.5 and 2.0 min
(respectively)
Space conditions: 22°C, 50% saturation (from psychrometric tables, $h_r = 43.34\,\text{kJkg}^{-1}$, $g_r = 0.0084\,\text{kgkg}^{-1}$)
Chilled water cooling coil: 20 kW (sensible), 8 kW (latent) (chilled water maximum flow rate $1.3\,\text{kgs}^{-1}$)
Hot water heating coil: 15 kW (heating water maximum flow rate $0.4\,\text{kgs}^{-1}$)
Steam humidifier: $4.87 \times 10^{-3}\,\text{kgs}^{-1}$ (steam at 1.0 bar dry saturated; $h_{st} = 2675.8\,\text{kJkg}^{-1}$)
Properties of air: $\rho_a = 1.2\,\text{kgm}^{-3}$, $c_{pa} = 1.005\,\text{kJkgK}^{-1}$

Solution

First, we derive the system matrix transfer function for the cool–humidify case. Neglecting the dynamics of the cooling coil, this will take the following form:

$$
\mathbf{G_{(c-h)}}(s) = \begin{bmatrix} \dfrac{K_{\theta_r}}{(\tau_{\theta_r}s + 1)(\tau_{cv}s + 1)} & \dfrac{K_{\theta_r g_r}}{(\tau_{\theta_r}s + 1)(\tau_{sv}s + 1)} \\[3mm] \dfrac{K_{g_r\theta_r}}{(\tau_{g_r}s + 1)(\tau_{cv}s + 1)} & \dfrac{K_{g_r}}{(\tau_{g_r}s + 1)(\tau_{sv}s + 1)} \end{bmatrix} \times \begin{bmatrix} K_{c\theta_r} & 0 \\ 0 & K_{cg_r} \end{bmatrix}
$$

$$
\times \begin{bmatrix} \dfrac{1}{(\tau_{d\theta_r}s + 1)} & 0 \\[3mm] 0 & \dfrac{1}{(\tau_{dg_r}s + 1)} \end{bmatrix} \tag{E6.3.1}
$$

Note that, for convenience in this example, we will adopt a positive sign convention for heat extracted (thus a positive gain for the cooling coil will follow).

Now, each transfer function is derived in turn. In $G_{\theta_r}(s)$ (equation (6.12)), K_{cc} will be the cooling coil temperature gain, i.e.

$$
K_{cc} = \frac{\Delta\theta a}{\Delta m_w} \quad \text{and} \quad \Delta\theta_a = \frac{q_{cc}(\text{sensible})}{m_a c_{pa}} = \frac{20}{2 \times 1.005} = 9.95\text{K}
$$

Thus,

$$K_{cc} = \frac{9.95}{1.3} = 7.654 \text{K} (\text{kgs}^{-1})^{-1}$$

K_{cv} will be the chilled water control valve gain, the water flow through the valve per unit control signal. If we adopt a control signal in the interval $0 \rightarrow 1$, the valve gain will simply be equal to the rated water flow rate if the valve can be assumed to give a linear response across its entire range, i.e. 1.3kgs^{-1}.

From equation (6.12),

$$K_{\theta_r} = \frac{2 \times 1.005 \times 7.654 \times 1.3}{\left[2 \times 1.005 + 2.7 + (3 \times 1000)/(3 \times 10^3)\right]} = 3.5 \text{K}$$

From equation (3.1),

$$\tau_{\theta_r} = \frac{1000 \times 1.005 \times 1.2}{\left[2 \times 1.005 + 2.7 + (3 \times 1000)/(3 \times 10^3)\right]} = 211.2 \text{s} = 3.52 \text{min}$$

and therefore

$$G_{\theta_r}(s) = \frac{K_{\theta_r}}{(\tau_{\theta_r} s + 1)(\tau_{cv} s + 1)} = \frac{3.5}{(3.5s + 1)(s + 1)} \qquad \text{(E6.3.2)}$$

For $G_{g,\theta_r}(s)$, K'_{cc} is the cooling coil air moisture content gain. The moisture content change across the cooling coil will be

$$\Delta g = \frac{q_{cc}(\text{latent})}{h_{fg} \times m_a}$$

where h_{fg} is the latent heat of evaporation of water at atmospheric pressure, i.e. $h_{fg} = 2501 \text{kJkg}^{-1}$. Hence

$$\Delta g = \frac{8}{2501 \times 2} = 1.6 \times 10^{-3} \text{kgkg}^{-1}$$

and

$$K'_{cc} = \frac{1.6 \times 10^{-3}}{1.3} = 1.23 \times 10^{-3} \text{kgkg}^{-1} (\text{kgs}^{-1})^{-1}$$

With the chilled water valve gain as before, then from equation (6.17),

$$K_{g,\theta_r} = \frac{2 \times 1.23 \times 10^{-3} \times 1.3}{(2 + 3 \times 1000 \times 1.2/3600)} = 1.07 \times 10^{-3} \text{kgkg}^{-1}$$

and τ_{g_r} is obtained from equation (6.14):

$$\tau_{g_r} = \frac{1000 \times 1.2}{(2 + 3 \times 1000 \times 1.2/3600)} = 400 \text{s} = 6.7 \text{min}$$

Therefore

$$G_{g,\theta_r}(s) = \frac{K_{g,\theta_r}}{(\tau_{g_r}s + 1)(\tau_{cv}s + 1)} = \frac{1.07 \times 10^{-3}}{(6.7s + 1)(s + 1)} \tag{E6.3.3}$$

For G_{θ,g_r}, K'_{sv} is the temperature gain of the steam humidifier, the temperature rise across the injection process per unit control signal. Again, for a control signal in the interval $0 \to 1$, K'_{sv} will simply be the temperature gain calculated from equation (6.15). Using the data supplied,

$$K'_{sv} = \Delta\theta_{r(g)} = 43.34 \times 2 + 4.87 \times 10^{-3} \times 2675.8$$

$$\frac{-2501(8.4 \times 10^{-3} \times 2 + 4.87 \times 10^{-3})}{2(1.805 \times 8.4 \times 10^{-3} + 1) + 1.805 \times 4.87 \times 10^{-3}} - 22 = 0.32\text{K}$$

K_{θ,g_r} is now found from equation (6.18):

$$K_{\theta,g_r} = \frac{2 \times 1.005 \times 0.32}{[2 \times 1.005 + 2.7 + (3 \times 1000)/(3 \times 10^3)]} = 0.11\text{K}$$

and

$$G_{\theta,g_r} = \frac{K_{\theta,g_r}}{(\tau_{\theta_r}s + 1)(\tau_{sv}s + 1)} = \frac{0.11}{(3.5s + 1)(s + 1)} \tag{E6.3.4}$$

Finally for G_{g_r}, K_{sv} is the steam-injection gain of the steam valve, the moisture rise across the steam humidifier per unit control signal. The moisture content rise across the steam humidifier will be the injection rate divided by the air mass flow rate. Thus for a control signal in the interval $0 \to 1$,

$$K_{sv} = \frac{4.87 \times 10^{-3}}{2} = 2.435 \times 10^{-3}\text{kgkg}^{-1}$$

and from equation (6.16),

$$K_{g_r} = \frac{2 \times 2.435 \times 10^{-3}}{(2 + 3 \times 1000 \times 1.2/3600)} = 1.62 \times 10^{-3}\text{kgkg}^{-1}$$

and

$$G_{g_r}(s) = \frac{K_{g_r}}{(\tau_{g_r}s + 1)(\tau_{sv}s + 1)} = \frac{1.62 \times 10^{-3}}{(6.7s + 1)(s + 1)} \tag{E6.3.5}$$

The sensor transfer functions will be

$$G_{d\theta_r} = \frac{1}{(0.5s + 1)} \qquad \text{(temperature)}$$

and

$$G_{dg_r} = \frac{1}{(2s + 1)} \qquad \text{(humidity)}$$

We can now write down the system transfer function matrix for the cool–humidify case. Substituting equations (E6.3.2)–(E6.3.5) into equation (E6.3.1),

$$\mathbf{G}_{(c-h)}(s) = \begin{bmatrix} \dfrac{3.5}{(3.5s+1)(s+1)} & \dfrac{0.11}{(3.5s+1)(s+1)} \\ \dfrac{1.07 \times 10^{-3}}{(6.7s+1)(s+1)} & \dfrac{1.62 \times 10^{-3}}{(6.7s+1)(s+1)} \end{bmatrix} \times \begin{bmatrix} K_{c\theta_r} & 0 \\ 0 & K_{cg_r} \end{bmatrix}$$

$$\times \begin{bmatrix} \dfrac{1}{(0.5s+1)} & 0 \\ 0 & \dfrac{1}{(2s+1)} \end{bmatrix} \qquad \text{(E6.3.6)}$$

We now turn to the dehumidify–heat case. The system transfer function matrix for this case will take the following general form (based on equation (6.21)):

$$\mathbf{G}_{(d-ht)}(s) = \begin{bmatrix} \dfrac{K'_{\theta_r}}{(\tau_{\theta_r}s+1)(\tau_{hv}s+1)} & \dfrac{K'_{\theta_r g_r}}{(\tau_{\theta_r}s+1)(\tau_{cv}s+1)} \\ 0 & \dfrac{K'_{g_r}}{(\tau_{g_r}s+1)(\tau_{cv}s+1)} \end{bmatrix} \times \begin{bmatrix} K_{c\theta_r} & 0 \\ 0 & K_{cg_r} \end{bmatrix}$$

$$\times \begin{bmatrix} \dfrac{1}{(\tau_{d\theta_r}s+1)} & 0 \\ 0 & \dfrac{1}{(\tau_{dg_r}s+1)} \end{bmatrix} \qquad \text{(E6.3.7)}$$

For $G'_{\theta_r}(s)$ we now insert the heating coil and valve gains. In a manner similar to that of the sensible cooling path of the previous case, we find that $K_{hc} = 18.7\,\mathrm{K}$ and $K_{hv} = 0.4\,\mathrm{kgs}^{-1}$. Hence $K'_{\theta} = 2.6\,\mathrm{K}$ and

$$G'_{\theta_r}(s) = \dfrac{2.6}{(3.5s+1)(s+1)} \qquad \text{(E6.3.8)}$$

$G'_{\theta_r g_r}(s)$ will be the same as $G_{\theta_r}(s)$ in the previous case and $G'_{g_r}(s)$ is the same as G_{g,θ_r} in the previous case. Hence,

$$\mathbf{G}_{(d-ht)}(s) = \begin{bmatrix} \dfrac{2.6}{(3.5s+1)(s+1)} & \dfrac{3.5}{(3.5s+1)(s+1)} \\ 0 & \dfrac{1.07 \times 10^{-3}}{(6.7s+1)(s+1)} \end{bmatrix} \times \begin{bmatrix} K_{c\theta_r} & 0 \\ 0 & K_{cg_r} \end{bmatrix}$$

$$\times \begin{bmatrix} \dfrac{1}{(0.5s+1)} & 0 \\ 0 & \dfrac{1}{(2s+1)} \end{bmatrix} \qquad \text{(E6.3.9)}$$

To assess the stability of these two interacting systems, we require to determine $|[\mathbf{I} + \mathbf{G}]| = 0$ for each case.

First, for the cool–humidify case, equation (E6.3.6) reduces to

$$G_{(c-h)}(s) = \begin{bmatrix} \dfrac{3.5K_{c\theta_r}}{(3.5s+1)(s+1)(0.5s+1)} & \dfrac{0.11K_{cg_r}}{(3.5s+1)(s+1)(2s+1)} \\[4mm] \dfrac{1.07\times10^{-3}K_{c\theta_r}}{(6.7s+1)(s+1)(0.5s+1)} & \dfrac{1.62\times10^{-3}K_{cg_r}}{(6.7s+1)(s+1)(2s+1)} \end{bmatrix} \qquad \text{(E6.3.10)}$$

from which

$$[\mathbf{I}+\mathbf{G}(s)]_{c-h} = \left[\begin{array}{c} \dfrac{(3.5s+1)(s+1)(0.5s+1)+3.5K_{c\theta_r}}{(3.5s+1)(s+1)(0.5s+1)} \\[4mm] \dfrac{1.07\times10^{-3}K_{c\theta_r}}{(6.7s+1)(s+1)(0.5s+1)} \end{array} \right.$$

$$\left. \begin{array}{c} \dfrac{0.11K_{cg_r}}{(3.5s+1)(s+1)(2s+1)} \\[4mm] \dfrac{(6.7s+1)(s+1)(2s+1)+1.62\times10^{-3}K_{cg_r}}{(6.7s+1)(s+1)(2s+1)} \end{array} \right]$$

Thus, $|[\mathbf{I}+\mathbf{G}(s)]_{c-h}| = 0$ implies that

$$\left[(3.5s+1)(s+1)(0.5s+1)+3.5K_{c\theta_r}\right]\times\left[(6.7s+1)(s+1)(2s+1)+1.62\times10^{-3}K_{cg_r}\right]$$
$$+\,1.18\times10^{-4}K_{c\theta_r}K_{cg_r} = 0$$

which leads to

$$as^6 + bs^5 + cs^4 + ds^3 + es^2 + fs + l = 0 \qquad \text{(E6.3.11)}$$

in which

$a = 23.5$
$b = 115.7$
$c = 211.1$
$d = 181.4 + 46.9K_{c\theta_r} + 2.835\times10^{-3}K_{cg_r}$
$e = 76.4 + 77.4K_{c\theta_r} + 9.315\times10^{-3}K_{cg_r}$
$f = 14.7 + 34K_{c\theta_r} + 8.1\times10^{-3}K_{cg_r}$
$l = 1 + 3.5K_{c\theta_r} + 1.62\times10^{-3}K_{cg_r} + 5.55\times10^{-3}K_{c\theta_r}K_{cg_r}$

That is, the multivariable system considered, consisting individually of third-order transfer functions, results in a sixth-order model when the various interactions are included.

In a similar manner, we can obtain the following characteristic equation for the dehumidify–heat case:

$$d's^6 + b's^5 + c's^4 + d's^3 + e's^2 + f's + l' = 0 \qquad \text{(E6.3.12)}$$

in which

$a' = 23.5$
$b' = 115.7$
$c' = 211.1$
$d' = 181.4 + 34.8K_{c\theta_r} + 1.873\times10^{-3}K_{cg_r}$
$e' = 76.4 + 57.5K_{c\theta_r} + 6.153\times10^{-3}K_{cg_r}$
$f' = 14.7 + 25.2K_{c\theta_r} + 5.35\times10^{-3}K_{cg_r}$
$l' = 1 + 2.6K_{c\theta_r} + 1.07\times10^{-3}K_{cg_r} + 2.782\times10^{-3}K_{c\theta_r}K_{cg_r}$

Stability can now be investigated by identifying the combinations of controller gain values which lead to a critically stable condition; for example, a Routh array might be used.

Initially, if we put $K_{cg_r} = 0$, then using a Routh array as set out in section 4.2, we find critical stability at $K_{c\theta_r} = 4.4\,\mathrm{K}^{-1}$ for the cool–humidify case, and $6.0\,\mathrm{K}^{-1}$ for the dehumidify-heat case. Similarly, when we put $K_{c\theta_r} = 0$ we obtain results for critical stability when humidifying control is active and these are $K_{cg_r} = 9250\,(\mathrm{kgkg}^{-1})^{-1}$ for cool–humidify and $14\,000\,(\mathrm{kgkg}^{-1})^{-1}$ for dehumidify–heat. The reader may verify these results but will find the algebra involved rather tedious, involving the solution of a number of polynomials in $K_{c\theta_r}$ and K_{cg_r}; one way to ease this is to 'guess' values of $K_{c\theta_r}$, K_{cg_r} and use trial and error. Inevitably, computational tools such as simulation modelling will be the best way forward.

Now, these results merely tell us what the critical controller gain values need to be when interaction is suppressed (resulting from setting the accompanying controller gain value to zero). They do not tell us whether these values will be safe when interaction is 'switched on'. But if we now substitute the pairs of critical values we have just calculated into the respective characteristic equations and express the Routh arrays, we find that both systems are stable – *just*. In other words, this system will be stable *with interaction* at controller gain values calculated for the individual loops when free of interaction.

So we could have saved ourselves all the trouble we have gone to and simply found the critical controller gains using the much simpler third-order system models of the individual control loops! However, this is a risky strategy as we will see a little later. Meanwhile, let us see how well our closed-loop system performs when disturbed. Actual settings for these simple proportional controllers may be taken as 50% of the critical values according to Ziegler–Nichols rules – see section 7.3 (Table 7.2).

Setting up the system with these values, Figure E6.3.1 shows a Simulink model (Simulink, 1996) for the cool–humidify case in which the cooling coil is active to achieve temperature control in the space and the humidifier is active to compensate for the resulting humidity depressions. Figure E6.3.2 shows the disturbance in temperature arising from a set point change in the humidity loop, whilst Figure E6.3.3 shows the disturbance in space moisture content arising from a set point change in the temperature control loop.

For the humidity disturbance, the set point was arbitrarily changed by $0.002\,\mathrm{kgkg}^{-1}$. This is equivalent to the entire capacity of the steam humidifier and we see that the effect on space temperature is negligible (Figure E6.3.2). This is to be expected since the temperature gain of the steam humidifier is very low; indeed this disturbance path can be neglected in practical systems involving steam humidifiers.

However, when an arbitrary temperature set point change of 2K is applied (Figure E6.3.3), the disturbance in space moisture content is initially significant, rising to $0.0004\,\mathrm{kgkg}^{-1}$. This is equivalent to about 25% of

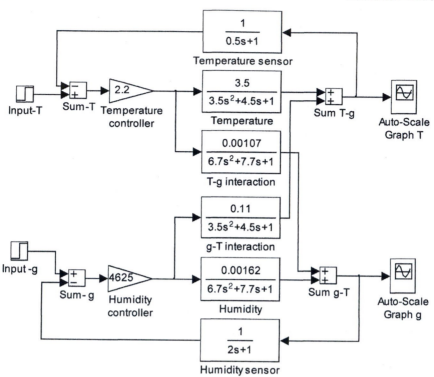

Figure E6.3.1 Simulink model of the cool–humidity case.

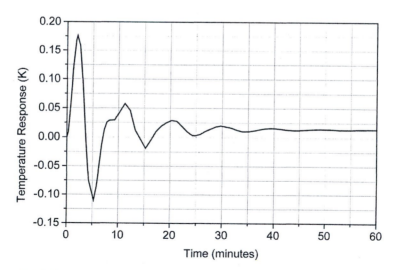

Figure E6.3.2 Temperature response to humidity loop set point change.

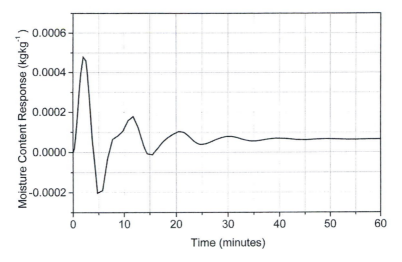

Figure E6.3.3 Humidity response to temperature set point change (cooling).

the overall range of space moisture content which the plant in this example is designed to cater for. The response eventually settles with an offset of about 4% of this overall range as a result of action by the steam humidifier.

In the dehumidify–heat case, the cooling coil now appears in the humidity control loop and a heating coil compensates for temperature disturbances, thus there will be no humidity disturbances arising from temperature control action in this situation. Figure E6.3.4 shows the Simulink model for this case and Figure E6.3.5 shows the disturbance pattern in temperature arising from a step change (again arbitrarily set at 0.002kgkg^{-1}) in space moisture content.

Evident from Figure E6.3.5, temperature disturbances when the cooling coil is active in the humidity control loop will be substantial. Here, the dramatic initial changes in temperature eventually settle but leave a steady-state offset of a little under 1K. Once again the humidity set point producing this disturbance is equivalent to the full range of humidity control in the space so this can be interpreted as the maximum likely disturbance pattern for this particular interaction.

To summarise for the temperature–humidity control example we have considered, interaction will mainly tend to present disturbance problems only when the cooling coil is active in the humidity control loop. Stability of the interacting system can be expected simply by tuning the individual control loops without the need to account for the interaction.

Inevitably, the linear methods we have used here restrict the robustness of the conclusions arrived at. To explore performance over a much wider range of plant operation requires the use of computer simulation, in turn involving wider plant participation in the modelling process – see for example Zaheer-uddin (1993).

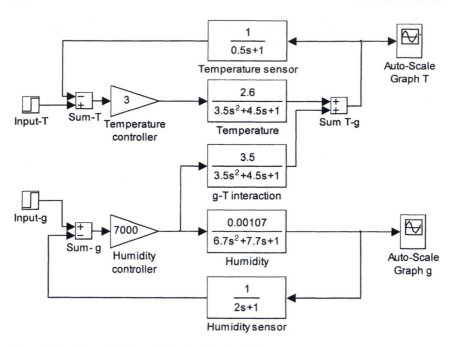

Figure E6.3.4 `Simulink` model for the dehumidify–heat case.

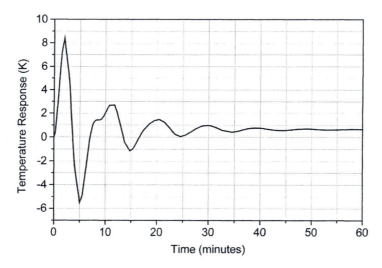

Figure E6.3.5 Temperature response to humidity set point change (dehumidification).

6.4 Design of multivariable compensators

The key to minimising interaction in multivariable systems lies in achieving *diagonal dominance* in the system transfer function matrix. In many cases, this may be assessed to an approximate degree through steady-state characteristics of the system which is especially valuable where the development of transfer function models for the various paths may not be possible.

Steady-state criteria

Seborg *et al.* (1989) discuss a method for assessing multivariable system stability based on *steady-state criteria only*. They show that a relative gain in the 1–1 path (see Figure 6.5) can be expressed as follows for a two-loop system:

$$\lambda_{11} = \frac{1}{1 - K_{12}K_{21}/K_{11}K_{22}} \tag{6.22}$$

Further elements forming a *relative gain array* are established as follows:

$$\lambda_{22} = \lambda_{11} \quad \text{and} \quad \lambda_{12} = \lambda_{21} = 1 - \lambda_{11}$$

which results in relative gain matrix (RGM), Λ,

$$\Lambda = \begin{bmatrix} \lambda & 1-\lambda \\ 1-\lambda & \lambda \end{bmatrix} \tag{6.23}$$

Note that the RGM rows and columns will always sum to unity, and that similar arrays can be derived for systems with greater than two loops. When $\lambda = 1$ or $\lambda = 0$ there is either one-way interaction or no interaction at all and at least one of the gains must be zero. For other conditions of λ there will be interaction to varying degrees but especially where $\lambda \to \infty$ or $\lambda \to -\infty$. In general when multivariable control design can be based on steady-state criteria, λ should be positive and as close to unity as possible.

Example 6.3 (Continued)

In Example 6.3 for instance, applying the existing gains for the cool–humidify case gives, from equation (6.22),

$$\lambda = \frac{1}{1 - (0.11 \times 1.07 \times 10^{-3})/(3.5 \times 1.62 \times 10^{-3})} = 1.021$$

which confirms that, as far as a steady-state criterion is concerned, the system in the example should be reasonably interaction-free which is what we found.

Now consider arbitrarily fixing the interaction path gains such that $K_{\theta_g r} K_{g_r \theta_r} = 3 \times K_{\theta_r} K_{g_r}$ (that is, in equation (6.22), $\lambda = -0.5$). We find the following:

- At critical stability, for $K_{cg_r} = 0$ we have $K_{c\theta_r} = 4.4 K^{-1}$, and for $K_{c\theta_r} = 0$ we have $K_{cg_r} = 9250 (\text{kgkg}^{-1})^{-1}$, as before.
- However, in a departure from the previous case, when $K_{c\theta_r} = 4.4 K^{-1}$ and $K_{cg_r} = 9250 (\text{kgkg}^{-1})^{-1}$ the system is now unstable (readers should be able to verify these results for themselves).

Dynamic considerations

Using steady-state data to establish the danger of interaction in multivariable systems as set out above can be very helpful because these data are quite easy to calculate or measure in practical systems. However, they do not necessarily reveal the truth about the likelihood of interaction in situations where the dynamics of the multi-loop system varies.

Example 6.3 (*Continued*)

Consider for instance Example 6.3 once more. For the cool–humidify case, we note that the product of the dynamic portions of the diagonal and off-diagonal elements of the system transfer function matrix are identical (equation (E6.3.10)). Suppose that the larger of the two time constants in the interaction (off-diagonal) path is dropped. The plant is now dynamically out of balance and we find that it enjoys far less stability at the existing controller settings than existed before (Figure E6.3.6). After progressively reducing the two controller gains to $K_{c\theta_r} = 1.5\,\mathrm{K}^{-1}$ and $K_{cg_r} = 3150\,(\mathrm{kg\,kg}^{-1})^{-1}$ a stable and reasonably damped response to a disturbance in either of the main control paths was obtained.

In summary, we find that for a multivariable system to be free of instability induced by loop interaction, there must be *diagonal dominance* in both steady-state and dynamic characteristics of the system transfer function matrix.

Decoupling multivariable control systems

We have seen that the open-loop matrix transfer function for the general 2×2 case can be expressed as follows (Figure 6.5):

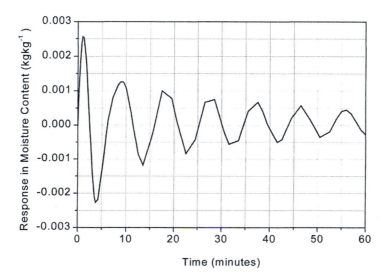

Figure E6.3.6 Temperature-humidity interaction when dynamically unbalanced.

$$\mathbf{G} = \mathbf{G_p G_c G_d} = \begin{bmatrix} G_{11} & G_{12} \\ G_{21} & G_{22} \end{bmatrix} \times \begin{bmatrix} G_{c1} & 0 \\ 0 & G_{c2} \end{bmatrix} \times \begin{bmatrix} G_{d1} & 0 \\ 0 & G_{d2} \end{bmatrix} = \begin{bmatrix} G_{11}G_{c1}G_{d1} & G_{12}G_{c2}G_{d2} \\ G_{21}G_{c1}G_{d1} & G_{22}G_{c2}G_{d2} \end{bmatrix}$$

$$(6.24)$$

To achieve zero interaction or fully *decoupled* control, we need an open-loop matrix transfer function for the system in which the off-diagonal elements are zero. This can be achieved with the use of *process decouplers* or *cross-compensators*.

For the fairly simple but common 2×2 case, cross-compensators G_{cc1} and G_{cc2} are added as shown in Figure 6.7. Thus the combined controller and cross-compensator matrix transfer function becomes

$$\mathbf{G_c} = \begin{bmatrix} G_{c1} & G_{cc1} \\ G_{cc2} & G_{c2} \end{bmatrix}$$

and the overall open-loop matrix transfer function will now be

$$\mathbf{G} = \begin{bmatrix} G_{11}G_{c1}G_{d1} + G_{12}G_{cc2}G_{d1} & G_{11}G_{cc1}G_{d2} + G_{12}G_{c2}G_{d2} \\ G_{21}G_{c1}G_{d1} + G_{22}G_{cc2}G_{d1} & G_{21}G_{cc1}G_{d2} + G_{22}G_{c2}G_{d2} \end{bmatrix} \qquad (6.25)$$

For interaction-free control we need

$$\mathbf{G} = \begin{bmatrix} G_{11}G_{c1}G_{d1} + G_{12}G_{cc2}G_{d1} & 0 \\ 0 & G_{21}G_{cc1}G_{d2} + G_{22}G_{c2}G_{d2} \end{bmatrix}$$

from which

$$G_{11}G_{cc1}G_{d2} + G_{12}G_{c2}G_{d2} = 0$$

$$G_{cc1} = \frac{-G_{12}G_{c2}}{G_{11}} \qquad (6.26)$$

and

$$G_{21}G_{c1}G_{d1} + G_{22}G_{cc2}G_{d1} = 0 \qquad \text{(see over)}$$

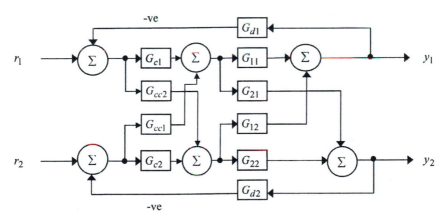

Figure 6.7 Application of cross-compensators in a 2×2 multi-loop system.

$$G_{cc2} = \frac{-G_{21}G_{c1}}{G_{22}} \qquad\qquad (6.27)$$

Note at this point the similarity between the results of the above cross-compensators, and the result for the feedforward compensator in section 4.3. They all share the following characteristic, which can be stated with respect to the compensator: *the (negative bias) disturbance transfer function divided by the compensated system transfer function.*

Example 6.4

Design cross-compensators for the system of Example 6.3.

Solution

First, for the cool–humidify case,

$$G_{11} = G_{\theta_i}(s) = \frac{3.5}{(3.5s + 1)(s + 1)} \qquad \text{(equation (E6.3.2))}$$

$$G_{12} = G_{\theta,g_i}(s) = \frac{0.11}{(3.5s + 1)(s + 1)} \qquad \text{(equation (E6.3.4))}$$

$$G_{21} = G_{g,\theta_i}(s) = \frac{1.07 \times 10^{-3}}{(6.7s + 1)(s + 1)} \qquad \text{(equation (E6.3.3))}$$

$$G_{22} = G_{g_i}(s) = \frac{1.62 \times 10^{-3}}{(6.7s + 1)(s + 1)} \qquad \text{(equation (E6.3.5))}$$

In Example 6.3, values of controller gains for this case were $G_{c1}(s) = K_{c1} = 4.4\,\mathrm{K}^{-1}$ and $G_{c2}(s) = K_{c2} = 4625\,(\mathrm{kgkg}^{-1})^{-1}$.

Cross-compensators for the cool–humidify case will therefore take the following form:

$$G_{cc1}(s) = \frac{-G_{12}(s)G_{c2}(s)}{G_{11}(s)} = \frac{-0.11}{(3.5s + 1)(s + 1)} \times 4625 \times \frac{(3.5s + 1)(s + 1)}{3.5}$$

$$= -145.4\left(\mathrm{kgkg}^{-1}\right)^{-1}$$

$$G_{cc2}(s) = \frac{-G_{21}(s)G_{c1}(s)}{G_{22}(s)} = \frac{-1.07 \times 10^{-3}}{(6.7s + 1)(s + 1)} \times 4.4 \times \frac{(6.7s + 1)(s + 1)}{1.62 \times 10^{-3}}$$

$$= -2.91\mathrm{K}^{-1}$$

In a similar manner, for the dehumidify–heat case, we find that $G_{cc1}(s) = K_{cc1} = -9423\,(\mathrm{kgkg}^{-1})^{-1}$ and $G_{cc2}(s) = K_{cc2} = 0$.

We find that, since the diagonal and off-diagonal dynamics of the system of Example 6.3 are identical, the necessary cross-compensators take the form of a simple set of gains, *steady-state decoupling*.

In the system of Example 6.3, full decoupling is achieved by applying a gain compensator with a gain of $-9423\,(\mathrm{kgkg}^{-1})^{-1}$ to the heating coil control path *when the cooling coil is called on to dehumidify.* A gain compensator with a gain of $-145.4\,(\mathrm{kgkg}^{-1})^{-1}$ is applied to the humidifier control path to compensate space humidity when the cooling coil is called on the provide temperature control. Finally, a gain compensator with a gain of $-2.9\,K^{-1}$ is applied to the heating coil control loop *when the humidifier is called*, to compensate room temperature for the sensible heating effect of the steam humidifier.

This form of compensation would be very easy to implement in practice, especially where a building energy management system or (certain types of) programmable digital controller is used. However, many multivariable control systems cannot be decoupled quite so easily. In particular, where the dynamics of the system paths and disturbance paths differ, or where the more usual PI, PID or other analytically derived controller is used in one or all system paths, the characteristics of the required compensation become dynamic. We will come across this situation in the next (and final) example of this chapter in which we will consider another increasingly common example of HVAC multivariable control – that of indoor air quality.

Example 6.5

Indoor air quality (IAQ) control is becoming increasingly popular as attention is given to improving the quality of the indoor environment (see section 1.2). Most commonly, control is achieved by varying the fresh air intake in order to dilute sources of pollution in the space. Multivariance results from variations in space temperature when the IAQ controller adjusts fresh air intake and is especially troublesome in winter. Clearly, actions in the temperature control loop do not affect air quality, so this system is an example of *one-way multivariance*. In fact, a simple model of this situation is identical in form to the temperature–humidity system of the previous case, where the earlier moisture content variable now becomes pollutant concentration.

Consider the case of an auditorium in winter. The space has a volume of $20\,000\,m^3$ and it is possible to ignore heat exchange with the surrounding fabric and due to fresh air infiltration, i.e. $\Sigma(AU_i) = 0$ and $n_v V_r/3 = 0$.

Air is delivered to the space at a rate of four air changes per hour and the fresh air and recirculating air proportions are fully variable by means of mixing dampers which can be represented by a first-order lag with a positioning time constant of 2 min. There is a heating coil and control valve which collectively can be represented by a first-order lag with a dead-time term (τ_{ch} and t_{dch} of 1.5 min and 0.2 min, respectively). The coil/valve and dampers can be assumed to give a linear response over a full range of their travels. In winter, when the coil is off and the fresh air damper is fully open, the auditorium will be at 0°C and, when the coil is full on, 21°C. Similarly, when the fresh air damper is fully open the concentration of CO_2 in the space is 0.05% and, at minimum position, the concentration can reach 1.8%. The concentration of CO_2 can be measured by a sensor with a first-order lag plus dead-time response (τ_{diaq} and t_{diaq} of 3 min and 0.5 min respectively), and temperature in the space can be measured with a sensor having a first-order lag response and a time constant of 0.5 min.

The damper and heating coil are to be controlled using digital feedback PI controllers using a sampling interval of one-tenth of the dominant time constant present in the system model.

Design a control system which *reduces* interaction.

Solution

Assigning a control signal range in the interval $0 \rightarrow 1$, the gain of the coil/valve and space will be equal to the output temperature range (i.e. $21 - 0 = 21°C$). The gain of the fresh air damper will, likewise, be equal to the range of pollutant concentration (i.e. $1.8 - 0.05 = 1.75\%$), making the assumption that both control systems offer a linear response across a full range of output.

The form of transfer function for the coil/valve–space will consist of a first-order lag for the space, with the first-order lag and dead time for the coil/valve itself. Since there is no heat exchange in the auditorium, the time constant for the space will be (based on equation (3.1))

$$\tau_r = \frac{V_r \rho_a}{m_a}$$

and

$$m_a = \frac{20\,000 \times 4 \times 1.2}{3600} = 26.67 \,\text{kgs}^{-1} \qquad \text{(i.e. four air changes per hour)}$$

and so

$$\tau_r = \frac{20\,000 \times 1.2}{26.67} = 899.9 \,\text{s} \,(15\,\text{min})$$

Comparing the time constants for coil/valve, fresh air damper and space, we see that the space has the dominant time constant (i.e. will place a pole closest to the imaginary axis in a root locus, with a mating zero on the imaginary axis in the right half-plane). Using the one-tenth dominant time constant criterion given for T, we can fix the sampling interval at 1.5 min.

The form of transfer function for the fresh air damper–space will be a second-order one; combining a first-order transfer function for the damper (as given) and a first-order transfer function for the space. The latter will take on the same form as that describing moisture content in a space (i.e. equation (6.13)) using the principle that a contaminant can be treated in the same manner as a moisture fraction. Note that the space transfer function resulting from this will have a time constant which is identical to that for the thermal loop in this example, since there is no infiltration.

For the air quality (i.e. damper) to temperature interaction path, the transfer function will be the same form as the air quality control path, but with a gain which is the same as that in the temperature control path. Note that there will be no interaction in the temperature control to air quality control paths.

Designating path 1–1 as temperature control and path 2–2 for air quality control, we can now write down transfer functions for this example corresponding to a 2×2 matrix structure. Since we are dealing in this case with discrete time control, we also translate each continuous-time transfer function to its discrete-time equivalent. A zero-order hold has been assumed for each path.

For convenience, discrete-time transfer functions are translated using the MATLAB control toolbox c2dm function (MATLAB, 1992) – Table E6.5.1. Readers without access to this should be able to translate these expressions using the appropriate tables in Chapter 5 (Tables 5.1 and 5.2). The form of transfer function for the discrete-time PI controllers in the two control paths will be as follows (see section 5.5):

$$G_c(z) = \frac{(1.5K_c + i_t K_c)z^2 - i_t K_c z}{i_t z^2 - i_t z} \tag{E6.5.6}$$

Next, we need to tune the controllers for each loop. In principle, we could do this by evaluating the critical controller gains by solving the denominator of the matrix closed-loop model of the system (equation (6.11)). But, as we found in Example 6.3, this is likely to prove to be intractable with a system of this complexity. Instead, a much less troublesome way to obtain reasonable tuning is to simulate step responses at a variety of controller parameter values. This was done as follows.

1 As a 'first guess', a Simulink simulation of the equivalent continuous-time system was set up and, with each partner controller gain respectively set to zero (with no integral action), the marginal controller gain of each loop was found from successive step response tests.

Table E6.5.1 Component transfer functions in continuous and discrete time

	$G(s)$	$G(z)$	
$G_{11}(s,z)$	$\dfrac{2\,\mathrm{lexp}(-0.2s)}{(15s+1)(1.5s+1)}$	$\dfrac{0.585z^2 + 0.672z + 0.0061}{z^3 - 1.273z^2 + 0.333}$	(E6.5.1)
$G_{12}(s,z)$	$\dfrac{21}{(15s+1)(2s+1)}$	$\dfrac{0.601z + 0.453}{z^2 - 1.377z + 0.427}$	(E6.5.2)
$G_{21}(s,z)$	0	0	
$G_{22}(s,z)$	$\dfrac{1.75}{(15s+1)(2s+1)}$	$\dfrac{0.05z + 0.038}{z^2 - 1.377z + 0.427}$	(E6.5.3)
$G_{d1}(s,z)$	$\dfrac{1}{(0.5s+1)}$	$\dfrac{0.95}{z - 0.05}$	(E6.5.4)
$G_{d2}(s,z)$	$\dfrac{\exp(-0.5s)}{(3s+1)}$	$\dfrac{0.284z + 0.109}{z^2 - 0.609z + 0.0015}$	(E6.5.5)

2 Using Ziegler–Nichols rules (Table 7.2, section 7.3), the critically stable
   criteria of each response were used to find values for each controller
   gain and integral time.
3 A simulation was then set up for the actual discrete-time system in which
   these controller parameters were used (Figure E6.5.1). The values required further fine tuning, and the following discrete-time controller
   transfer functions were finally arrived at:

$$G_{c1}(z) = \frac{0.0605z^2 - 0.0556z}{z^2 - z} \tag{E6.5.7}$$

$$G_{c2}(z) = \frac{0.76z^2 - 0.7z}{z^2 - z} \tag{E6.5.8}$$

Note the system matrix transfer function will be

$$\mathbf{G}(z) = \begin{bmatrix} \dfrac{0.585z^2 + 0.672z + 0.0061}{z^3 - 1.273z^2 + 0.333} & \dfrac{0.601z + 0.453}{z^2 - 1.377z + 0.427} \\ 0 & \dfrac{0.05z + 0.038}{z^2 - 1.377z + 0.427} \end{bmatrix}$$
$$\text{(system)}$$

$$\times \begin{bmatrix} \dfrac{0.0605z^2 + 0.672z + 0.0061}{z^2 - z} & 0 \\ 0 & \dfrac{0.76z^2 - 0.7z}{z^2 - z} \end{bmatrix}$$
$$\text{(controllers)}$$

$$\times \begin{bmatrix} \dfrac{0.95}{z - 0.05} & 0 \\ 0 & \dfrac{0.284z + 0.109}{z^2 - 0.609z + 0.0015} \end{bmatrix} \tag{E6.5.9}$$
$$\text{(sensors)}$$

Results of the interacting system are summarised in Figures E6.5.2–E6.5.4.
Figure E6.5.2 gives the response in indoor air quality for a +0.1% step
change in the IAQ set point. Figure E6.5.3 gives the corresponding disturbance in the temperature control path and reveals significant nuisance
disturbance which, through the action of the temperature controller, eventually eradicates. Figure E6.5.4 shows the response in auditorium temperature arising from a unit step change in temperature loop set point.

Decoupling

We now consider measures which might reduce or eliminate the one-way
interaction in this plant. Only one cross-compensator will be needed;
$G_{cc1}(z)$. One option will be to derive a discrete-time compensator directly
using equation (6.26). However, this will give very high-order compensation and may well be numerically ill-conditioned. A better approach will
often be to derive a compensator based on the continuous-time transfer

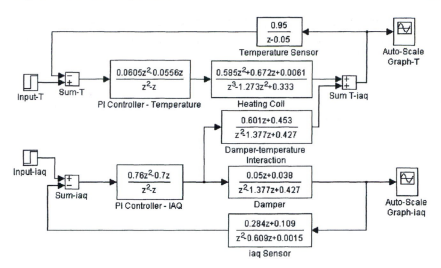

Figure E6.5.1 Simulink model of the interacting plant.

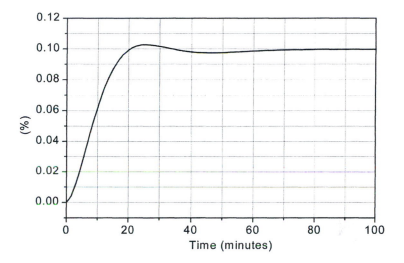

Figure E6.5.2 IAQ response to IAQ disturbance.

functions and *then* translate the resulting compensator to discrete time for implementation. This results in the following continuous-time compensator:

$$-\frac{G_{12}(s)}{G_{11}(s)} = \frac{2.25s^3 + 24.15s^2 + 16.6s + 1}{3s^3 - 28.3s^2 - 16.9s + 1}$$

which, using c2dm in the MATLAB control system toolbox, translates to

$$-\frac{G_{12}(z)}{G_{11}(z)} = \frac{2.54z^2 - 3.11z + 0.738}{-3.27z^2 + 4.5z - 1.4} \tag{E6.5.10}$$

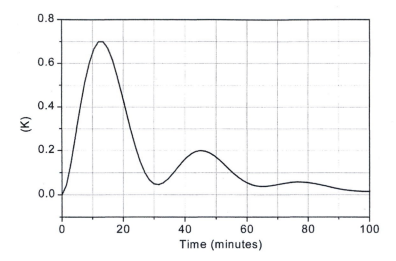

Figure E6.5.3 Temperature response to IAQ disturbance.

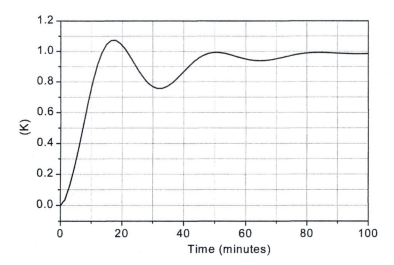

Figure E6.5.4 Temperature response to temperature disturbance.

Combining with the controller which lies in interaction path,

$$G_{cc1}(z) = -\frac{G_{12}(z)G_{c2}(z)}{G_{11}(z)} = \frac{1.93z^3 - 4.15z^2 + 2.74z - 0.517}{-3.27z^3 + 7.77z^2 - 5.9z + 1.4} \qquad (E6.5.11)$$

Figure E6.5.5 shows the modified Simulink block diagram with this compensation and Figure E6.5.6 confirms that this leads to improved, but by no means interaction-free, control (compare this with Figure E6.5.3). The peak deviation in temperature following the IAQ loop disturbance is now about one-half of the deviation without compensation. Note that to

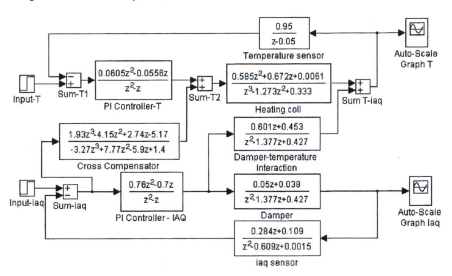

Figure E6.5.5 Addition of a dynamic cross-compensator.

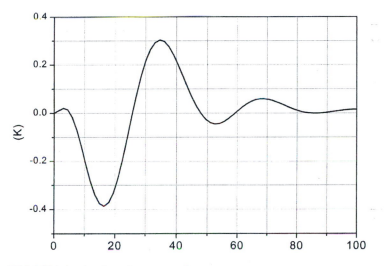

Figure E6.5.6 Third-order dynamic compensation.

implement this compensation in a practical on-line situation would require equation (E6.5.11) to be expressed in the following time-sequence form:

$$u_{ccl}(n) = 2.38u_{ccl}(n-1) - 1.81u_{ccl}(n-2) + 0.43u_{ccl}(n-3) - 0.59\varepsilon_{iaq}(n)$$
$$+ 1.27\varepsilon_{iaq}(n-1) - 0.84\varepsilon_{iaq}(n-2) + 0.16\varepsilon_{iaq}(n-3) \qquad \text{(E6.5.12)}$$

in which $u_{ccl}(n)$ and $\varepsilon_{iaq}(n)$ would be the control algorithm output and applied input at time instant n, respectively.

An altogether simpler option would be to opt for steady-state compensation. Putting $z = 0$ in equation (E6.5.11) results in a gain compensator with

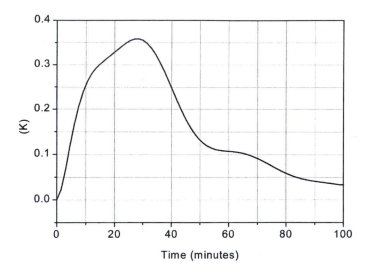

Figure E6.5.7 Gain compensation.

a value of about −0.4. Simulating this value of gain compensation leads to an improvement in control. The peak deviation is reduced to about the same level as can be achieved with third-order compensation but the lack of dynamic compensation means that the steady-state method is more protracted (Figure E6.5.7).

In summary, the need for any form of dynamic cross-compensation in HVAC multivariable systems will not always be overwhelming. However it will be a matter of interest when the dynamics in the diagonal and off-diagonal paths of the overall system matrix transfer function differ substantially or where the respective gains differ substantially. Often, steady-state compensation will provide sufficient improvement.

It is hoped that this chapter has identified the nature of the potential problem of interaction and considered some ways of dealing with it. Inevitably, a weakness with the methods discussed in this and earlier chapters of this book is that they are good for the precise set of conditions to which they are applied but tend to break down when these conditions vary. Control must be flexible in time as well as parametrically flexible in order to deal with the non-stationary control environment encountered in most practical situations. We will take a look at methods which provide the necessary degrees of flexibility in Chapters 8 and 9. Meanwhile, we need to take a look at methods for arriving at desirable controller parameters in the first instance – this forms the basis of Chapter 7.

References

Howell, J. (1988) Relative humidity in HVAC systems: improving control. *Building Services Engineering Research and Technology*, **9** (2), 55–61.

Howell, R.H. (1986) Variations in relative humidity in a conditioned space due to coil bypass control systems. *ASHRAE Transactions*, **92** (1B), 499–508.

Jones, W.P. (1995) *Air Conditioning Engineering*. Edward Arnold, London.

Letherman, K.M. (1988) Room air moisture content: dynamic effects of ventilation and vapour generation. *Building Services Engineering Research and Technology*, **9** (2), 49–53.

Loveday, D.L., Virk, G.S., Cheung, J.Y.M. (1993) Advanced modelling for control in buildings. *Proceedings of the CLIMA 2000 Conference*, London.

MATLAB (1992) *Control System Toolbox User's Guide*. The Mathworks Inc., Natick, MA.

MATLAB (1996) *MATLAB The Language of Technical Computing – Using MATLAB*. The Mathworks Inc., Natick, MA.

Seborg, D.E., Edgar, T.E., Mellichamp, D.A. (1989) *Process Dynamics and Control*. John Wiley, New York.

Simulink (1996) *SIMULINK 2 Dynamic Simulation for MATLAB*. The Mathworks Inc., Natick, MA.

Townsend, M.A., Chercas, D.B., Abdelmessih, A. (1986) Optimal control of a general environmental space. *Transactions of the American Society of Mechanical Engineers – Journal of Dynamic Systems, Measurement and Control*, **108**, 330–339.

Yik, F.W.H., Underwood, C.P., Chow, W.K. (1995) Simultaneous modelling of heat and moisture transfer and air conditioning systems in buildings. *Proceedings of the Fourth IBPSA Conference*, Madison, WI, 14–21.

Zaheer-uddin, M. (1992) Decentralised control systems for HVAC. *ASHRAE Transactions*, **98** (2), 114–126.

Zaheer-uddin, M. (1993) Temperature and humidity control of indoor environmental spaces. *Energy and Buildings*, **19**, 275–284.

7   System identification and controller tuning

7.1 System identification

Much of what we have looked at so far has depended on the ability to build
a mathematical representation of a system from first principles using established theory, backed up with a few judicious assumptions. There are of
course circumstances in which this is either not necessary (if, for instance,
the plant or system exists and it is possible to derive a much more accurate
representation of the plant by observing its actual performance) or not
possible (the derivation of a mathematical model from first principles is too
complex or requires too many questionable assumptions). *System identification* is the process by which mathematical representations of dynamic systems can be obtained from performance data collected from the actual
plant or physically similar plant. This technique is at the heart of *adaptive*
control which we will look at in Chapter 8. Identification applied to HVAC
plant has received considerable attention (Penman, 1990; Zaheer-uddin,
1990; So *et al.*, 1995; Coley & Penman, 1992).

The process–reaction curve

This is the simplest method of identification and leads to a transfer function representation of the plant which can subsequently be used to estimate
required controller parameters or can be used with any of the modelling,
stability and response methods considered in earlier chapters.

Sufficient data to derive a process–reaction curve can be obtained by
carrying out a simple test on-site. Any controller present is switched to
manual; the system is now under open-loop conditions. The positioning
device (e.g. valve or damper actuator) is manually adjusted by a small
amount, resulting in an open-loop step response by the system. This response is plotted, forming a *process–reaction curve.*

A response of the form shown will usually be obtained (Figure 7.1). Note
the dead-time component. This is the point at which the response will
have cut the time axis were the response of a purely exponential form. In
practice, a mathematical model can be easily fitted to the process–reaction
curve in the form of a first-order lag plus dead time; this form of representation has been dealt with in a number of places in this book.

The process–reaction curve can therefore be approximated by

$$G_p(s) = \frac{K_p \exp(-t_d s)}{(\tau s + 1)} \tag{7.1}$$

where K_p = the overall plant gain, t_d = the apparent system dead time and
τ = the overall system time constant. With a step input, we can express the
time-domain form as

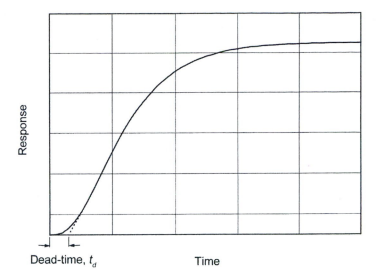

Figure 7.1 Process–reaction curve.

$$y(t) = y(0) + \Delta y\left[1 - \exp\left(\frac{-(t - t_d)}{\tau}\right)\right] \qquad (t \geq t_d) \qquad (7.2)$$

where Δy = the change in the response variable, $y(t)$ = the response variable value as a function of time, $y(0)$ = the initial response variable value and t = time.

All that remains is to determine values for t_d and τ. To do this, we solve equation (7.2) at two arbitrary points in time. The arbitrary points in time can be anywhere on the response curve above the dead time and below the final value. In most cases, the following arbitrary times will suffice:

$$t_1 = t_d + \tau/3 \qquad\qquad\qquad\qquad\qquad\qquad\qquad (7.3)$$

$$t_2 = t_d + \tau \qquad\qquad\qquad\qquad\qquad\qquad\qquad\qquad (7.4)$$

Substituting these arbitrary times for t in equation (7.2) leads to

$$y(t_1) = y(0) + 0.2835\Delta y \qquad\qquad\qquad\qquad\qquad (7.5)$$

$$y(t_2) = y(0) + 0.6321\Delta y \qquad\qquad\qquad\qquad\qquad (7.6)$$

Thus, having plotted the process–reaction curve, we note the values of time corresponding to 28.35% and 63.21% of Δy, the overall change in the response variable, from the curve. We then put these time results into equations (7.3) and (7.4) and solve them for t_d and τ.

Besides using the resulting estimated transfer function with many of the techniques we have explored earlier in this book, t_d and τ form the basis of a variety of controller parameter recommendations, some of which we will look at in section 7.3. Meanwhile, the following example illustrates this simple method of model fitting.

Example 7.1

An open-loop step response test has been conducted on a hot water to air heating coil by manually adjusting the heating control valve by 10% of its range and observing the resulting change in off-coil air temperature. The following results were obtained (given in $y(t)$ format in which y is in °C and t in seconds).

30.00(0)	30.03(20)	30.09(40)	30.18(60)	30.26(80)	30.36(100)
30.45(120)	30.52(140)	30.60(160)	30.66(180)	30.72(200)	31.00(300)
31.20(400)	31.32(500)	31.40(600)	31.45(800)	31.46(1000)	

Solution

The open-loop response data are plotted to produce a process–reaction curve (Figure E7.1.1). We note that our two arbitrary times (equations (7.3) and (7.4)) coincide with 28.35% and 63.21% of the overall change in response (i.e. 30.41°C and 30.92°C respectively) at 112 and 273 seconds respectively (t_1 and t_2 in Figure E7.1.1). Substituting these times into the simultaneous equations gives the following:

$$t_1 = 112 = t_d + \tau/3$$

$$t_2 = 273 = t_d + \tau$$

leading to

$$\tau = 241.5 \text{ seconds}$$

$$t_d = 31.5 \text{ seconds}$$

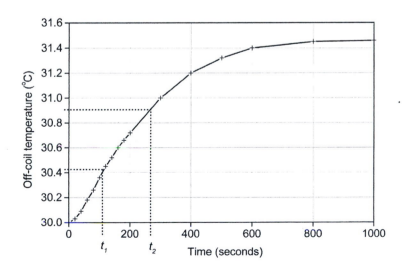

Figure E7.1.1 Process–reaction curve.

We are now in a position to fit the model to the data. For fractional changes in heating valve position, $u(s)$, the gain between the off-coil temperature, $\theta_{ao}(s)$, and valve position will be $1.46/0.1 = 14.6$. Thus,

$$G_p(s) = \frac{\theta_{ao}(s)}{u(s)} = \frac{14.6 \exp(-31.5s)}{(241.5s + 1)} \qquad \text{(E7.1.1)}$$

This is quite a useful result, for if it were necessary to carry out some further detailed analysis on the system (for example, to investigate in detail various options for improving control), we now have a transfer function representation of the system at our disposal.

Autoregressive models

Several structures exist for the fitting of models to much more detailed experimental data than that considered in the previous section, including the presence of disturbance or noise inputs. A summary of some of the leading methods is given below. A more substantial treatment of the subject can be found in a number of texts – see for example Ljung (1987) or Wellstead and Zarrop (1991). Note that t in the following refers to sampled time.

The autoregressive (AR) structure

The simplest model structure relates a time-series output, $y(t)$, to a noise or disturbance term, $e(t)$, as follows:

$$y(t) = \frac{e(t)}{A} \qquad \text{(7.7)}$$

where

$$A = 1 + a_1 z^{-1} + a_2 z^{-2} + \ldots + a_{na} z^{-a_{na}}$$

This forms the _autoregressive_ (AR) model structure, and A is a polynomial in z.

The ARX structure

For closed-loop control, there will be an input, $u(t)$, and the AR model is therefore inappropriate for a full description of the data. Hence an _autoregressive_ with independent or _exogenous input_ (ARX) model structure can be defined:

$$y(t) = \frac{B}{A} u(t) + \frac{e(t)}{A} \qquad \text{(7.8)}$$

where

$$B = b_0 + b_1 z^{-1} + b_2 z^{-2} + \ldots + b_{nb} z^{-b_{nb}}$$

The ARMAX structure

A further refinement for situations in which the disturbance can be measured is to introduce a time-series disturbance parameter which results in the _autoregressive moving average_ with _exogenous input_ structure:

$$y(t) = \frac{B}{A} u(t) + \frac{C}{A} e(t) \tag{7.9}$$

where

$$C = c_0 + c_1 z^{-1} + c_2 z^{-2} + \ldots + c_{nc} z^{-c_{nc}}$$

The ARIMAX structure

For models which include one or more disturbance inputs, the random nature of many disturbance patterns is not always adequately represented in the ARMAX structure. This situation can be improved by integrating the C time series with respect to time which can be conveniently achieved by introducing the discrete-time integrator $1/(1 - z^{-1})$. The _autoregressive integrated moving average_ with _exogenous input_ (ARIMAX) structure results:

$$y(t) = \frac{B}{A} u(t) + \frac{C}{IA} e(t) \tag{7.10}$$

where

$$I = 1 - z^{-1}$$

Note that the disturbance takes the form of _white noise_ in the special situation where $C = A = 1$.

Other model structures include the _output error_ (OE) structure, in which the disturbance is treated as white measurement noise and $e(t)$ is regarded as an error with respect to the undisturbed output, and a refinement to this, Box–Jenkins, in which the output error is parametrised (for a full treatment see Ljung, 1987).

Estimation

System identification using any one of the various model structures involves three main stages:

* Select a suitable model structure.
* Estimate values of the model parameters, A, B, etc.
* Verify the resulting model using, ideally, data which were not used in model estimation.

In many cases, the nature of any known disturbances will determine the best model structure and it might often be desirable to test the results from a number of different model structures in order to find the best one. For many applications, the ARX model structure will give satisfactory results;

indeed the ARX structure tends to be the most commonly used. Verification will therefore determine whether the results obtained from a given model structure are adequate, or whether an alternative model structure should be investigated. It will also enable a choice to be made as to the best combination of parameters in a given model structure.

The routine part of system identification is the estimation of the parameters themselves. In some applications, the input/output $(u(t) \ldots y(t))$ sequence will contain a dead-time delay of nk time samples, so if we consider the case of the ARX structure, the form of model offered is:

$$y(t) + a_1 y(t-1) + \ldots + a_{na} y(t-na) =$$
$$b_0 u(t) + b_1 u(t-nk) + \ldots + b_{nb} u(t-nk-nb) \tag{7.11}$$

Thus the current output, $y(t)$, is related to previous values of output and input, which may include a delay of nk time samples in the input sequence. Note that b_0 will always be zero so that the model will be physically realisable (i.e. based on known previous values) and $nk = 1$ signifies no dead-time delay between input and output sequences.

Estimation therefore involves the derivation of the parameter set which, since applicable to an evolving time series of input/output values, will be a column vector, $\hat{\theta}$,

$$\hat{\theta} = [a_1 \quad a_2 \quad \ldots \quad a_{na} \quad b_1 \quad b_2 \quad \ldots \quad b_{nb}]^T \tag{7.12}$$

By far the most common method of parameter estimation is the *least-squares algorithm* (for a proof, see Wellstead & Zarrop, 1991),

$$\hat{\theta} = [\mathbf{X}^T \mathbf{X}]^{-1} \times [\mathbf{X}^T \mathbf{y}] \tag{7.13}$$

where

$$\mathbf{y}^T = [y(t-1) \quad \ldots \quad y(t-n)]$$
$$\mathbf{X} = [(y(t-1) \quad \ldots \quad y(t-n)), \quad (u(t-1) \quad \ldots \quad u(t-n))]$$

Note that the above assumes that there is no disturbance sequence present (this would merely add a further set of elements to the $\mathbf{X}$ matrix after the $u(t)$ sequence).

We also note that

$$\hat{\mathbf{e}} = \mathbf{y} - \mathbf{X}\hat{\theta} \tag{7.14}$$

where $\hat{\mathbf{e}}$ is a vector of modelling errors.

We will now demonstrate the least-squares algorithm in the estimation of a model from some experimental data.

Example 7.2

The data given in Table E7.2.1 relate the room temperature, $\theta_r(t)$, to control signal, $u(t)$, for the feedback space heating of a room. The data have been sampled at uniform intervals in time during a morning period

Table E7.2.1 Data for day 1

Sample	$\theta_r(t)$	$u(t)$
0	16.1	1.000
1	18.6	1.000
2	19.6	1.000
3	20.3	1.000
4	20.9	0.997
5	21.4	0.852
6	21.8	0.437
7	21.5	0.027

Table E7.2.2 Data for day 2

Sample	$\theta_r(t)$	$u(t)$
0	17.0	1.000
1	19.1	1.000
2	19.9	1.000
3	20.4	1.000
4	20.8	1.000
5	21.3	0.905
6	21.7	0.572
7	21.6	0.167

(including the preheating period). The data in Table E7.2.2 represent the same observations made one day later. (These data are in fact segments of the more substantial data set used in Example 7.3 and are based on the performance monitoring of a campus building at the University of Northumbria at Newcastle (Hudson & Underwood, 1996).)

Fit a two-input two-output (four-parameter) ARX model with zero dead time to the data recorded on day 1 and verify the resulting model using the data recorded on day 2.

Solution

First, we note that the form of model will be ARX(n_a, n_b, n_k) $\rightarrow$ ARX(2, 2, 1) for two input parameters, two output parameters and zero dead-time, i.e.

$$\theta_r(t) = a_1\theta_r(t-1) + a_2\theta_r(t-2) + b_0u(t) + b_1u(t-1) + b_2u(t-2) \qquad \text{(E7.2.1)}$$

and b_0 is zero so that the model is feasible (i.e. based entirely on past values). We therefore require estimates of a_1, a_2, b_1 and b_2 from the day 1 data.

The parameter estimate will be, $\hat{\boldsymbol{\theta}}$,

$$\hat{\boldsymbol{\theta}} = \left[\mathbf{X}^T\mathbf{X}\right]^{-1} \times \left[\mathbf{X}^T\mathbf{y}\right] = \begin{bmatrix} a_1 & a_2 & b_1 & b_2 \end{bmatrix}^T \qquad \text{(E7.2.2)}$$

In most cases involving dynamic data, a more refined model results from fitting to deviations from the data mean values. This also produces a

Table E7.2.3 Removing the mean values from the day 1 data

T	$\theta_r(t)$	$\Delta\theta_r(t)$	$u(t)$	$\Delta u(t)$
0	16.1	−3.925	1.000	0.211
1	18.6	−1.425	1.000	0.211
2	19.6	−0.425	1.000	0.211
3	20.3	0.275	1.000	0.211
4	20.9	0.875	0.997	0.208
5	21.4	1.375	0.852	0.063
6	21.8	1.775	0.437	−0.352
7	21.5	1.475	0.027	−0.762
	$(\bar{\theta}_r(t) = 20.025)$		$(\bar{u}(t) = 0.789)$	

parameter set appropriate to transfer function representation of the model, if required. Removing the mean values from the day 1 data, the values in Table E7.2.3 are obtained.

X can now be formed from the input and output sequences from rows 1–6 and 0–5 in Table E7.2.3, representing parameter estimate bases for $\theta_r(t-1)$, $\theta_r(t-2)$ and $u(t-1)$, $u(t-2)$ respectively. Thus **X** will be a matrix of four columns and six rows,

$$\mathbf{X} = \begin{bmatrix} -1.425 & -3.925 & 0.211 & 0.211 \\ -0.425 & -1.425 & 0.211 & 0.211 \\ 0.275 & -0.425 & 0.211 & 0.211 \\ 0.875 & 0.275 & 0.208 & 0.211 \\ 1.375 & 0.875 & 0.063 & 0.208 \\ 1.775 & 1.375 & -0.352 & 0.063 \end{bmatrix}$$

and the transpose of **X** will be

$$\mathbf{X}^{\mathrm{T}} = \begin{bmatrix} -1.425 & -0.425 & 0.275 & 0.875 & 1.375 & 1.775 \\ -3.925 & -1.425 & -0.425 & 0.275 & 0.875 & 1.375 \\ 0.211 & 0.211 & 0.211 & 0.208 & 0.063 & -0.352 \\ 0.211 & 0.211 & 0.211 & 0.211 & 0.208 & 0.063 \end{bmatrix}$$

With some effort, we now obtain

$$[\mathbf{X}^{\mathrm{T}}\mathbf{X}]^{-1} = \begin{bmatrix} 10.625 & -5.675 & 22.693 & -51.218 \\ -5.675 & 3.115 & -11.693 & 27.366 \\ 22.693 & -11.693 & 56.270 & -113.525 \\ -51.218 & 27.366 & -113.525 & 254.457 \end{bmatrix}$$

The output, **y**, will be a column vector formed from the values $\theta_r(2 \ldots 7)$ i.e.

$$\mathbf{y} = \begin{bmatrix} -0.425 & 0.275 & 0.875 & 1.375 & 1.775 & 1.475 \end{bmatrix}^{\mathrm{T}}$$

and therefore

$$\mathbf{X}^T\mathbf{y} = [6.991 \quad 4.864 \quad 0.032 \quad 0.905]^T$$

The parameter estimate, $\hat{\boldsymbol{\theta}}$, is now given by

$$\hat{\boldsymbol{\theta}} = \left[\mathbf{X}^T\mathbf{X}\right]^{-1} \times \left[\mathbf{X}^T\mathbf{y}\right] = [a_1 \quad a_2 \quad b_1 \quad b_2]^T = [1.032 \quad -0.127 \quad 0.793 \quad 1.777]^T$$

We can now express our model, remembering that $\theta_r(t)$ and $u(t)$ are expressed as deviations from their mean values,

$$\theta_r(t) = 20.025 + 1.032\Delta\theta_r(t-1) - 0.127\Delta\theta_r(t-2)$$
$$+ 0.793\Delta u(t-1) + 1.777\Delta u(t-2) \tag{E7.2.3}$$

Note that the matrix algebra above could be handled with the following MATLAB operations (MATLAB, 1996). Given the matrix $\mathbf{X}$, to obtain the transpose,

```
>xt=rot90(x,-1)
```

```
>xt=fliplr(xt)
```

For the inverse of $\mathbf{X}^T\mathbf{X}$,

```
>xtx=xt*x
```

```
>xtxinv=inv(xtx)
```

For the $\mathbf{X}^T\mathbf{y}$ term,

```
>y=[-0.425 0.275 0.875 1.375 1.775 1.475]
```

```
>y=rot90(y,-1)
```

```
xty=xt*y
```

Hence the parameter estimate,

```
>theta=xtxinv*xty
```

Note also that the time-domain model form of equation (E7.2.3) can be expressed in the form of a discrete-time transfer function, $G_r(z)$,

$$G_r(z) = \frac{\theta_r(z)}{u(z)} = \frac{0.793z^{-1} + 1.777z^{-2}}{-1.032z^{-1} + 0.127z^{-2}} = \frac{14 + 6.25z}{1 - 8.14z} \tag{E7.2.4}$$

We can now apply the estimated model (equation (E7.2.3)) to the verification data from the day 2 measurements in order to establish how adequate our model is. The results are shown in Table E7.2.4. Evidently, modelling error is initially quite high due to the lack of initial model results at $t = 1$, -2 from which to construct model predictions at $t = 0$, 1. Nevertheless, the error does recede and a reasonable mean error across the six-sample prediction of 0.38K is obtained.

Manual autoregressive model fitting of this type, even to small groups of data such as that considered in this example, is likely to prove intractable in

Table E7.2.4 Application of estimated model for day 1 to verification data for day 2

n	$\theta_r(t)$ (measured)	$u(t)$ (measured)	$\Delta u(t)$ (measured)	$\Delta\theta_r(t)$ (model)	$\theta_r(t)$ (model)	Error
0	17.0	1.000	0.169	0	—	—
1	19.1	1.000	0.169	0	—	—
2	19.9	1.000	0.169	0.434	20.5	0.6
3	20.4	1.000	0.169	0.882	20.9	0.5
4	20.8	1.000	0.169	1.300	21.3	0.5
5	21.3	0.905	0.074	1.665	21.7	0.4
6	21.7	0.572	−0.259	1.913	21.9	0.2
7	21.6	0.167	−0.664	1.690	21.7	0.1

$\bar{u}(t) = 0.831$ mean error $= 0.38K$

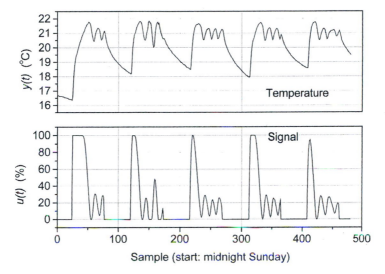

Figure E7.3.1 Heating system monitored data.

practice without the help of computational methods. Example 7.3 illustrates the potential of autoregressive identification for a much more substantial family of data using such methods.

Example 7.3

Hudson has monitored the space air temperature and control valve signal in a number of zones of a heated and naturally ventilated campus building at the University of Northumbria at Newcastle (Hudson & Underwood, 1996). Figure E7.3.1 gives a five-day (Monday to Friday) sample of the data for a typical zone from December 1995. The data were recorded at 15 min intervals. The heating system consists of a feedback loop using a thermistor in the space, with a control signal generated from a building management system PID control algorithm which sampled at 5 s intervals and was tuned with a gain of $0.1\mathrm{K}^{-1}$, integral action time of 10min and derivative action time of 5min.

From a control point of view, we can identify some interesting features in these data. Each day commences with the heating system preheating at full capacity – note the protracted preheating on the first day (Monday), following the weekend shutdown. From mid-morning (mid-day in the case of the Monday cycle), the heating reverts to modulating control and there is clear evidence of low-frequency oscillatory behaviour during these periods of modulating control.

Model fitting

A four-parameter ARX model (with zero dead time) has been fitted to the data using the MATLAB system identification toolbox (Ljung, 1988) – indeed most statistical or spread-sheet programs offering autoregressive modelling functions could be used for this. The procedure adopted is as follows:

- The data were first modified by removing their mean values.
- The first half of the data (240 samples) were then used for model fitting.
- The second half of the data were then used for model verification.

A 'best-fit' model was found to be an ARX(4, 6, 1), which had a mean fitting error of 0.00267. However, a good fit was also obtained with the following ARX(2, 2, 1) structure (mean fitting error 0.00305):

$$a_1 = 1.6145, \qquad a_2 = -0.6213, \qquad b_1 = 0.0089, \qquad b_2 = -0.0079$$

and thus

$$\begin{aligned}
\theta_r(t) = \overline{\theta}_r &+ 1.6145\Delta\theta_r(t-1) - 0.6213\Delta\theta_r(t-2) \\
&+ 0.0089\Delta u(t-1) - 0.0079\Delta u(t-2)
\end{aligned} \tag{E7.3.1}$$

Evidently, the estimated parameters here are quite different to those obtained from the much smaller segment of data used in the previous example (equation (E7.2.3)). This is not to say that the above model is better; merely that it reflects a more substantial set of observations.

Figure E7.3.2 shows how the model-predicted room temperature compares with the second segment of measured values. Certainly, the middle days reproduce quite well but the first day predictions are not good. Once again, this is due to the need to set initial predicted values of zero (i.e. the first pair of input/output values used to get the model started are not known). The error associated with this can be appreciated from Figure E7.3.3 in which the first pair of values from the measured data set are used for the first predictions of the model (this is of course technically cheating, because these values are not known to the model). Clearly then, there will always tend to be a certain amount of front-loaded error in identified models of this type.

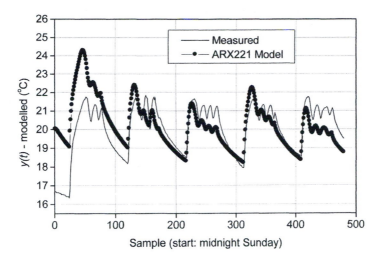

Figure E7.3.2 ARX(2, 2, 1) model performance.

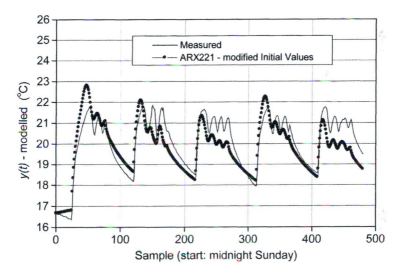

Figure E7.3.3 ARX(2, 2, 1) model performance with modified initial values.

Though two models were identified (ARX(4, 6, 1) and ARX(2, 2, 1)), Figure E7.3.4 shows that there is little to choose between them; and the four-parameter model is simpler.

One final point: there are two dominant frequencies in these data. One is due to the daily system switching pattern whilst the other is due to system dynamics. It will of course be difficult to fit a model with excellent robustness in terms of reproducing all features of such a complex pattern of data. One possible remedy where a higher-quality model is needed would be to

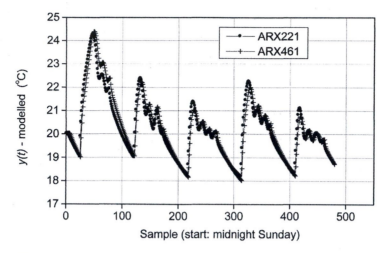

Figure E7.3.4 ARX(2, 2, 1) and ARX(4, 6, 1) model comparison.

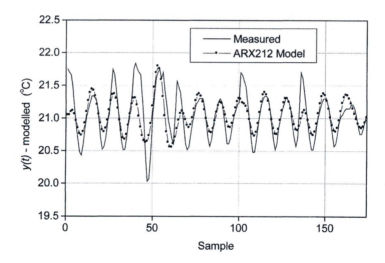

Figure E7.3.5 ARX(2, 1, 2) model fitted to daytime cyclic data only.

identify separately the daytime data only. Figure E7.3.5 shows the performance of an ARX(2, 1, 2) fitted to a segment of daytime data only for one of the daily patterns forming Figure E7.3.1. Again, a 'best fit' was found to be an ARX(2, 6, 1) but an ARX(2, 1, 2) gave results which were almost as good. This ARX(2, 1, 2) for this confined data set is as follows:

$$a_1 = 1.3845, \qquad a_2 = -0.6119, \qquad b_1 = 0.0046$$

and thus

$$\theta_r(t) = \overline{\theta}_r + 1.3845\Delta\theta_r(t-1) - 0.6119\Delta\theta_r(t-2) + 0.0046\Delta u(t-1) \qquad \text{(E7.3.2)}$$

Figure E7.3.5 shows that the cyclic pattern of activity is much better represented by this model, even though the model does not represent the peaks and troughs especially well. In many cases then, the best resort will often be to fit a combination of models, especially where a multi-modal pattern of data is evident.

Autoregressive system identification is a straightforward enough business provided that suitable computational tools are available though the quality of model fit can vary considerably. In Chapter 9 we will discuss the use of artificial neural networks for identification and will see that these have the potential to produce excellent representations of complex data patterns. We will also return to a more practical application of ARX methods a little later in Chapter 8.

7.2 Assessing the quality of response

Stable operation is the primary concern in control system design. However, once stable operation has been established, further adjustment will lead to the 'optimum' response of a plant as judged by a number of performance quality criteria. We will consider some of the basic response quality criteria in the following then go on to see how we might achieve them with reference to *controller tuning.*

Synoptic criteria

In the response of Figure 7.2, we can identify a number of features which constitute measures of how well a plant is behaving under control. The figure shows the response of a feedback control system to a unit change in set point. When tuning a controller, minimising the following response features will lead to some kind of an 'optimum' dynamic response.

- *Settling time, t_s:* the time taken for the response to reach a steady state.
- *Rise time, t_r:* the time taken for the response to first reach the required target condition.
- *Offset:* which we are aware is a feature of purely proportional control or ill-tuned proportional plus integral control (see section 4.1).

There are some response features which do not necessarily require to be minimised, but adjusted to within some specified tolerance. The *percentage overshoot, $(a/r) \times 100$* (where r is the target response required), and the *decay ratio, a/b,* give indications of the degree of damping present. A common 'rule of thumb' for process control is to target a decay ratio of about one-quarter.

Error-integral criteria

Lopez *et al.* (1967) propose error-integral criteria as a more rigorous way of evaluating the quality of response of a control system. These criteria are

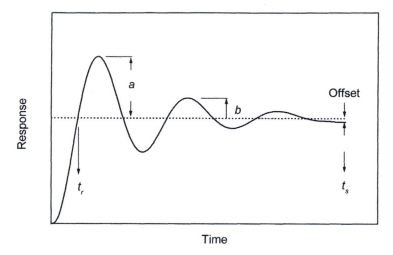

Figure 7.2 Synoptic response quality criteria.

more difficult to evaluate than synoptic methods, other than in on-line situations. There are several ways of applying error-integral criteria but all are based on the time integral of the error experienced by a control system following some predefined disturbance (usually a small step change in set point).

Consider the typical set point response of Figure 7.3. The area of the hatched regions represents the integral of the error with respect to time. Discriminating between positive contributions to error (above the desired response value) and negative contributions (below the desired response) is unnecessary so we can define an *integral of absolute error* $(I_{(|\varepsilon|)})$,

$$I_{(|\varepsilon|)} = \int_0^\infty |\varepsilon| \mathrm{d}t \tag{7.15}$$

Clearly, a perfect response is satisfied at $I_{(|\varepsilon|)} = 0$, but this will never be achievable in practice due to intrinsic system dynamics. Instead, a 'good' response is obtained by minimising $I_{(|\varepsilon|)}$ in some way.

For conventional stand-alone controllers with fixed settings, this minimising procedure will be a matter of hands-on trial and error, but for programmable controllers it should be possible to use this criterion as a basis for on-line controller parameter adjustment. Nominally fixed-parameter self-tuning controllers work in this way (for examples, see Kamimura *et al.*, 1994; Underwood, 1989; Nesler, 1986; Pinnella *et al.*, 1986) – note that automated tuners of these types should not be confused with adaptive controllers which we consider in Chapter 8.

In some circumstances, it may be desirable to weight the integral error in some way. The *integral of error squared* $(I_{(\varepsilon^2)})$ ensures that large values of error contribute more heavily to the integral than smaller values,

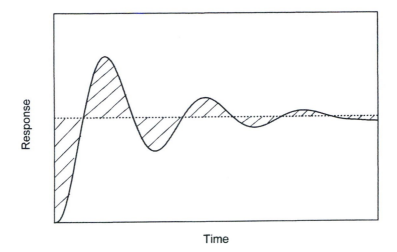

Figure 7.3 Integral-error criteria.

$$I_{(\varepsilon^2)} = \int_0^\infty \varepsilon^2 \mathrm{d}t \qquad\qquad (7.16)$$

Large error tends to occur early in a response, immediately following some disturbance, and take the form of overshoot/undershoot. Hence, minimising this criterion will tend to result in minimised oscillation in a response.

Another possibility is the *integral of time multiplied by absolute error* ($I_{(|\varepsilon|t)}$),

$$I_{(|\varepsilon|t)} = \int_0^\infty |\varepsilon| t \mathrm{d}t \qquad\qquad (7.17)$$

Because time accumulates independently after a disturbance, this criterion weights more heavily errors that occur late in time (such as offset), putting less emphasis on errors which occur immediately after the disturbance (i.e. overshoot).

7.3. Fixed-parameter controller tuning methods

The process of adjusting a control system or control loop to achieve its best response characteristics is often referred to as *tuning the control loop*. Essentially, this is the process of commissioning and optimisation of system performance. Once a system has been installed, then the tuning process involves adjusting the various controller parameters or settings in order to achieve the best possible response (such as a minimum error-integral value). In traditional fixed-parameter control systems, this is usually achieved not by scientific reasoning but by trial and error – backed up by some initial intelligent guess work, though some work has been done on systematic methods for obtaining optimum controller parameters (see for example Kelly *et al.*, 1991). We will look at this later but first we need to consider what degrees of freedom we have in the pre-commissioned system.

Controller parameters

The 1980s saw a dramatic shift in the UK from the use of electronic and pneumatic analogue control systems for HVAC applications, to digital control systems and, with it, the expansion of distributed control and building management systems.

Overwhelmingly, the HVAC control algorithm in use in these systems is the discretised form of the three-term proportional–integral–derivative (PID) controller. We looked at this in Chapter 5 (section 5.5) and the equations are reproduced below in sampled time form:

$$u(n) = a\varepsilon(n) - b\varepsilon(n-1) + c\varepsilon(n-2) + u(n-1) \tag{7.18}$$

where $u(n)$ and $\varepsilon(n)$ are the control signal and control error at the *n*th sampling instant, respectively, $u(n-1)$ and $\varepsilon(n-1)$ are the control signal and error at the previous sampling instant, $\varepsilon(n-2)$ is the control error at the previous but one sampling instant, and the coefficients are

$$a = \frac{K_c}{Ti_t} \times \left[T^2 + Ti_t + i_t d_t \right] \tag{7.19}$$

$$b = \frac{K_c}{T} \times (T + 2d_t) \tag{7.20}$$

$$c = \frac{K_c}{T} \times d_t \tag{7.21}$$

Thus control system 'degrees of freedom' are the parameters: K_c = the controller gain, i_t = the integral time (or reset time), d_t = the derivative time (or rate time) and T = the sampling time interval.

Equation (7.18) is therefore used to calculate the required control signal at time, $t = nT$, from previous values of control signal and control error. Practical algorithms take this form though there are some minor variations. For example, some make the feedback variable value the object of the derivative action instead of control error. Integral and derivative parameters are sometimes expressed as an *integral gain* K_c/i_t and a *derivative gain* $K_c d_t$, respectively.

There may also need to be a provision for *integral wind-up*. Wind-up occurs when, during a protracted period in which the error is high, the control signal saturates (i.e. settles at its maximum value) due to the accumulated integral. A sustained period of negative error would be needed to cause a reduction in the control signal. Instead, a mechanism can be included in the algorithm to reset the saturated signal rapidly.

Note that setting $i_t = \infty$ and $d_t = 0$ effectively eliminates any integral and derivative control action, leaving us with purely proportional control.

We will come back to these parameters in a later section but first let us look at two other aspects of controller set-up.

Proportional band and proportional gain

For largely historical reasons, many of the controllers and control algorithms used in HVAC applications use a *proportional band setting* as opposed to proportional gain, though it appears that this term is gradually being replaced by the latter.

The proportional band, p, is defined as the range of the controlled variable to which the control signal is directly proportional. It can be expressed simply as a band of the controlled variable value. As a controller setting, the proportional band is usually expressed as a percentage of the corresponding overall range of the system,

$$p = \frac{\delta y}{\Delta y} \times 100\% \qquad (7.22)$$

where δy represents the band of the controlled variable within which control is to be maintained and Δy is the maximum range of the controlled variable which is determined by the limits of the capacity of the plant. Thus, the percentage proportional band is also sometimes called the *throttling range* (CIBSE, 1985).

The following example illustrates the relationship between the proportional band and gain.

Example 7.4

A proportional fan speed controller is used to maintain a critical static pressure in a space. The fan when operating at top speed can maintain a static pressure of $300\mathrm{Nm}^{-2}$ in the space, and at its minimum speed, the static pressure will be $25\mathrm{Nm}^{-2}$. The speed controller has been set up to achieve a proportional band of $\pm 10\mathrm{Nm}^{-2}$. What is the gain and percentage proportional band of the controller?

Solution

The proportional band in units of the controlled variable is, therefore, $10 - (-10) = 20\mathrm{Nm}^{-2}$. The gain of this controller will therefore be

$$\frac{\text{signal range}}{\delta y} = \frac{1}{20} = 0.05\mathrm{m}^2\mathrm{N}^{-1} \quad \left(\text{or } 5\%\mathrm{m}^2\mathrm{N}^{-1}\right)$$

and the percentage proportional band will be

$$p = \frac{\delta y}{\Delta y} = \frac{20}{(300 - 25)} \times 100\% = 7.27\%$$

Now suppose that the controller proportional band setting is adjusted to 15%. The range (or proportional band) of the control variable will be

$15\% \times (300 - 25) = 41.25\mathrm{Nm}^{-2}$ $\left(\text{i.e. } \pm 20.625\mathrm{Nm}^{-2}\right)$

and the modified gain will be

$$\frac{1}{41.25} = 0.024\mathrm{m}^2\mathrm{N}^{-1} \left(\text{or } 2.4\%\mathrm{m}^2\mathrm{N}^{-1}\right)$$

Note that the proportional band and proportional gain settings of a controller are inversely related.

Identifying PID terms

In many applications, there will be no advantage in adopting the derivative term in a PID controller; in fact, the PI controller is quite common in HVAC applications, and in a number of these cases, purely proportional control might suffice with adequate results. The reasoning is as follows. If the system has sufficient intrinsic stability to allow a purely proportional controller to be used with a high gain (i.e. low proportional band) without fear of instability, then as we know from section 4.1 the high gain setting will be reasonably free from offset and satisfactory control will result. If the high gain is not possible due to the danger of an unstable response, then a low-gain integral PI controller will be necessary and it becomes necessary to fix a sufficiently low value of integral time to eliminate the offset. This then prompts the need for derivative action to compensate for the resulting slowness of response caused by the integral action; full PID control is necessary.

Though most current generation controllers will offer full PID control, there are clearly advantages in set-up time if the necessary combination of controller term(s) is known prior to commissioning. Borresen (1990) has proposed a simple method which is summarised below.

Using an open-loop process–reaction test, the system gain, K_p, and time constant and dead time (τ, t_d), are obtained (see section 7.1). Borresen argues that the controllability of a system depends on the system gain, offset, and the ratio of the dead time to time constant of the system. He defines the *relative control difficulty*, D_{REL}, as

$$D_{REL} + \frac{K_p}{\Delta y_{ss}} \times \frac{t_d}{\tau} \tag{7.23}$$

where Δy_{ss} is the required maximum offset (equation (7.23) implies that there must be some tolerance to offset, even after applying integral control action).

The interpretation of D_{REL} is summarised in Table 7.1. We note that D_{REL} = 1 when the offset is equal to the proportional band and it would therefore be unnecessary to specify other than simple proportional control. At $D_{REL} > 1$, offset becomes a problem and integral action becomes necessary. The introduction of derivative action at the higher D_{REL} can be thought of as compensation for system dead-time effects.

Table 7.1 Interpretation of D_{REL} (Borresen, 1990)

Relative difficulty	Controllability	Controller choice
$D_{REL} < 1$	Simple	P
$1 < D_{REL} < 2.5$	Medium difficult	PI
$D_{REL} > 2.5$	Difficult	PID

As a general guide, a maximum offset of around 1–2K for air temperature control systems, 10–15% for relative humidity control and 5–10K for water temperature control (boiler and chiller plant) should be aimed for.

Example 7.5

For the system parameters determined in Example 7.1, find the relative control difficulty and, hence, determine which of the three term control actions should be specified. The control system should operate within a maximum offset of 1.0K.

Solution

We use the system time constant and dead-time characteristics in equation (7.23) to determine the relative control difficulty, noting that the maximum acceptable offset is to be 1.0K:

$$D_{REL} = \frac{K}{\Delta\theta_{oss}} \times \frac{t_d}{\tau} = \frac{14.6}{1.0} \times \frac{31.5}{241.5}$$

$$D_{REL} = 1.9$$

From this, we conclude that the system will experience 'medium control difficulty' (Table 7.1), since $D_{REL} > 1$. We therefore specify a PI controller.

Methods of establishing controller parameters

There are several methods for establishing the required controller parameters or settings to achieve the 'best' control for a given control system. With conventional fixed-parameter controllers, these parameters can often only be determined by trial and error. The methods we will look at in the following can be used to establish 'first guess' approximations to inform this trial-and-error process.

We noted that there are up to four controller parameters for traditional P, PI or PID control, giving up to four degrees of freedom for system tuning. These are the sampling interval, T, and up to three controller settings. We will look at the latter in the following sections, but it is appropriate to reflect once more on the sampling interval, which we first considered in section 5.4.

For a system which can be adequately described by a first-order lag plus a dead-time term, we note from section 3.5 that the dead-time term can be treated in exactly the same way as a transport lag and can be represented by a Padé approximation. This results in our simple lag plus dead-time model having two poles and a zero in the right half of the *s*-plane; thus the dominant pole is the inverse of the time constant. One of the 'rules of thumb' discussed in section 5.4 was to fix T (as a maximum) of one-tenth of the dominant time constant. Therefore, Example 7.1 under proportional control might be expected to operate satisfactorily with a sampling interval of about 24s. We also noted in Chapter 5 (Example 5.5) that with the sampling interval low enough, our system might be expected to behave like a continuous-time system but with a need for a lower proportional gain and, as a consequence, a tendency for offset which must be checked using integral control. On the other hand, a high sampling interval tends to result in a response which resembles a continuous system with high dead time, a combination which equation (7.23) confirms will lead to reduced stability.

In summary for the sampling interval, unless we expect high-frequency disturbances (i.e. rapid load disturbances or measurement noise, not common in HVAC applications), a very low sampling interval is desirable. Proportional control alone is unlikely to be adequate in most cases. In any event, the sampling interval should be less than any dead time and within one-tenth of the dominant time constant if known.

Ziegler and Nichols tuning based on closed-loop criteria

The earliest work on controller tuning with particular reference to process control was due to Ziegler & Nichols (1942). In the first of two methods, the controller participates in the loop (i.e. the control system is maintained in its closed-loop mode) and the controller is set for proportional control only. The controller gain is gradually increased (i.e. proportional band decreased) as small step disturbances are introduced until a point is reached where the control loop is marginally stable. At this point, the controlled variable will be seen to oscillate at uniform amplitude (Figure 7.4).

The following characteristics at marginal stability are noted: K_c', the controller gain at marginal stability; and p_ω, the period of oscillation at marginal stability. From these, settings can be estimated using the expressions given in Table 7.2. These estimates are based on achieving one-quarter decay ratio to a step change in input to the controller.

This method is very simple to carry out under site conditions. However, there are two disadvantages which restrict its use:

1   Operation at or close to marginal stability for some systems may be dangerous (e.g. boiler plant, high-pressure systems, etc.).
2   In some of the highly damped control loops often found in HVAC applications (e.g. underfloor heating), it may be difficult to force a marginally stable condition, or to do so may take a considerable time.

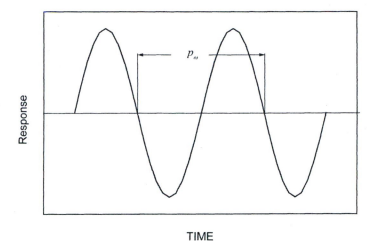

TIME

Figure 7.4 Closed-loop response at marginal stability.

Table 7.2 Ziegler and Nichols tuning estimates – closed-loop criteria (Ziegler & Nichols, 1942)

Controller	K_c	i_t	d_t
P	$0.5K'_c$	—	—
PI	$0.45K'_c$	$p_\omega/1.2$	—
PID	$0.6K'_c$	$p_\omega/2$	$p_\omega/8$

Tuning based on open-loop response criteria

To overcome the disadvantages imposed by extracting the closed-loop data, alternative estimates for controller settings have evolved based on the identified parameters, τ and t_d, from the open-loop process–reaction curve (section 7.1).

Ziegler and Nichols estimates As well as the work they did based on closed-loop performance characteristics, Ziegler & Nichols (1942) also offer an estimate based on open-loop process–reaction curve information, again based on the one-quarter decay ratio criterion (Table 7.3).

These classical parameter estimates have been used widely in process control for many years, though Geng & Geary (1993) found that they can lead to conservative results when used in some HVAC applications and propose a modified procedure.

Cohen and Coon settings Cohen & Coon (1953) pointed out that the Ziegler & Nichols (1942) work did not consider system 'self-regulation' resulting from dead-time effects and they derived alternative setting esti-

Table 7.3 Ziegler and Nichols tuning estimates – open-loop criteria (Ziegler & Nichols, 1942)

Controller	K_c	i_t	d_t
P	$\dfrac{\tau}{K_p t_d}$	—	—
PI	$0.9 \times \dfrac{\tau}{K_p t_d}$	$3t_d$	—
PID	$1.2 \times \dfrac{\tau}{K_p t_d} \to 2.0 \times \dfrac{\tau}{K_p t_d}$	$2t_d$	$0.5t_d$

Table 7.4 Cohen and Coon tuning estimates (Cohen & Coon, 1953)

Controller	K_c	i_t	d_t
P	$\dfrac{\tau}{K_p t_d} \times \left[\dfrac{3\tau + t_d}{3\tau}\right]$	—	—
PI	$\dfrac{\tau}{K_p t_d} \times \left[\dfrac{10.8\tau + t_d}{12\tau}\right]$	$t_d \left[\dfrac{10\tau + t_d}{3\tau + 6.7 t_d}\right]$	—
PID	$\dfrac{\tau}{K_p t_d} \times \left[\dfrac{5.3\tau + t_d}{4\tau}\right]$	$t_d \left[\dfrac{32\tau + 6t_d}{13\tau + 8t_d}\right]$	$t_d \left[\dfrac{2\tau}{55\tau + t_d}\right]$

mates based on the open-loop process reaction curve data. These are summarised in Table 7.4.

Lopez *et al.* estimates Lopez *et al.* (1967) make recommendations for estimating controller settings which seek to minimise the various error-integral criteria (see section 7.2). They also make recommendations for achieving either good set point response or good disturbance-rejection response. The following summarises their recommendations for disturbance rejection control:

- for the controller gain setting,

$$K_c = \frac{a}{K_p} \times \left[\frac{t_d}{\tau}\right]^b \qquad (7.24)$$

- for the integral action time,

$$i_t = \frac{\tau}{c} \times \left[\frac{t_d}{\tau}\right]^d \qquad (7.25)$$

- for the derivative action time,

$$d_t = e\tau \times \left[\frac{t_d}{t}\right]^f \qquad (7.26)$$

Table 7.5 Constants for Lopez et al. tuning estimates (Lopez et al., 1967)

Integral-error criterion	Controller	a	b	c	d	e	f
$I_{(\|e\|)}$	P	0.437	−1.098	—	—	—	—
$I_{(e^2)}$	P	0.666	−1.027	—	—	—	—
$I_{(\|e\|t)}$	P	0.362	−1.119	—	—	—	—
$I_{(\|e\|)}$	PI	0.984	−0.986	0.608	0.707	—	—
$I_{(e^2)}$	PI	1.305	−0.960	0.492	0.739	—	—
$I_{(\|e\|t)}$	PI	0.859	−0.977	0.674	0.680	—	—
$I_{(\|e\|)}$	PID	1.435	−0.921	0.878	0.749	0.482	1.137
$I_{(e^2)}$	PID	1.495	−0.945	1.101	0.771	0.560	1.006
$I_{(\|e\|t)}$	PID	1.357	−0.947	0.842	0.738	0.381	0.995

The constants, a–f, are given in Table 7.5.

One of the main advantages of the open-loop process–reaction curve methods is that they allow a simple identified model to be fitted to the system. This, in turn, enables a much wider analysis to be carried out on an existing system if required, using, for example, simulation methods. The following example compares the results of the various methods.

Example 7.6

Using the results of Examples 7.1 and 7.5, select and compare suitable controller parameters using the various methods discussed in this section.

Solution

From Example 7.5, we recall that a PI controller should be used for this case.

- **Ziegler and Nichols estimates**

Based on the open-loop process–reaction curve of Example 7.1, we note that the plant gain, K_p, is 14.6K^{-1} and the dead time and time constant are 31.5s and 241.5s respectively. Referring to Table 7.3 for a PI controller,

$$K_c = 0.9 \times \frac{\tau}{K_p t_d} = 0.9 \times \frac{241.5}{14.6 \times 31.5} = 0.473\text{K}^{-1}$$

and

$$i_t = 3t_d = 3 \times 31.5 = 94.5\text{s}$$

As a matter of interest, it is also possible to find the setting estimates based on Ziegler and Nichols' closed-loop performance criteria. We note from Example 7.1 that the closed-loop model for this system is

$$G(s) = \frac{14.6 \exp(-31.5s)}{(241.5s + 1)} \times K_c$$

Using a Routh array based on the characteristic equation of the above, we can obtain the controller gain and period of the response based on the system natural frequency, ω_n, as described in Example 4.2 (Chapter 4). The results are $K'_c = 1.119$ and $p_\omega = 93.1$.

Referring to the estimates of Table 7.2, the alternative Ziegler and Nichols tuning estimates will therefore be

$$K_c = 0.45 \times K'_c = 0.45 \times 1.119 = 0.503\text{K}^{-1}$$

and

$$i_t = \frac{p_\omega}{1.2} = \frac{93.1}{1.2} = 77.6\text{s}$$

- **Cohen and Coon estimates**

Based on Table 7.4,

$$K_c = \frac{\tau}{K_p t_d}\left[\frac{10.8\tau + t_d}{12\tau}\right] = \frac{241.5}{14.6 \times 31.5} \times \left[\frac{10.8 \times 241.5 + 31.5}{12 \times 241.5}\right] = 0.478\text{K}^{-1}$$

and

$$i_t = t_d\left[\frac{10\tau + t_d}{3\tau + 6.7t_d}\right] = 31.5 \times \left[\frac{10 \times 241.5 + 31.5}{3 \times 241.5 + 6.7 \times 31.5}\right] = 82.4\text{s}$$

- **Lopez et al. estimates**

We refer to equations (7.24) and (7.25) and the associated constants in Table 7.5. For the absolute error-integral criterion:

$$K_c = \frac{a}{K_p}\left[\frac{t_d}{\tau}\right]^b = \frac{0.984}{14.6} \times \left[\frac{31.5}{241.5}\right]^{-0.986} = 0.502\text{K}^{-1}$$

and

$$i_t = \frac{\tau}{c}\left[\frac{t_d}{\tau}\right]^d = \frac{241.5}{0.608} \times \left[\frac{31.5}{341.5}\right]^{0.707} = 94.1\text{s}$$

Similarly, for the squared error-integral criterion:

$$K_c = \frac{1.305}{14.6} \times \left[\frac{31.5}{241.5}\right]^{-0.960} = 0.632\text{K}^{-1}$$

and

$$i_t = \frac{241.5}{0.492} \times \left[\frac{31.5}{241.5}\right]^{0.739} = 109.0\text{s}$$

For the time multiplied by error-integral criterion:

$$K_c = \frac{0.859}{14.6} \times \left[\frac{31.5}{241.5}\right]^{-0.977} = 0.430\text{K}^{-1}$$

and

$$i_t = \frac{241.5}{0.674} \times \left[\frac{31.5}{241.5}\right]^{0.680} = 89.7s$$

Table E7.6.1 Tuning estimates

Method	Controller gain	Integral action time
Ziegler–Nichols (open-loop criteria)	0.473	94.5
Ziegler–Nichols (closed-loop criteria)	0.503	77.6
Cohen–Coon	0.478	94.1
Lopez et al. ($I_{(|e|)}$)	0.502	94.1
Lopez et al. ($I_{(e^2)}$)	0.632	109.0
Lopez et al. ($I_{(|e|t)}$)	0.430	89.7

Table E7.6.1 summarises the various results. We see that most of the estimates are in reasonably close agreement, but we can pick out two of the Lopez *et al.* results for closer examination since they stand apart from the other results – specifically those estimates based on squared error integral ($I_{(e^2)}$) and time multiplied by error integral ($I_{(|e|t)}$). We note that the latter fixes a lower integral time evidently to force the offset (i.e. long-term error) to zero, but in doing so, a lower gain is necessary for stability. The opposite pattern is evident for the squared error-integral criterion.

We should note a particularly important point about the estimates in Example 7.6; they are entirely based on the performance optimisation of a continuous-time process system though we are aware that most modern control is conducted in discrete time. How will these estimates behave when applied to the equivalent discrete-time system when the sampling-time degree of freedom is also introduced? We shall consider this for the two Lopez *et al.* estimates in a simulation in Example 7.7.

Example 7.7

For the system described by the model of Example 7.1, apply the Lopez *et al.* controller setting estimates based on ($I_{(e^2)}$) and ($I_{(|e|t)}$) to the discrete-time control case with the initial sampling interval taken to be one-tenth of the dominant time constant of the system.

Compare the resulting unit step time response simulations of the continuous-time and discrete-time cases.

Solution

The sampling interval will be 24.15s (one-tenth of the system time constant which in this case is essentially the dominant time constant).

Firstly, we need to discretise the continuous-time system model. If we assume a zero-order hold, then the modified z-transform of equation (5.14) (section 5.3) can be used. Representing the case in which the dead time is a non-integer multiple of sampling intervals (i.e. $t_d = (315/24.15) \times T = 1.304T$), equation (5.14) gives

$$G(z,md) = \frac{K_c K_p}{z^i}\left[\frac{[\exp(-B) - \exp(-A)]z^{-2} + [1 - \exp(-B)]z^{-1}}{1 - \exp(-A)z^{-1}}\right]$$

where $A = T/\tau$, $B = (T/\tau) \times (1 - \lambda)$, $t_d = 1.304T$, $\lambda = 0.304$, $m = 1 - \lambda = 0.696$, $i = 1$, $K_p = 14.6$ and $\tau = 241.5$. We can therefore express the discrete-time form of system model with zero-order hold as

$$G(z,md) = \frac{0.407z^{-2} + 0.982z^{-1}}{z - 0.905} = \frac{0.982z + 0.407}{z^3 - 0.905z^2}$$

Now we consider the controller. For a PI controller, we recall the discrete-time (z-domain) form as follows (see section 5.5):

$$G_c(z) = \frac{az^2 - bz}{z^2 - z}$$

where

$$a = \frac{K_c}{Ti_t} \times \left(T^2 + K_c Ti_t\right) \qquad \text{and} \qquad b = K_c$$

For the first of the two Lopez et al. controller settings estimates, $K_c = 0.430\,\mathrm{K}^{-1}$ and $i_t = 89.7\mathrm{s}$ imply

$$a = \frac{0.430 \times 24.15^2 + 0.430 \times 24.15 \times 89.7}{24.15 \times 89.7} = 0.546 \qquad \text{and} \qquad b = 0.430$$

Therefore the discrete-time controller transfer function will be:

$$G_c(z)\big|_{(K_c=0.430, i_t=89.7)} = \frac{0.546z^2 - 0.430z}{z^2 - z}$$

and, similarly, for the other,

$$G_c(z)\big|_{(K_c=0.632, i_t=109.0)} = \frac{0.772z^2 - 0.632z}{z^2 - z}$$

A comparative simulation model which generates side-by-side results for the discrete-time and continuous-time cases is now generated using a Simulink model (Simulink, 1996). The model is shown in Figure E7.7.1 for the first of the two estimated controllers considered above and is based on a unit step change in set point.

Results

Simulation results from the continuous-time system confirm good responses for either set of settings estimates (Figure E7.7.2). In particular, the $I_{(|\varepsilon|t)}$ criterion has resulted in settings which give minimal oscillation and (slightly) faster settling time than for the $I_{(\varepsilon^2)}$-based estimates.

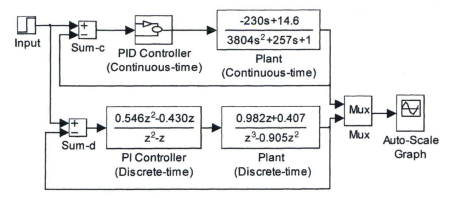

Figure E7.7.1 Simulink model of discrete- and continuous-time systems.

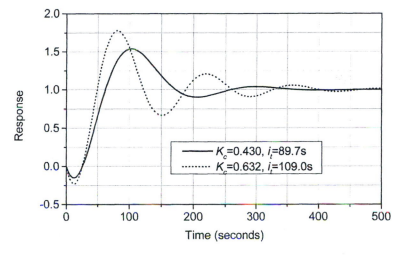

Figure E7.7.2 Results for the continuous-time case.

However, the discrete-time equivalent system at the second of the two controller settings estimates exhibits instability (Figure E7.7.3) and for the first of the two settings estimates the discrete-time plant was found to show considerable oscillation prior to settling (not shown).

Clearly we must either conclude for this example that the settings estimates, whilst satisfactory for a continuous-time situation, are inadequate for the discrete-time case; or that the one-tenth dominant time constant sampling interval for the latter is unsound when using these settings estimates.

If we modify the settings estimates and leave the sampling interval as it was, our choices are confined to reducing the gain or increasing the integral time or both. In any event, this is likely to lead to offset and protracted response times (as we found in Example 5.5, Chapter 5). The

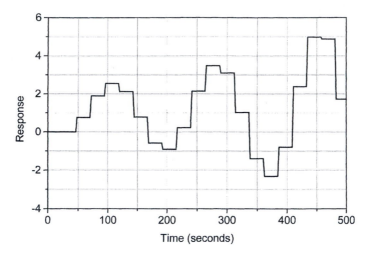

Figure E7.7.3 Discrete-time case ($T = 24.15$s and $I_{(|\epsilon|t)}$-based estimates.

correct option therefore appears to be a decrease in the sampling interval. Indeed decreasing the sampling interval, at least for systems which are free of high-frequency disturbances, leads to the discrete-time system approaching a response which is closer to the equivalent continuous-time system.

It is possible to show that, with a sampling interval of one-twentieth of the dominant time constant ($T = 12.0575$s), the discrete-time system continues to give an unstable response for the $I_{(\epsilon^2)}$-based estimates though a satisfactory response is achieved for the $I_{(|\epsilon|t)}$-based estimates.

At a sampling interval of one-fortieth of the dominant time constant ($T = 6.0375$s), the system and controller transfer functions become

$$G(z,md) = \frac{0.283z^2 + 0.077}{z^7 - 0.975z^6}$$

$$G_c(z)\big|_{(K_c=0.430, i_i=89.7)} = \frac{0.459z^2 - 0.43z}{z^2 - z}$$

and

$$G_c(z)\big|_{(K_c=0.632, i_i=109.0)} = \frac{0.667z^2 - 0.632z}{z^2 - z}$$

Figure E7.7.4 shows that stable offset-free control can now be expected for either set of settings estimates.

Clearly then, for systems which exhibit significant dead time at least, the various settings estimates considered in this section will work, but only when care is taken over the choice of sampling interval, T. It seems clear that the one-tenth time constant 'rule of thumb' for T will be too irresolute. Indeed, as a general rule, provided that care is taken with regard to the

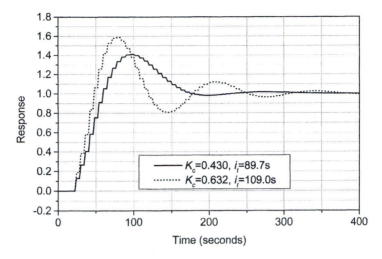

Figure E7.7.4 Discrete-time case at $T = \tau/40$.

information loss effect (see section 5.4), it might be concluded for HVAC control that a choice of T as low as is practicable will usually be the best course of action.

In this chapter, we have given extensive consideration to the estimation of controller parameters and a reasonable amount of work has been done in this field of relevance to HVAC applications. Bekker *et al.* (1991) look at the use of a pole–zero cancellation tuning method (pole–zero cancellation is applied to adaptive controller design in Chapter 8). Kasahara *et al.* (1997) introduce a two-degrees-of-freedom controller which acts separately on feedback and reference signals and go on to propose a tuning method for this, whilst Nesler & Stoecker (1984) evaluate PI settings for HVAC plant experimentally. Pinnella *et al.* (1986) describe the development of a self-tuning algorithm for a fixed-parameter purely integral controller. A first-order plus dead-time model is fitted to open-loop step response data collected on-line and the results are used to identify an integral gain using Ziegler–Nichols tuning rules. The method is tested on a fan speed regulator and a steam heat exchanger having fast- and slow-response dynamics respectively. Both systems exhibited critically damped responses. Ho (1993) develops a search algorithm for the optimised selection of fixed PID controller parameters. The search uses energy consumption as the objective function as well as a stability criterion. However, the search times are protracted (it is debatable whether the method could be used on-line) and though results seem good, they are compared with simulated reference results only.

Another important aspect of controller performance is performance and fault detection in use – aspects which are beyond the scope of this book. However, several workers have developed methods of fault detection in control systems – see for instance Pape *et al.* (1991), Solberg & Teeters (1992) or Fasolo & Seborg (1995).

Increasingly, the advantages of *adaptive* control in which controller parameters are adjusted on-line to accommodate changing circumstances and operating conditions by the plant are being realised. We will therefore give special consideration to this in the next chapter.

References

Bekker, J.E., Meckl, P.H., Hittle, D.C. (1991) A tuning method for first-order processes with PI controllers. *ASHRAE Transactions*, **97** (2), 19–23.

Borresen, B.A. (1990) Controllability: back to basics. *ASHRAE Transactions*, **14** (1), 817–819.

CIBSE (1985) *Automatic Controls and Their Implications for System Design*. Chartered Institution of Building Services Engineers, London.

Cohen, G.H., Coon, G.A. (1953) Theoretical considerations of retarded control. *Transactions of the American Society of Mechanical Engineers*, **75**, 827–834.

Coley, D.A., Penman, J.M. (1992) Second order system identification in the thermal response of real buildings. Paper II: Recursive formulation for on-line building energy management and control. *Building and Environment*, **27** (3), 269–277.

Fasolo, P.S., Seborg, D.E. (1995) Monitoring and fault detection for an HVAC control system. *ASHRAE Transactions*, **1** (3), 177–193.

Geng, G., Geary, G.M. (1993) On performance and tuning of PID controllers in HVAC systems. *Proceedings of the Second IEEE Conference on Control Applications*, Vancouver, 819–824.

Ho, W.F. (1993) Development and evaluation of a software package for self-tuning of three-term DDC controllers. *ASHRAE Transactions*, **99** (1), 529–534.

Hudson, G., Underwood, C.P. (1996) Effect on energy usage of over-capacity and emission characteristics in the space heating of buildings with high thermal capacity. *Proceedings of the CIBSE National Conference*, Harrogate.

Kamimura, K., Yamada, A., Matsuba, T., Kimbara, A., Kurosu, S., Kasahara, M. (1994) CAT (computer-aided tuning) software for PID controllers. *ASHRAE Transactions*, **100** (1), 180–190.

Kasahara, M., Matsuba, T., Murasawa, I., Hashimoto, Y., Kamimura, K., Kimbara, A., Kurosu, S. (1997) A tuning method of two degrees of freedom PID controller. *ASHRAE Transactions*, **103** (1), 278–289.

Kelly, G.E., Park, C., Barnett, J.P. (1991) Using emulators/testers for commissioning EMCS software, operator training, algorithm development, and tuning local control loops. *ASHRAE Transactions*, **97** (1), 669–678.

Ljung, L. (1987) *System Identification: Theory for the User*. Prentice-Hall, Englewood Cliffs, NJ.

Ljung, L. (1988) *System Identification Toolbox for Use with MATLAB*. The Mathworks Inc., Natick, MA.

Lopez, A.M., Miller, J.A., Smith, C.L., Murrill, P.W. (1967) Tuning controllers with error-integral criteria. *Instrumentation Technology*, **14** (11), 57–62.

MATLAB (1996) *MATLAB The Language of Technical Computing*. The Mathworks Inc., Natick, MA.

Nesler, C.G. (1986) Automated controller tuning for HVAC applications. *ASHRAE Transactions*, **92** (2B), 189–201.

Nesler, C.G., Stoecker, W.F. (1984) Selecting the proportional and integral constants in the direct digital control of discharge air temperature. *ASHRAE Transactions*, **90** (2B), 834–845.

Pape, F.L.F., Mitchell, J.W., Beckman, W.A. (1991) Optimal control and fault detection in heating, ventilating and air conditioning systems. *ASHRAE Transactions*, **97** (1), 729–736.

Penman, J.M. (1990) Second order system identification in the thermal response of a working school. *Building and Environment*, **25** (2), 105–110.

Pinnella, M.J., Wechselberger, E., Hittle, D.C., Pedersen, C.O. (1986) Self-tuning digital integral control. *ASHRAE Transactions*, **92** (2B), 202–210.

Simulink (1996) *SIMULINK 2 Dynamic System Simulation for MATLAB*. The Mathworks Inc., Natick, MA.

So, A.T.P., Chan, W.L., Chow, T.T., Tse, W.L. (1995) New HVAC control by system identification. *ASHRAE Transactions*, **30** (3), 349–357.

Solberg, D.P.W., Teeters, M.D. (1992) Specification of spreadsheet trend log sheets for DDC/EMCS and HVAC system commissioning, energy monitoring, life safety cycles, and performance based service contracts. *ASHRAE Transactions*, **98** (2), 553–571.

Underwood, D.M. (1989) Response of self-tuning single loop digital controllers to a computer simulated heating coil. *ASHRAE Transactions*, **95** (2), 424–430.

Wellstead, P.E., Zarrop, M.B. (1991) *Self-tuning Systems: Control and Signal Processing*. John Wiley, Chichester.

Zaheer-uddin, M. (1990) Combined energy balance and recursive least squares method for the identification of system parameters. *ASHRAE Transactions*, **96** (2), 239–244.

Ziegler, J.G., Nichols, N.B. (1942) Optimum settings for automatic controllers. *Transactions of the American Society of Mechanical Engineers*, **64**, 759–768.

8 Adaptive control

8.1 Adaptive controllers

Successful control owes everything to the care that is given at the design and commissioning stages. There are numerous cases, upon which it would not be appropriate to dwell here, where well-intentioned design and installation are compromised by such things as cost cutting, tight timescales and misunderstandings between procurer and supplier and, very often, because control is often the after-thought of an HVAC project, things go wrong. Many control suppliers use factory-preset controller parameters so that the installation team make no site adjustments unless manifestly necessary – resulting control may be generally satisfactory but sub-optimal. Indeed until recently, many electronic HVAC controllers had fixed tuning which could not be adjusted on-site, often leaving the maintenance team with the problem of making the best out of a bad job in situations where the controller is not suited to an application.

Thus, development of current and future generations of HVAC control has begun to focus on in-built intelligence at the controller, in particular, adaptive control. Many of the ideas presently being explored are not new. John & Dexter (1989) first discussed the potential for intelligence in building services control in 1989, identifying progress and ongoing work in fields such as self-tuning, adaptive and rule-based control. Zaheer-uddin (1993) also reviews progress in adaptive and optimal control and gives illustrative examples of their applications in the HVAC field, whilst Wittenmark & Åström (1984) review some of the general basic methods in relation to practical difficulties associated with their implementation. See also Clarke & Gawthrop (1979) who summarise early progress in self-tuning control, including implementation problems, and Dexter & Haves (1989) who discuss implementation issues with reference to HVAC applications.

To some, *self-tuning* control and *adaptive* control are the same and represent control systems capable of adjusting themselves to changes in their operational domain as well as changing requirements. To others, the self-tuning control system makes these adjustments in search of optimality over an initial period only – then freezes (these have also been mentioned in section 7.2), whilst the adaptive controller continues to make adjustments throughout life. A general consensus is that with self-tuning controllers it is assumed that the process under control has initially unknown but constant parameters which can be estimated on-line whilst no such assumptions about parametric constancy can be made with adaptive control. The early heating optimisers are examples of *self-tuning systems* in so far as a finite adjustment timespan is concerned and considerable work has been done on these over the years (Levermore, 1992; Coley & Penman, 1996;

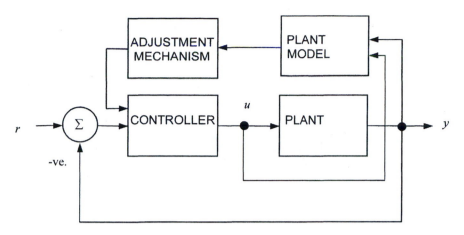

Figure 8.1 Structure of an adaptive controller.

Murdoch *et al.*, 1990; Birtles & John, 1985). Since both self-tuning and adaptive control systems tend to use the same or similar methods, the differences between the two are often subtleties, but from here on, we will discuss adaptive control with reference to lifelong adaptivity.

Key elements of adaptive and self-tuning control systems are an *adjustment mechanism* and a predictive model of the process (Figure 8.1). The latter predicts future values of the process output and will most commonly use a recursive least-squares estimation method.

The function of the adjustment mechanism is to determine the parameters of the controller, using the coefficients of an estimated process model. There are many possibilities for this, and we will take a brief look at some of them in the following section.

8.2 Direct and indirect adaptive control algorithms

The plant model structure may take one of a number of forms. A common approach is to use the discrete form of the ARX structure (see section 7.1),

$$Ay(t) = Bu(t) + e(t) \tag{8.1}$$

where

$$A = 1 + a_1 z^{-1} + \ldots + a_{na} z^{-na} \quad \text{and} \quad B = b_0 + b_1 z^{-1} + \ldots + b_{nb} z^{-nb}$$

Here $y(t)$ and $u(t)$ are the process output and input and $e(t)$ is a disturbance sequence (Figure 8.2). t refers to sampled time.

It is possible to write down an expression for the controller. A general form of linear two-degrees-of-freedom controller with tunable parameter sequences, K_1, K_2, K_3, is frequently used:

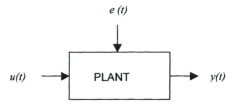

Figure 8.2 Plant discrete-time variables.

$$K_1 u(t) = K_2 r(t) - K_3 y(t) \tag{8.2}$$

where

$$K_1 = 1 + k_1 z^{-1} + \ldots + k_{nk_1} z^{-nk_1}$$

$$K_2 = k_{20} + k_2 z^{-1} + \ldots + k_{nk_2} z^{-nk_2}$$

$$K_3 = k_{30} + k_3 z^{-1} + \ldots + k_{nk_3} z^{-nk_3}$$

Adaptation or self-tuning now amounts to two operations:

1 System identification; essentially estimating the sequences A, B.
2 Controller adaptation; using the parameters of the system model and an adjustment mechanism, determine the coefficients of the controller, K_1, K_2, K_3.

These two operations are carried out on-line during each sampling interval. There are several methods of controller parameter adjustment but they generally fall into two categories, *direct* and *indirect*. We will take a look at a leading method in each category – the direct minimum-variance method and the indirect pole-placement method a little later. First, the problem of on-line identification is considered.

Recursive estimation

In section 7.1 the least-squares algorithm for estimating the parameters of a plant model from experimental data (i.e. equation (7.13)) was discussed. This method is used widely for system model parameter estimation but for adaptive control applications the plant variables can be observed periodically in real time and it is therefore possible (and desirable) to obtain a refreshed estimate of the plant model parameters at each time step. Of essence is the need to carry out the computations associated with adaptive control comfortably within each sampling interval. To save computational effort, adaptive controllers make use of *recursive least-squares* (RLS) estimation such that measurement observations made at time t are used to obtain plant parameter estimates for time $t + 1$. Thus computational efficiency is achieved by requiring the storage of data from the previous time step only and further computational efficiency is achieved by avoiding the need for a complicated matrix inversion operation through the use of a 'matrix inversion lemma'. The method is summarised below. For a detailed treatment, readers are referred to Åström and Wittenmark (1989) or Ljung (1988).

The RLS estimate at time t can be expressed as

$$\hat{\theta}(t) = \hat{\theta}(t + 1) + \mathbf{K}(t)\left[\mathbf{y}(t) - \mathbf{X}^{\mathrm{T}}(t)\hat{\theta}(t - 1)\right] \tag{8.3}$$

in which $\mathbf{K}(t)$ sometimes referred to as the Kalman gain, is given by

$$\mathbf{K}(t) = \frac{\mathbf{P}(t - 1)\mathbf{X}(t)}{\lambda + \mathbf{X}^{\mathrm{T}}(t)\mathbf{P}(t - 1)\mathbf{X}(t)} \tag{8.4}$$

and

$$\mathbf{P}(t) = \left[\mathbf{I} - \mathbf{K}(t)\mathbf{X}^{\mathrm{T}}(t)\right] \times \mathbf{P}(t - 1)\lambda^{-1} \tag{8.5}$$

The method assumes that parameters, once estimated, are constant. In many cases a time weighting of earlier data is desirable. A typical need for this is where parameters change slowly and estimation then becomes unresponsive to a subsequent rapid change. A popular and simple way of dealing with this is to introduce the constant, λ, as shown in equations (8.4) and (8.5) which causes current data to have unit weighting but previous sampled data to be diminished exponentially. This is called *forgetting* or *discounting* and thus λ is an exponential forgetting factor. Though $0 \leq \lambda \leq 1$, typical practical values for λ lie in the range 0.95–0.99.

One point of caution here is that exponential forgetting only works when the system variables receive continuous excitement. Where the system or process spends long periods without excitement (for example systems which are designed mainly to accommodate periodic set point changes only) then discarding old data tends to lead to uncertainty in the estimated parameters – a condition sometimes referred to as *estimator wind-up* (Åström & Wittenmark, 1989).

An RLS algorithm can therefore be summarised as follows:

1 *Step 1.* At time $= t$, form $\mathbf{X}(t)$ based on measurements of $y(t)$, $u_1(t - k_1), \ldots, u_m(t - k_m), \ldots$ (i.e. for up to m inputs where k_1, k_m are input–output delays in sampling instants).
2 *Step 2.* Form $\mathbf{P}(t)$ using equation (8.5).
3 *Step 3.* Form $\mathbf{K}(t)$ using equation (8.4).
4 *Step 4.* Update $\hat{\theta}$ using equation (8.3).
5 *Step 5.* At time $= t + 1$, go back to Step 1.

Demonstrating the RLS method off-line, though academic, can be insightful. Often in these situations, the big question is what to do with $e(t)$. In practice, disturbances imposed on the system will be a complex mix of load disturbances and measurement 'noise' and will rarely be fully understood *a priori*. In the simulation of systems a random but uniform disturbance pattern can be very helpful and a zero-mean *white noise* source will often be considered giving the essential qualification of randomness. In the following example, we will consider the application of RLS estimation to a simple feedback system response excited with a white noise disturbance. For a practical discussion of measurement noise and a method of dealing with it using Kalman filtering, see Diderrich & Kelly (1984).

Example 8.1

The damper control system of Example 5.6 has been simulated with the external temperature disturbance path represented by a zero-mean white noise source of unit variance, sampled at the same interval as the control system option with $T = 1.5$s. The damper model, based on the mid-season gain and corrected to unit controller range, is

$$G_p(z) = \frac{0.578}{z^3 - 0.928z^2} \tag{E8.1.1}$$

The results are given in Figure E8.1.1.

Note that, though quite suitable from the point of view of a step response analysis, the controller arrived at in Example 5.6 is insufficiently responsive for the high-frequency disturbance dynamics we have arbitrarily imposed here.

The RLS estimation (equations (8.3), (8.4), (8.5)) will need to be carried out by computer due to the number of operations involved. The estimated parameters shown in Figure E8.1.2 are based on a first-order ARX structure ($n_a = n_b = 1$) and were arrived at using the MATLAB system identification toolbox command (Ljung, 1987),

```
>theta=rarx(z,[111],'ff',1)
```

In the above, the parameter estimates at each time step are returned in theta and the arguments of the recursive ARX estimator (rarx) are z

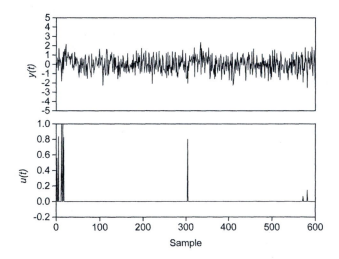

Figure E8.1.1 Simulated damper control system when excited with *white noise*.

(which contains the column vectors of $\mathbf{y}(t)$, $\mathbf{u}(t)$ from Figure E8.1.1). The square bracket contains the model order details and the remaining data signify exponential forgetting factor fixed, in this case, at 1. Figure E8.1.2 shows the asymptotic convergence of the parameter estimates of the first-order model (in this case $f = 1$).

Figure E8.1.3 shows the estimated responses compared with the original simulated sequence. For comparison, a fourth-order estimation is included together with results using an exponential forgetting factor of 0.95.

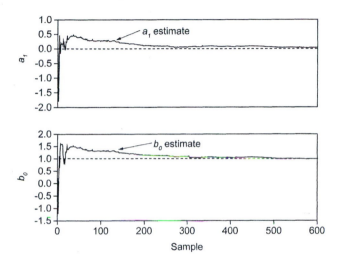

Figure E8.1.2 Parameter estimates for the first-order RLS model.

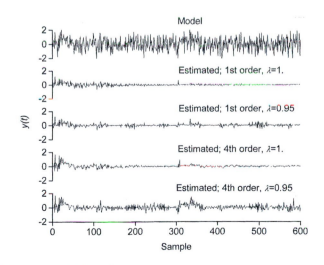

Figure E8.1.3 Model and RLS predictions.

Clearly, the ARX structure (which does not parametrise the noise term) does not give a good representation of the system response though slightly improved results occur with exponential forgetting. Only the fourth-order model with exponential forgetting factor (in this case 0.95) begins to reproduce the actual system output with some measure of agreement – at least in the general pattern of the results if not the variance.

When an ARMAX structure is adopted (see section 7.1), which does parametrise the noise term, considerably improved results are obtained (Figure E8.1.4) in that the random nature of $y(t)$ is now much better represented. Here we see little point in using a forgetting factor when the noise term appears in the model directly. This was fitted using the `rarmax` function in the MATLAB system identification toolbox (very similar to the previous function). In many practical situations however, it is not possible to measure the noise term which leads to complications regarding its inclusion or otherwise in the model structure.

In many cases, the main features of the plant might be adequately traced using a low-order model especially where disturbances are periodic but well defined and random noise influences are not significant. As is generally the case with these methods, considerable trial and error will often be necessary before a satisfactory compromise on model structure, model order and choice of data weighting are arrived at.

Direct adaptation – the minimum-variance method

A controller which minimises the cost function, J,

$$J = E\left(y^2(t + k)\right) \tag{8.6}$$

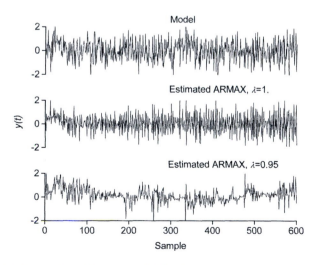

Figure E8.1.4 Influence of model structure (ARMAX).

where k is the input–output delay in sampling intervals and E is a polynomial in the delay operator z, is termed a *minimum-variance controller* (MVC).

Let the predicted output from the system be $\hat{y}(t)$ and the corresponding prediction error be $y_\varepsilon(t)$, i.e. $y(t) = \hat{y}(t) + y_\varepsilon(t)$. Then

$$J = E(\hat{y}(t+k) + y_\varepsilon(t+k))^2 = E\hat{y}^2(t+k) + Ey_\varepsilon^2(t+k) + 2E(\hat{y}(t+k) + y_\varepsilon(t+k))$$

$$(8.7)$$

For the linear controller of the form of equation (8.2) the final (linear) term in the above disappears since only the non-linear terms will contribute to J. Also, $\hat{y}_\varepsilon(t+k)$ is not influenced by the controller, merely by the model predictor. Hence for minimum-variance control we require that

$$J = E\hat{y}^2(t+k) = 0$$

or

$$\hat{y}(t+k) = 0 \tag{8.8}$$

So that for the general ARMAX case,

$$y(t+k) = \frac{B}{A} u(t) + \frac{C}{A} e(t+k) \tag{8.9}$$

it is possible to show that

$$\hat{y}(t+k|t) = BFu(t) + Gy(t) = 0$$

(the vertical bar notation implies that the estimate of $y(t+k)$ is based on data received up to and including time t).

From this, the minimum-variance controller is

$$u(t) = -\frac{G}{BF} y(t) \tag{8.10}$$

where the polynomial coefficients, F and G, can be found from the polynomial identity (or Diophantine equation)

$$C = AF + z^{-k}G \tag{8.11}$$

and take the form

$$F = 1 + f_1 z^{-1} + \ldots + f_{k-1} z^{-(k-1)}$$
$$G = g_0 + g_1 z^{-1} + \ldots + g_{n_g} z^{-n_g} \qquad n_g = \max(n_a - 1, n_c - k)$$

The result is a controller which gives the following output:

$$y(t) = Fe(t) \tag{8.12}$$

and minimum output variance

$$J_{min} = \left(1 + f_1^2 + \ldots + f_{k-1}^2\right) \times \sigma_e^2 \qquad (8.13)$$

where σ_e^2 is the disturbance variance.

For further details, see Wellstead & Zarrop (1991). Note also that the use of Diophantine equations features frequently in this and related branches of control algorithm design and a detailed treatment of these can be found in Kučera (1979).

Note that when implemented with respect to an *error* signal (i.e. with a reference set point), the MVC output becomes

$$u(t) = \frac{G}{BF}[r(t) - y(t)] \qquad (8.14)$$

Because the controller output, $u(t)$, is calculated directly at each time step from the estimated parameters of the system model, this is a direct or implicit method of adaptation. Note however that k must be known and in some instances the inverse of B is unstable, thus a basic minimum-variance controller can lack robustness. This can however be remedied by adopting a more computationally demanding general minimum-variance (GMV) controller. A treatment of GMV control is beyond the scope of this book – interested readers are referred to Wellstead & Zarrop (1991) or Harris & Billings (1981).

Example 8.2

We can derive a 'snapshot' MVC (i.e. based only on a current system model) for the damper system of Examples 5.6 and 8.1 as an illustration of the method.

Solution

We recall the (theoretical) system model for this case,

$$G(z) = \frac{y(z)}{u(z)} = \frac{0.578}{z^3 - 0.928z^2}$$

Adding a noise sequence, which we assume for the purpose of this example to be truncated after the first term, we can express the system model in the time domain:

$$y(t) = 0.928\,y(t-1) + 0.578u(t-3) + e(t)$$

from which $A = 1 - 0.928z^{-1}$, $B = 0.578z^{-3}$, $C = 1$, $k = 3$, $n_a = n_b = 1$ and $n_c = 0$; therefore $n_g = 0$.

Applying the polynomial identity of equation (8.11) implies that $C = AF + z^{-k}G$. Noting that $F = 1 + f_1z^{-1} + f_2z^{-2}$ and $G = g_0$ we get

$$1 = \left(1 - 0.928z^{-1}\right)\left(1 + f_1z^{-1} + f_2z^{-2}\right) + g_0z^{-k}$$

Multiplying out and equating the coefficients of z^{-1}, z^{-2}, z^{-3} respectively yields

$f_1 = 0.928$ $\quad f_2 = 0.861$ $\quad$ and $\quad g_0 = 0.799$

The MVC output will therefore be, for a controller with a reference set point (equation (8.14)),

$$u(t) = \frac{G}{BF}[r(t) - y(t)] = \frac{0.799}{0.578(1 + 0.928z^{-1} + 0.861z^{-2})}[r(t) - y(t)] \qquad \text{(E8.2.1)}$$

or, in time domain,

$$u(t) = 1.382[r(t) - y(t)] - 0.928u(t - 1) - 0.861u(t - 2)$$

The closed-loop response (with constant set point) will be (equation (8.12))

$$y(t) = e(t) + 0.928e(t - 1) - 0.861e(t - 2)$$

and the minimum output variance will be (equation (8.13))

$$J_{min} = \sigma_e^2(1 + 0.928^2 + 0.861^2) = 2.6\sigma_e^2$$

Note that if $k = 1$ then $F = 1$ (equation (8.11)) in which case the minimum variance reduces to σ_e^2. Thus any time delay, k, present in the system model critically determines the variance of the adaptive system output.

The system may be summarised as in Figure E8.2.1.

In summary, a simple adaptive MVC algorithm can be stated as follows:

1   *Step 1.* At time $= t$, estimate the system model using the RLS algorithm (k must be known) based on current and past measurements of $y(t, t - 1, \ldots)$ and $u(t - 1, \ldots)$.

2   *Step 2.* Solve the polynomial identity of equation (8.11) to obtain parameter sequences F and G.

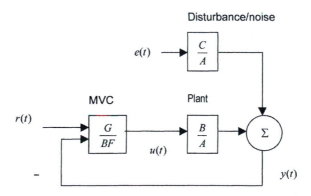

Figure E8.2.1 Summary of MVC system.

3   *Step 3.* Calculate the required controller output using equation (8.10)
    or (8.14) using the current and previous measurements of
    $y(t, t-1, \ldots)$ and $u(t-1, \ldots)$
4   *Step 4.* Wait for time $= t + 1$, then go back to Step 1.

Indirect adaptation – the pole-placement method

In this method, the key objective is to allocate or *place* open-loop system
poles to some desired set of closed-loop pole locations (recall our earlier
use of pole–zero terminology in connection with the root locus stability
method – section 4.2). Thus in an adaptive pole-placement controller
(PPC) the controller design adapts to achieve a pre-defined system output
characteristic. As we shall see, it has the advantage over the adaptive MVC
in that the input–output delay, k, need not be known.

As a simple illustration of this concept, suppose a system can be described by the following continuous-time open-loop model:

$$y(s) = \frac{K_p}{(\tau s + 1)} u(s) \tag{8.15}$$

and a proportional controller is used,

$$u(s) = K_c[r(s) - y(s)] \tag{8.16}$$

The system open-loop dynamic response is characterised by the time constant, τ (i.e. by a pole at $-1/\tau$).

Now, combine equations (8.15) and (8.16) to form the closed-loop
model of the system,

$$y(s) = \frac{K_c K_p}{\tau s + (1 + K_c K_p)} = \frac{K_c K_p (1 + K_c K_p)^{-1}}{[\tau/(1 + K_c K_p)]s + 1} \tag{8.17}$$

We see that the *closed-loop* dynamic response is characterised by the time
constant $\tau/(1 + K_c K_p)$ or a pole located at $-(1 + K_c K_p)/\tau$. In this closed-
loop system, K_c gives us a degree of freedom so that we can place or assign
the closed-loop pole where we want. For example, a high positive K_c reduces
the time constant, thus speeding response; a low negative K_c increases the
time constant which slows response. Furthermore, the steady-state response
will be $K_c K_p/(1 + K_c K_p)$ so that if we introduce a further parameter into the
numerator of the closed-loop model, K_c^*, we can achieve an offset-free
response through

$$\frac{K_c^* K_p}{(1 + K_c K_p)} \tag{8.18}$$

Modifying the standard proportional controller of equation (8.16) to the
following two-parameter form achieves precisely this:

$$u(s) = K_c r(s) - K_c^* y(s) \qquad (8.19)$$

However, it is also possible to exercise pole-placement design through a three-term PID controller as an alternative, using the integral gain to achieve a specified steady-state response (Wellstead & Zarrop, 1991).

We now turn to the use of the pole-placement method for adaptive control. A summary of the basis for the method is given in the following; readers are referred to Wellstead & Zarrop (1991) or Åström & Wittenmark (1989) for a more detailed treatment of this subect. The objective is to achieve a controller which tracks the reference signal, $r(t)$, in some desired way and, coincidentally, rejects any random disturbance sequence in $e(t)$ which may corrupt $y(t)$. Consider once again the ARMAX structure for the system model and note that since b_0 in B must be zero (since $u(t)$ is the subject of our controller), we can put

$$Ay(t) = Bu(t)z^{-1} + Ce(t) \qquad (8.20)$$

Combining equation (8.20) with the general linear controller of equation (8.2) leads to the closed-loop system model:

$$y(t) \times \left(AK_1 + BK_3z^{-1}\right) = BK_2 r(t)z^{-1} + CK_1 e(t) \qquad (8.21)$$

We now set the following polynomial identity, or *Diophantine* equation,

$$AK_1 + BK_3z^{-1} = XC \qquad (8.22)$$

in which the poles of the closed-loop model can be placed at their desired locations through the setting of X and $X = 1 + x_1 z^{-1} + \ldots + x_{n_x} z^{-n_x}$. Combining equations (8.21) and (8.22) leads to

$$y(t) = \frac{BK_2}{XC} r(t-1) + \frac{K_1}{X} e(t) \qquad (8.23)$$

Thus the polynomial sequences of the linear controller, K_1, K_3, are calculated in this method indirectly in contrast to the direct approach of the previous method. Note that K_2 is selected to achieve some pre-defined outcome; frequently, to cancel C in equation (8.23) and to apply a constant to the reference set point ($r(t-1)$) such that

$$K_2 = C\left[\frac{X}{B}\right]_{z=1} \qquad (8.24)$$

Where the noise term, $e(t)$, is unmodelled, then $C = 1$ and K_2 will be a constant.

Solving the polynomial identity of equation (8.22) requires some consideration especially for computationally efficient on-line calculation in real time. First, to achieve a unique solution we set

$$n_{k_1} = n_b \qquad (8.25a)$$

$$n_{k_3} = n_a - 1 \qquad (8.25b)$$

$$n_x \leq n_a + n_b - n_c \qquad (8.25c)$$

(though there are instances where over-parametrisation is called for as we shall see later). Provided that the polynomials A, B have no common zeros, the identity can be solved as follows:

1   Set n_{k_1}, n_{k_3}, n_x using the above criteria.
2   Write down the identity in full (equation (8.22)) and multiply out the brackets.
3   Form a set of linear equations by equating coefficients of z^{-m} ($m = 1$, $2, \ldots$) and express in stacked matrix–vector notation.
4   Inverting the resulting matrix of the A, B, C polynomials leads to the controller parameter polynomials, K_1, K_3.

As an illustration, consider the general case of the ARX system model structure ($C = 1$) in which $n_a = n_b = 2$. Suppose we set $n_x = 1$. The polynomial identity of equation (8.22) will be

$$AK_1 + BK_3 z^{-1} = X$$

Then $n_{k1} = 2$ (equation (8.25a) implies
$$K_1 = 1 + k_{11}z^{-1} + k_{12}z^{-2}$$
$$n_a = 2 \text{ implies}$$
$$A = 1 + a_1 z^{-1} + a_2 z^{-1}$$
$$n_b = 2 \text{ implies}$$
$$B = b_0 + b_1 z^{-1} + b_2 z^{-2}$$
$$n_{k_3} = 1 \text{ (equation (8.25b)) implies}$$
$$K_3 = k_{30} + k_{31} z^{-1}$$
and $n_x = 1$ implies
$$X = 1 + x_1 z^{-1}$$

The full polynomial identity will therefore be

$$\left(1 + a_1 z^{-1} + a_2 z^{-1}\right)\left(1 + k_{11}z^{-1} + k_{12}z^{-2}\right) + z^{-1}\left(b_0 + b_1 z^{-1} + b_2 z^{-2}\right)\left(k_{30} + k_{31}z^{-1}\right)$$
$$= 1 + x_1 z^{-1}$$

Multiplying out and equating the coefficients of z^{-m} ($m = 1, 2, \ldots$) leads to the following set of linear equations:

$$k_{11} + b_0 k_{30} = x_1 - a_1$$
$$k_{12} + a_1 k_{11} + b_0 k_{31} + b_1 k_{30} = -a_2$$
$$a_1 k_{12} + a_2 k_{11} + b_1 k_{31} + b_2 k_{30} = 0$$
$$a_2 k_{12} + b_2 k_{31} = 0$$

or, in matrix–vector form,

$$\begin{bmatrix} 1 & 0 & b_0 & 0 \\ a_1 & 1 & b_1 & b_0 \\ a_2 & a_1 & b_2 & b_1 \\ 0 & a_2 & 0 & b_2 \end{bmatrix} \times \begin{bmatrix} k_{11} \\ k_{12} \\ k_{30} \\ k_{31} \end{bmatrix} = \begin{bmatrix} x_1 - a_1 \\ -a_2 \\ 0 \\ 0 \end{bmatrix}$$

AK = b

The matrix **A** is a *Sylvester* structure, quite commonly encountered in matrix algebra, due in this case to the distinctive column banding formed by the A, B coefficients. Provided that A and B are co-prime (i.e. contain no common factors), the matrix **A** may be inverted to reveal a solution for the controller parameters in the column vector, **K**, i.e.

$$\mathbf{K} = \mathbf{A}^{-1}\mathbf{b} \qquad\qquad (8.26)$$

When A, B are not co-prime other methods must be resorted to (see for example Kučera's method (Kučera, 1979), described in Wellstead & Zarrop (1991)). The matrix–vector format illustrated gives a convenient and efficient basis for on-line processing of the data, especially where high-order estimation is to be used. Clearly, for 'snapshot' controller design using manual calculations, low-order models will yield a linear equation set that will be more conveniently solved using simple substitution as illustrated in the following example.

Example 8.3

Revisit the damper control system considered in Example 8.2. The closed-loop response is desired to consist of a first-order lag with a time constant of $4T$ (where T is the sampling interval). Specify a controller using the pole-placement method.

Solution

The sampling interval for the system model considered in Example 8.1 is $1.5\,\mathrm{s}$, i.e. $4T = 6\,\mathrm{s}$. From Table 5.1, a first-order closed-loop response with a time constant of $6\,\mathrm{s}$ gives

$$\frac{z}{z - \exp(-T/\tau)} = \frac{z}{z - \exp(-1.5/6)} = \frac{1}{1 - 0.78z^{-1}}$$

from which we can set $X = 1 - 0.78z^{-1}$ (i.e. $x_1 = 0.78$).

We recall that the system model in linear time-domain form from Example 8.2 is

$$y(t) + 0.928\,y(t-1) + 0.578u(t) + e(t)$$

from which we note that

$$n_a = 1 \quad\Rightarrow\quad A = 1 - 0.928z^{-1}$$
$$n_b = 1 \quad\Rightarrow\quad B = 0.578z^{-1} \qquad \text{(i.e. } b_0 = 0)$$
$$n_c = 0 \quad\Rightarrow\quad C = 1$$

We therefore need to solve the polynomial identity,

$$AK_1 + BK_3z^{-1} = X \qquad \text{with} \quad n_{k_1} = n_b = 1 \quad \text{and} \quad n_{k_3} = n_a - 1 = 0$$

i.e.

$$\left(1 - 0.928z^{-1}\right)\left(1 + k_{11}z^{-1}\right) + \left(0.578z^{-2}\right)(k_{30})z^{-1} = 1 - 0.78z^{-1}$$

Multiplying out

$$(k_{11} - 0.928)z^{-1} - 0.928z^{-2} + 0.578k_{30}z^{-3} = 1 - 0.78z^{-1}$$

Equating the coefficients of z^{-1} and z^{-3} yields $k_{11} = 0.148$ and $k_{30} = 0$.

But this eliminates the participation of $y(t)$ in the controller (i.e. $k_{30} = 0$), thus leading to an unfeasible controller design. The reason for this is noted to be the significant input–output delay in this particular system example.

One solution is to over-parametrise one or both of the controller polynomials so that we obtain a balance of z^{-m} coefficients for all m. An obvious way to do this is to try $n_{k_1} = k - n_a = 2$ (k is the input–output time delay which is 3 in this example), i.e.

$$\left(1 - 0.928z^{-1}\right)\left(1 + k_{11}z^{-1} + k_{12}z^{-2}\right) + \left(0.578z^{-2}\right)(k_{30})z^{-1} = 1 - 0.78z^{-1}$$

which leads to $k_{11} = 0.148$, $k_{12} = 0.137$, $k_{30} = 0.238$.

Finally, K_2 is obtained from equation (8.24). Since, in this case, $C = 1$ then $K_2 = k_{20}$, a simple constant,

$$k_{20} = C\left[\frac{X}{B}\right]_{z=1} = 1 \times \left[\frac{1 - 0.78}{0.578}\right] = 0.381$$

From equation (8.2), a possible PPC emerges,

$$u(t) = -0.148u(t - 1) - 0.137u(t - 2) + 0.381r(t) - 0.238y(t) \qquad \text{(E8.3.1)}$$

or, in z-domain,

$$u(z) = \frac{0.381[r(z) - 0.625y(z)]}{\left(1 + 0.148z^{-1} + 0.137z^{-2}\right)} \qquad \text{(E8.3.2)}$$

Figure E8.3.1 gives a comparison of the simulated results using the above PPC, the MVC of Example 8.2 (equation (E8.2.1)) and the PID controller from the original Example 5.6. Here an (arbitrary) square-wave variation in set point is used to excite the controllers and random noise has been suppressed. Evident from these results is that, though the MVC and PPC are more stable and much faster acting than the PID, the steady-state performance is not as good – a point which has already been mentioned and can be remedied through pole-placement design in harness with an integral gain. Overall for the three controllers, the PPC clearly gives much the better response for this example and with this disturbance sequence. It is of course emphasised that these simulated results give a performance 'snapshot'; they do not represent adaptive performance of the MVC and PPC.

In summary, a simple adaptive pole-placement control algorithm can be realised as follows:

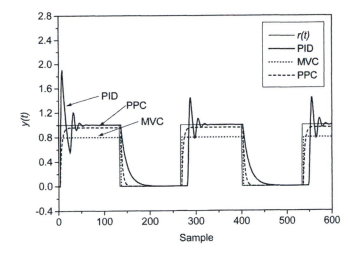

Figure E8.3.1 Comparative performance of PID, MVC and PPC control.

1 *Step 1.* At time $= t$, estimate the system model using the RLS algorithm (see section 8.2) based on the current and past measurements of $y(t, t-1, \ldots)$ and $u(t-1, t-2, \ldots)$.

2 *Step 2.* Solve the polynomial identity of equation (8.22) for the pre-defined closed-loop pole locations, in X. Extract the two controller parameter sets (from K_1, K_3) from this and calculate the remaining controller parameter(s) (K_2) from equation (8.24).

3 *Step 3.* Calculate the required controller output using equation (8.2), based on the current and previous measurements of $y(t, t-1, \ldots)$ and $u(t-1, t-2, \ldots)$.

4. *Step 4.* Wait for time $= t+1$, then go back to step 1.

8.3 Model-reference adaptive control

In model-reference adaptive control (MRAC) a fixed model of the plant under control is applied, generally taking the form of one (or a set of) difference equation(s) which calculate(s) desired plant response from on-line measurements of current and previous plant input and output signals. The error between the model-predicted response and actual measurement is used as a basis for adjusting the controller parameters. In this way, the controller attempts to force the plant or system to follow the response of the model. This approach has its place especially where a reliable model of the plant can easily be expressed and to which the actual plant response can conceivably be matched.

The desired closed-loop response is specified by a model, as in the case of pole-placement design. Suppose the model generates an output, y_m, then the error between the model output and system output, y, will be

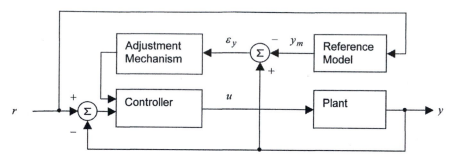

Figure 8.3 The MRAC system.

$$\varepsilon_y = y - y_m \tag{8.27}$$

and the general MRAC system will be realised as in Figure 8.3. The essence of the MRAC adjustment mechanism is the minimisation of ε_y through the setting of adjustable controller parameters, K.

To illustrate the principles, early MRAC systems made use of a simple approach known as the *gradient method* (for a detailed treatment, see Åström & Wittenmark (1989) or Landau (1979)), in which the controller parameters are adjusted according to a *sensitivity derivative*, $\partial\varepsilon_y/\partial K$, such that

$$\frac{\mathrm{d}K}{\mathrm{d}t} = -\gamma\varepsilon_y\frac{\partial\varepsilon_y}{\partial K} \tag{8.28}$$

This assumes that the controller parameters can be changed according to the negative gradient of ε_y^2, scaled by a constant, γ, which is called the *adaptation gain*. This is sometimes called the *MIT rule* (based on work done at MIT in the USA) and adaptation based on the MIT rule is referred to as the gradient method since $\partial\varepsilon_y/\partial K$ represents the gradient of the output-model error with respect to the controller parameters, K.

If it can be assumed that K will vary more slowly than other system variables, then $\partial\varepsilon_y/\partial \mathbf{K}$ can be found quite easily with the assumption that K is 'constant' (or at least for the duration of a sampling interval).

The method is developed in the following illustrative example.

Example 8.4

Consider the familiar case of a system which is known to have the dynamic characteristics of a first-order lag with dead time, i.e.

$$y(s) = \frac{K_p\exp(-t_d s)}{(\tau s + 1)}u(s)$$

or, if a first-order Padé approximation is used for the dead-time term,

$$y(s) = \frac{K_p(1 - 0.5t_d s)}{0.5\tau \times t_d s^2 + (\tau + 0.5t_d)s + 1}u(s) \qquad \text{(E8.4.1)}$$

Suppose, for the sake of this illustration, that we require that the closed-loop system responds according to a first-order lag, then we can express a reference model for the closed loop as

$$y_m(s) = \frac{K_p'}{(\tau's + 1)}r(s) \qquad \text{(E8.4.2)}$$

Suppose also that a two-parameter controller is to be used,

$$u(s) = k_r r(s) - k_y y(s) \qquad \text{(E8.4.3)}$$

The interpretation of an MRAC based on the MIT rule is now developed as follows.

Combining equations (E8.4.1) and (E8.4.3) leads to

$$y(s) = \frac{K_p k_r(1 - 0.5t_d s)}{0.5\tau \times t_d s^2 + (\tau + 0.5t_d - 0.5K_p k_y t_d)s + (1 + K_p k_y)}r(s) \qquad \text{(E8.4.4)}$$

Based on equation (8.28), the sensitivity derivative with respect to the first controller parameter will be

$$\frac{\partial \varepsilon_y}{\partial k_r} = \frac{\partial}{\partial k_r}\left[\frac{K_p k_r(1 - 0.5t_d s)}{0.5\tau \times t_d s^2 + (\tau + 0.5t_d - 0.5K_p k_y t_d)s + (1 + K_p k_y)}r(s) \right]$$

which yields,

$$\frac{\partial \varepsilon_y}{\partial k_r} = \left[\frac{K_p(1 - 0.5t_d s)}{0.5\tau \times t_d s^2 + (\tau + 0.5t_d - 0.5K_p k_y t_d)s + (1 + K_p k_y)}r(s) \right] \qquad \text{(E8.4.5)}$$

Similarly, the sensitivity derivative with respect to the second controller parameter is obtained as

$$\frac{\partial \varepsilon_y}{\partial k_y} = \frac{k_r(K_p - 0.5K_p t_d s)^2}{\left[0.5\tau t_d s^2 + (\tau + 0.5t_d - 0.5K_p k_y t_d)s + (1 + K_p k_y)\right]^2}r(s)$$

which, after substituting equation (E8.4.4), reduces to

$$\frac{\partial \varepsilon_y}{\partial k_y} = -\frac{K_p(1 - 0.5t_d s)}{0.5\tau t_d s^2 + (\tau + 0.5t_d - 0.5K_p k_y t_d)s + (1 + K_p k_y)}y(s) \qquad \text{(E8.4.6)}$$

The snag is that the performance of the actual system, expressed here in terms of the system parameters K_p, τ, t_d, will not normally be known. One way out of this dilemma is to argue that, at optimum control, a model-following equality must exist between the dynamic characteristics of equations (E8.4.2) – our desired closed-loop response – and (E8.4.4) – the actual response – i.e.

$$\frac{1}{(\tau's + 1)} = \frac{(1 - 0.5t_d s)}{0.5\tau t_d s^2 + (\tau + 0.5t_d - 0.5K_p k_y t_d)s + (1 + K_p k_y)}$$

which, applying the general adaptation rule of equation (8.28), allows us to express the following adaptation rules for the controller parameters:

$$\frac{dk_r}{dt} = -\gamma \varepsilon_y r(s)\left[\frac{K_p}{(\tau's + 1)}\right]$$

and

$$\frac{dk_y}{dt} = \gamma \varepsilon_y y(s)\left[\frac{K_p}{(\tau's + 1)}\right]$$

which now leaves only one unknown, K_p, but this can conveniently be absorbed into γ, the adaptation gain, whose value will be set to achieve stable model-following, giving

$$k_r(s) = -\frac{\gamma}{s(\tau's + 1)}\varepsilon_y r(s) \qquad\qquad\qquad (E8.4.7)$$

and

$$k_y(s) = \frac{\gamma}{s(\tau's + 1)}\varepsilon_y y(s) \qquad\qquad\qquad (E8.4.8)$$

Generally, a low value of γ will need to be selected for stable model-following. As an illustration, suppose our actual system has a time constant of 10 units of time, a dead time of 2 units and a gain of unity. To force a reasonably quick response, we set the reference model time constant at half the value of the actual system (i.e. 5 units of time) and fix the reference model gain at unity also.

Figure E8.4.1 gives a `Simulink` model of the resulting MRAC system (Simulink, 1996). Simulating with an arbitrary square-wave set point sequence having unit amplitude and a period of 100 units of time produces the results shown in Figure E8.4.2. These results required a little trial and error with the setting of γ, which proved to give good results when set at 0.1, but unstable results at higher values.

The results are good – just one period is taken for adaptation to give near-perfect model-following for this simple illustration. In practice of course, the model of the actual system or plant will not normally be known unless some on-line estimation has been carried out. Indeed this is one of the advantages of the MRAC system – a model of the actual system, though helpful in tuning γ, is not essential. The sensitivity derivatives can be calculated using a suitable differencing scheme based on measured values

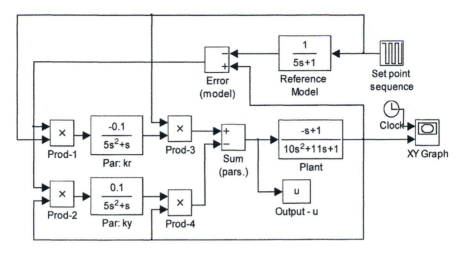

Figure E8.4.1 Simulink model for the MRAC system.

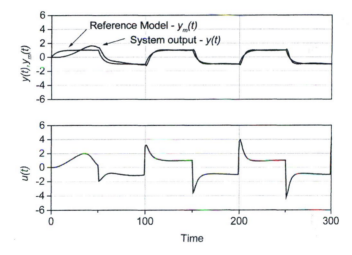

Figure E8.4.2 MRAC adaptation to a square-wave excitation in set point.

of the output-model error and the previous controller parameter values. Nevertheless, using the MIT rule as a basis for gradient-based adaptation can give unstable results and, because some approximation is needed due to the absence of the system parameters, the algorithm will attempt to drive the output-model error to zero but does not guarantee zero error model-following.

As an alternative, the design of MRAC adaptation rules based on Lyapunov functions can successfully be used to generate stable adaptation rules which have structural similarities to MIT-based adaptation rules

(Åström & Wittenmark, 1989). The use of *hyperstability theory* in which a system is split into linear time-invariant and non-linear time-varying 'component blocks' (Landau, 1979) and adaptation laws obtained based on the linear and non-linear blocks can also be used to guarantee stability. Another possibility is the use of *variable structure systems* in which controller structure changes on reaching a 'switching plane' (see for example Balestrino *et al.* (1984) for a comparison of hyperstability and variable-structure methods applied to MRAC).

8.4 Gain scheduling

Gain scheduling (GS) is a special case of adaptive control in that adaptation takes place without necessarily having any reference to the controlled condition. GS is useful (and easy to apply) in situations where the plant non-linearity results from some known and measurable condition. In these circumstances, it can lead to good improvement over conventional control (see for instance Tödtli, 1985).

A common application is the compensation of a control system which contains a component whose non-linear behaviour depends on its position (i.e. the input control signal). For example, Åström & Wittenmark (1989) describe the application of GS to compensate for a non-linear valve in which the valve characteristic is a known function of the incoming control signal. They show how two simple linear equations can be used to impose an approximate inverse valve characteristic on the plant thereby compensating for the non-linearity of the valve. In an HVAC application, Zaheeruddin (1989, 1991) uses a 'dynamic gain factor' to update the controller of a space heating system by adapting to space temperature and heating system dynamics.

Another application is where GS is carried out with reference to some measurable external variable which is known to influence plant behaviour in some way. A good example of this is to return once more to the time-variant control over mixing dampers discussed in Examples 5.6 and 8.1. We found that the damper gain was different at different parts of the operating year, specifically due to variations in inlet air temperature. In mid-season conditions when the inlet air entered the plant at 12°C, the damper gain was $0.08\,\text{K}(\%)^{-1}$ (control signal defined as a percentage scale). Then in winter conditions, when the entering air temperature had fallen to $-3°C$, the damper gain increased to $0.23\,\text{K}(\%)^{-1}$ causing the control system, which was stable at the mid-season condition, to become unstable. We will consider a simple GS remedy for this in the following example.

Example 8.5

Incorporate a simple compensating gain schedule in the damper control system of Examples 5.6 and 8.1.

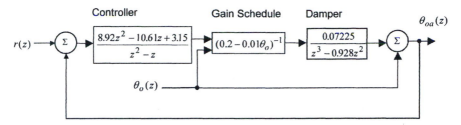

Figure E8.5.1 Gain schedule applied to the damper controller.

Solution

For the case where $T = 1.5$s, the discrete-time transfer functions for controller and damper adjusted to unity damper gain were found to be

$$G_p(z) = K_d \times \frac{0.07225}{z^3 - 0.928z^2} \qquad \text{(damper)}$$

and

$$G_c(z) = \frac{1}{K_d} \times \frac{8.92z^2 - 10.61z + 3.15}{z^2 - z} \qquad \text{(controller)}$$

where K_d is the (variable) damper gain. Thus as K_d increases, the controller gain requires to be commensurately reduced. GS can therefore be accomplished by applying the inverse of the external temperature-dependent damper gain to the control signal. Provided that the damper itself is linear, it is reasonable to suppose that K_d will vary linearly with external air temperature, θ_0. Within the range of temperature considered we can therefore express

$$K_d = 0.2 - 0.01\theta_o \qquad (-3 \leq \theta_o \leq 12) \qquad \text{(E8.5.1)}$$

and form a possible gain schedule for the damper controller as illustrated in Figure E8.5.1.

To illustrate the nature of the original problem and its GS remedy, Figure E8.5.2 gives a Simulink model of the system based on a ramp decay in external air temperature across the range considered (i.e. 12°C falling to −3°C). The decay period is 120 min for the purpose of the illustration and the damper gain equation (i.e. equation (E8.5.1)) and resulting gain schedule have been adjusted to these conditions. Also simulated is the case in which the gain schedule is disabled and the controller parameters are fixed at the values determined in Example 5.6 (i.e. the existing situation).

Figure E8.5.3 shows the results in terms of the original problem – which can be seen by the lapse into instability with increasing damper gain (reducing external temperature) – and confirms that the GS gives excellent compensation for the external temperature effect.

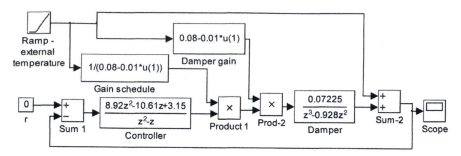

Figure E8.5.2 Simulink model of the GS implementation.

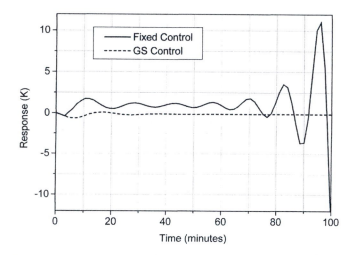

Figure E8.5.3 Damper control and gain scheduling.

In many cases, GS is capable of giving good improvement to systems experiencing instability as a result of plant non-linearity as this simple example has shown. However, there are instances where the precise nature of the non-linearity is difficult to establish and more robust methods then need to be resorted to.

8.5 Progress in HVAC adaptive control

It is appropriate at this point to consider areas in which relevant progress has been made in some of the areas discussed above.

Farris & McDonald (1980) describe one of the first reported applications of adaptive control to HVAC, based on a 'systems' approach in which the entire HVAC system and plant participates (i.e. energy generation and distribution, secondary plant at the space, and the space itself). The entire plant is represented by an RLS estimator and all controller parameters are generated in an attempt to minimise a cost function, seeking essentially to minimise energy use besides achieving acceptable comfort. Simulations which are based on an application involving a solar water heating system, air handling plant and space show that this 'adaptive whole system' approach can save substantial energy over conventional controllers whilst maintaining acceptable comfort.

Jota & Dexter (1985) have developed *general minimum-variance* (GMV) and *general predictive control* (GPC) algorithms for the heating and cooling coils of an air handling system. These have been applied directly to valve actuators and also 'cascaded' in which the algorithm modifies the set point of a conventional fixed-parameter PI controller. Direct application was found to produce unacceptable oscillation whilst the cascaded method gave a substantial improvement.

Virk & Loveday (1991) and Loveday *et al.* (1992) compare conventional on:off and fixed-parameter PI control with a minimum-variance on:off algorithm with the latter achieving superior stability performance and energy saving.

In work on cascaded self-tuning control, Graham & Dexter (1985) describe the development and application of a cascaded self-tuning control system to the control of a large heating system employing a mixing valve with both external temperature feedforward and room temperature feedback. The system uses two controllers. The first is an outer loop controller which receives a room set point, space temperature and external temperature signals and generates a desired mixed water temperature value at the mixing valve. The second controller, which is an inner loop controller, receives a signal from the outer loop controller (in effect, a floating set point) and a feedback signal from the actual mixed water temperature, thus generating a valve positioning signal based on any deviation between the two. A simple parameter estimator predicts the output of the system several steps ahead and the difference between the predicted and desired plant output is used to tune the parameters of the outer loop controller using a search technique which minimises a cost function. Based on field trials, the method works well and is robust, though this is no doubt assisted by the innate stability enjoyed by the plant. Dexter & Haves (1989) also consider robustness in *general predictive control* algorithms through the use of 'jacketing software'.

Brandt (1986) developed a pole-cancellation algorithm with 'bracketing software' and applied it to two (small and large) VAV systems with the bracketing software applied to the large test case only. The algorithm was applied to cooling coil and mixing damper control in each system. The bracketing software was used to suspend adaptation during periods when the system failed to achieve control objectives (i.e. became unstable or failed to meet set point requirements due, for example, to insufficient plant capacity). Brandt concluded that self-tuning control can work as well as well-tuned PID fixed-parameter control without the need for lengthy and repetitive tuning of the latter, but the bracketing software proved essential for robustness. Other work on HVAC adaptive controllers using the pole-placement method are reported by MacArthur *et al.* (1989), Zaheer-uddin (1993) and Wallenborg (1991). The issue of robustness in adaptive control is also considered in work reported by MacArthur & Woessner (1993) who develop a 'receding horizon controller' designed to force plant to stability in a finite number of control 'moves' (robust control is given detailed consideration in Chapter 9).

Nesler (1986) reports on the use of fitted first-order plus dead-time models to observable plant data and the use of this to calculate parameters for a PI control algorithm. He adopted Ziegler–Nichols tuning rules (see section 7.3) to determine the controller parameters, leading to an initial tuning. An RLS algorithm with exponential forgetting factor is then used to update the process model after this initial tuning and this is then used for the ongoing adjustment of the PI controller parameters again using the Ziegler–Nichols criteria. The algorithm is tested on an existing 10-zone dual-duct air conditioning system with reasonable results. Nesler identifies set point changes and the existence of severe non-linearities as areas requiring special measures when developing self-tuning control algorithms of this type.

Seem (1997) considers a novel method of adaptive HVAC control based on recognising patterns of closed-loop plant behaviour, in particular addressing 'sluggish' or oscillatory behaviour in PI controlled systems. This was developed initially as a self-tuning controller but is also claimed to offer promise for adaptive control. The use of 'look-ahead' adaptive control algorithms which predict future energy demands and use these to produce trajectories of required control behaviour are reported by Athienitis (1988), Zaheer-uddin (1994) and Oestreicher *et al.* (1996). These methods have significance in high-thermal-capacity buildings for predicting set-back control strategies, night cooling switching patterns, etc.

References

Åström, K.J., Wittenmark, B. (1989) *Adaptive Control.* Addison-Wesley, Reading, MA.

Athienitis, A.K. (1988) A predictive control algorithm for massive buildings. *ASHRAE Transactions,* **94** (2), 1050–1068.

Balestrino, A., De Maria, G., Zinober, A.S.I. (1984) Nonlinear adaptive model-following control. *Automatica,* **20** (5), 559–568.

Birtles, A.B., John, R.W. (1985) A new optimum start control algorithm. *Building Services Engineering Research and Technology,* **6** (3), 117–122.

Brandt, S.G. (1986) Adaptive control implementation issues. *ASHRAE Transactions,* **92** (2B), 211–219.

Clarke, D.W., Gawthrop, P.J. (1979) Self-tuning control. *IEE Proceedings: Part D – Control Theory and Applications,* **126** (6), 633–640.

Coley, D.A., Penman, J.M. (1996) Simplified thermal response modelling in building energy management. Paper III: Demonstration of a working controller. *Building and Environment,* **31** (2), 93–97.

Dexter, A.L., Haves, P. (1989) A robust self-tuning predictive controller for HVAC applications. *ASHRAE Transactions*, **95** (2), 431–438.

Diderrich, G.T., Kelly, R.M. (1984) Estimating and correcting sensor data in a chiller subsystem: an application of Kalman filter theory. *ASHRAE Transactions*, **90** (2B), 511–517.

Farris, D.R., McDonald, T.E. (1980) Adaptive optimal control – an algorithm for direct digital control. *ASHRAE Transactions*, **86** (1), 880–893.

Graham, W.J., Dexter, A.L. (1985) Self-tuning control of the heating plant in a large building. *Proceedings of the IEE International Conference – Control '85*, Institute of Electrical Engineers, London. IEE Conf. Publ. no. 252, vol. 2, 575–580.

Harris, C.J., Billings, S.A. (1981) *Self-tuning and Adaptive Control: Theory and Applications*. Peter Peregrinus, London.

John, R.W., Dexter, A.L. (1989) Intelligent controls for building services. *Building Services Engineering Research and Technology*, **10** (4), 131–141.

Jota, F.F., Dexter, A.L. (1985) Self-tuning control of an air handling plant. *Proceedings of the IEE International Conference – Control '88*, Institute of Electrical Engineers, London. IEE Conf. Publ. no. 285.

Kučera, V. (1979) *Discrete Linear Control: The Polynomial Equation Approach*. John Wiley, Chichester.

Landau, Y.D. (1979) *Adaptive Control – The Model Reference Approach*. Marcel Dekker, New York.

Levermore, G.J. (1992) *Building Energy Management Systems – An Application to Heating Control*. E&FN Spon, London.

Ljung, L. (1987) *System Identification: Theory for the User*. Prentice-Hall, Englewood Cliffs, NJ.

Ljung, L. (1988) *System Identification Toolbox for Use with MATLAB*. The Mathworks Inc., Natick, MA.

Loveday, D.L., Virk, G.S., Cheung, J.M. (1992) Advanced control for BEMS: a model-based predictive approach. *Building Services Engineering Research and Technology*, **13** (4), 217–233.

MacArthur, J.W., Woessner, M.A. (1993) Receding horizon control: a model-based policy for HVAC applications. *ASHRAE Transactions*, **99** (1), 139–148.

MacArthur, J.W., Grald, E.W., Konar, A.F. (1989) An effective approach for dynamically compensated adaptive control. *ASHRAE Transactions*, **95** (2), 415–423.

Murdoch, N., Penman, J.M., Levermore, G.J. (1990) Empirical and theoretical optimum start algorithms. *Building Services Engineering Research and Technology*, **11** (3), 97–103.

Nesler, C.G. (1986) Automated controller tuning for HVAC applications. *ASHRAE Transactions*, **92** (2B), 189–201.

Oestreicher, Y., Bauer, M., Scartezzini, J.-L. (1996) Accounting free gains in a non-residential building by means of an optimal stochastic controller. *Energy and Buildings*, **24**, 213–221.

Seem, J.E. (1997) Implementation of a new pattern recognition adaptive controller developed through optimisation. *ASHRAE Transactions*, **103** (1), 494–506.

Simulink (1996) *SIMULINK 2 Dynamic System Simulation for MATLAB*. The Mathworks Inc., Natick, MA.

Tödtli, J. (1985) Adaptive control of nonlinear plants. *Landis & Gyr Review*, **32** (2), 3–7.

Virk, G.S., Loveday, D.L. (1991) A comparison of predictive, PID, and on/off techniques for energy management and control. *ASHRAE Transactions*, **97** (2), 3–10.

Wallenborg, A.O. (1991) A new self-tuning controller for HVAC systems. *ASHRAE Transactions*, **97** (1), 19–25.

Wellstead, P.E., Zarrop, M.B. (1991) *Self-tuning Systems – Control and Signal Processing*. John Wiley, Chichester.

Wittenmark, B., Åström, K.J. (1984) Practical issues in the implementation of self-tuning control. *Automatica*, **20** (5), 595–605.

Zaheer-uddin, M. (1989) Sub-optimal controller for a space heating system. *ASHRAE Transactions*, **95** (2), 201–208.

Zaheer-uddin, M. (1991) Dynamic performance of a sub-optimal controller. *Energy and Buildings*, **17**, 117–130.

Zaheer-uddin, M. (1993) Optimal, sub-optimal and adaptive control methods for the design of temperature controllers for intelligent buildings. *Building and Environment*, **28** (3), 311–322.

Zaheer-uddin, M. (1994) Temperature control of multizone indoor spaces based on forecast and actual loads. *Building and Environment*, **29** (4), 485–493.

9 Advanced methods

9.1 Robust control

Plant sensitivity and robustness

Consider the simple single-input single-output (SISO) feedback system with proportional controller and unity feedback, shown in Figure 9.1. The plant model, $G(t)$, is of the general form

$$G(t) = \frac{A}{B} u(t) \tag{9.1}$$

where A and B (the parameters of the system model) contain coefficients of a time series. In practical systems, these coefficients generally vary with time, constituting non-linear system characteristics.

The closed-loop model of Figure 9.1 will be $G'(t)$, where

$$G'(t) = \frac{K_c G(t)}{1 + K_c G(t)} \tag{9.2}$$

and the change in closed-loop response with respect to changes in the open-loop system response, $dG'(t)/dG(t)$, represents the *sensitivity* of the closed-loop system with respect to changes at the plant. That is,

$$\frac{dG'(t)}{dG(t)} = \frac{d}{dG(t)} \left[\frac{K_c G(t)}{1 + K_c G(t)} \right] = \frac{1}{(1 + K_c G(t))^2}$$

which can be expressed as a normalised sensitivity function, S, if we multiply by $K_c G(t)/G'(t)$, which gives

$$S = \frac{1}{1 + K_c G(t)} \tag{9.3}$$

From this we conclude that, so long as the loop gain (i.e. the controller gain, K_c here) is high, the closed-loop system response will be insensitive to plant parameter variations.

Of course we have also concluded in a number of places throughout this book that a high controller gain might imply unstable dynamic response of the closed-loop system. Hence the basis of robust fixed-parameter control is to identify conditions for which the closed-loop system will be stable at high loop gain and, therefore coincidentally, insensitive to disturbances and non-linearities experienced by the plant. Similarly, it is possible to envisage robust–adaptive control in which robustness is designed to ensure that adaptation is insensitive to uncertainties in the estimated plant model. In general then, a robust control system is one in which control is insensitive

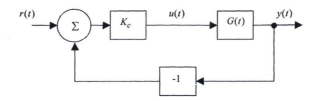

Figure 9.1 Basic SISO system.

to uncertainties in the plant model or plant behaviour, plant disturbances and measurement noise. Whilst many of the techniques for robust design have mainly been associated with fixed-parameter controller design (very often for aerospace applications) there is considerable potential for these methods in situations where a system model forms the basis of controller design. For example in adaptive and self-tuning systems, and where the model contains uncertainties.

Inevitably, as we shall see, achieving robustness in the case of fixed-parameter control will often prove, at least to some extent, to be at the expense of performance which is why robust methods are increasingly paired with adaptive control. Nevertheless, some protagonists of robust fixed-parameter control argue that adaptation is unnecessary provided that the robust design is based on the maximum conceivable range of uncertainty imposed on the plant.

Perturbations and uncertainty

When dealing with robust control, we concern ourselves with *structured* or *unstructured* perturbations (disturbances). These will result from uncertainties in the plant model, disturbances imposed on the plant (both along the forward path) and, possibly, measurement noise received in the feedback path. It is possible to envisage these as *additive* or *multiplicative* perturbations as shown in Figure 9.2 (here, the usual domain notation is neglected to imply generality).

Unstructured perturbations or uncertainties arise from plant disturbances or unmodelled plant behaviour whilst structured perturbations are generally taken to arise from uncertainties in specific plant pole–zero locations and gains.

Two-degree-of-freedom structure

Early work on robust control is due to Horowitz & Sidi (1972). In this, frequency-plane methods are used to design two controllers in which command inputs, r, are controlled independently of plant sensitivity control according to the *canonic* structure of Figure 9.3.

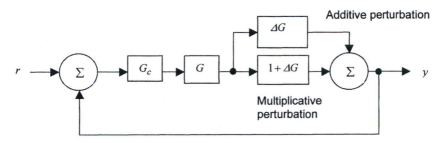

Figure 9.2 Additive and multiplicative uncertainty.

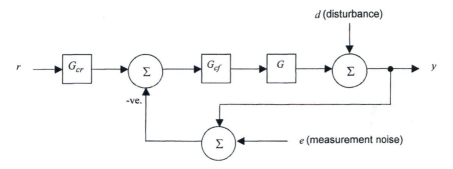

Figure 9.3 *Canonic two-degree-of-freedom structure.*

In this approach, acceptable tolerances of the time-domain closed-loop model are identified and translated to equivalent frequency response tolerances. That is, time-domain bounds on $y(t)$ (such as step response criteria) are expressed as frequency bounds on $G'(j\omega)$. These bounds are in turn used to express bounds on the open-loop plant model, $G_{cf}(j\omega)\,G(j\omega)$, from which a specification for the loop controller, G_{cf}, is obtained. The command signal 'prefilter', G_{cr}, is then designed such that closed-loop performance criteria are met (e.g. response time, offset, etc.). The method requires some trial and error on account of the two controllers to be determined.

Robust controller design using H^{∞} methods

The robust control 'problem' is to find a controller, G_c, which is *stabilising* when subject to a combination of disturbance, measurement noise and model uncertainty. A standard robust system configuration can therefore be summarised as in Figure 9.4. For this configuration

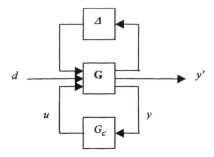

Figure 9.4 System configuration for robust control.

$$\mathbf{G} = \begin{bmatrix} G_{11} & G_{12} \\ G_{21} & G_{22} \end{bmatrix}$$

where G_{11} = the transfer function mapping disturbance d to output y', G_{12} = the transfer function mapping control signal u to output y', G_{21} = the transfer function mapping disturbance d to feedback y and G_{22} = the transfer function mapping control signal u to feedback y.

Thus the closed-loop transfer function of the standard robust configuration of Figure 9.4 as far as disturbance d and feedback y are concerned will be

$$G' = G_{11} + \frac{G_{21}G_cG_{12}}{(1 + G_cG_{22})} \tag{9.4}$$

Now, the magnitude of a frequency response, $x(j\omega)$, or a time series, $x(t)$ (etc.), can be measured by its *norm*, $\|x(j\omega)\|$ or $\|x(t)\|$. One powerful method in robust control design is to minimise the H^∞ norm of the plant. For example, for the plant transfer function matrix, $\mathbf{G}(s)$, the H^∞ norm, $\|\mathbf{G}\|_\infty$, can be defined in terms of frequency-dependent *singular values*, $\sigma(j\omega)$ of $\mathbf{G}(j\omega)$ as

$$\|\mathbf{G}\|_\infty \overset{\Delta}{=} \sup_\omega \bar{\sigma}(\mathbf{G}(j\omega)) \tag{9.5}$$

(in which $\sup_\omega \bar{\sigma}$ denotes the supremum of the frequency-dependent singular values – the maximum across the full spectrum of values).

Classical work on the subject can be found in Zames (1981) (who introduced the H^∞ principle for robust control design) and Kwakernaak (1985). Here, we confine ourselves to a practical overview of the subject.

For a practical interpretation of the H^∞ method, we require to find a stabilising controller, G_c, which minimises the H^∞ norm of G' (equation (9.4)), that is $\min_{G_c} \|G'\|_\infty$.

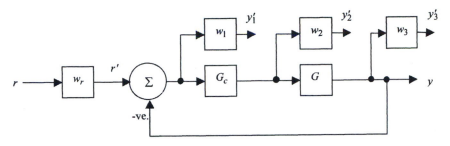

Figure 9.5 Weighting functions in robust control design.

In practice, this is achieved by introducing one or more *weighting functions* w_1, w_2, . . . , as illustrated in Figure 9.5. From this, we seek a stabilising controller which satisfies

$$\min_{G_c} \left\| \begin{array}{c} w_1 S w_r \\ w_2 G_c S w_r \\ w_3 G G_c S' w_r \end{array} \right\|_{\infty} \tag{9.6}$$

where S' is the complementary sensitivity function (i.e. $S' = 1 - S$).

The choice of weights is far from straightforward. For SISO cases, $w_r = 1$ will often be adequate and putting $w_3 = 0$ will lead to reasonable results in many cases, which limits the choice to w_1 and w_2. Some guidance on choice of weights is given by Poslethwaite (1991) and Lundström *et al.* (1991).

Generally, weight selection is problem-specific and difficult to establish procedurally. When w_1 and w_2 are used, reasonable results are often achieved using stable first-order lead–lag combinations such that w_1 is high-gain low-pass and w_2 is low-gain high-pass, i.e.

$$w_1(s) = \frac{K_{w_1}\left(\tau_{w_{1a}} s + 1\right)}{\left(\tau_{w_{1b}} s + 1\right)} \qquad \tau_{w_{1b}} \gg \tau_{w_{1a}}, \; K_{w_1} \to \text{high} \tag{9.7}$$

$$w_2(s) = \frac{K_{w_2}\left(\tau_{w_{2a}} s + 1\right)}{\left(\tau_{w_{2b}} s + 1\right)} \qquad \tau_{w_{2a}} > \tau_{w_{2b}}, \; K_{w_2} \to \text{low} \tag{9.8}$$

The weights are therefore generally designed complementarily, using Bode singular-value plots in which the singular magnitude of the function is plotted against frequency. Figure 9.6 shows a typical pattern for w_1 and w_2. Weighting functions of the typical forms shown hence ensure that, in minimising the path weighted by w_1, the plant will be insensitive to low-frequency disturbances. Minimising the path weighted by w_2 ensures that high-frequency plant noise (due to measurement noise and the high plant gain in the ultimate robust controller result) is desensitised.

The choice of gains in the weighting functions can be made to address steady-state plant performance criteria. For instance, the choice of K_{w_1} can

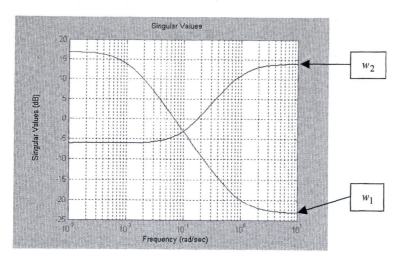

Figure 9.6 Typical Bode singular-value patterns for weights w_1 and w_2.

be made to limit the steady-state offset, whilst fixing $K_{w_2} < 1$ ensures that plant actuators will not saturate.

The following example makes use of the MATLAB robust control toolbox (Chiang & Safonov, 1992) to illustrate the design of an H^∞ robust controller for a situation involving a structured uncertainty.

Example 9.1

Consider the plant of Example 8.4 which consists of a first-order lag with dead time and suppose that the time units are minutes. The time constant is 10 min and the dead time is 2 min. Though the nominal plant gain is unity, suppose that this value is uncertain.

A PID controller can be designed for this application based on the nominal plant gain, using the methods set out in section 7.3. For example, using the Ziegler–Nichols tuning estimates of Table 7.2, one obtains the following for a PID controller:

$$K_c = 6K^{-1}, \qquad i_t = 4\text{min}, \qquad d_t = 1\text{min}$$

Using a Routh array with the above controller specification and the following plant transfer function with uncertain gain, K_p,

$$G(s) = \frac{K_p(1-s)}{(10s + 1)(s + 1)}$$

(a first-order Padé approximation has been used for the dead-time term), one establishes that the system will become unstable under fixed-parameter PID control when the plant gain is 1.66 or higher.

Consider now a robust design. The MATLAB robust control toolbox function HINFOPT can be used to find a stabilising controller which satisfies the condition of expression (9.6) and, coincidentally, searches for an optimum value for the weighting function gain K_{w_1}. The following weights were selected after shaping the Bode singular values to give profiles which are similar to those shown in Figure 9.6 (these can be generated by the MATLAB control toolbox using the SIGMA function):

$$w_1(s) = \frac{K_{w_1}(s+1)}{(100s+1)} \qquad w_2(s) = \frac{0.5(10s+1)}{(s+1)}$$

The third weight, w_3, was not applied (i.e. $w_3 = 0$).

The following summarises the commands used in the MATLAB robust control toolbox:

```
1  [a,b,c,d]=tf2ss([0 -1 1],[10 11 1]);
2  Gss=mksys(a,b,c,d);
3  W1=[[1 1];[100 1]];
4  W2=[0.5*[10 1];[1 1]];w35[];
5  Ass=augtf(Gss,w1,w2,w3);
6  [Kw1,Gcss]=hinfopt(Ass);
7  [Gca,Gcb,Gcc,Gcd]=branch(Gcss);
8  [num,den]=ss2tf(Gca,Gcb,Gcc,Gcd);
9  Gc=tf(num,den);
```

which yields the stabilising robust control specification,

$$G_c(s) = \frac{37.71s^3 + 79.18s^2 + 45.25s + 3.771}{s^4 + 190.3s^3 + 243.5s^2 + 54.75s + 0.5233} \qquad \text{(E9.1.1)}$$

The following points should be noted for the above steps:

- Step 1 – the plant model is converted to state-space form.
- Step 2 – **mksys** converts the state-space plant model into a singular MATLAB variable.
- Steps 3, 4 – the weights are interpreted as 2×2 matrices (w_3 is not used).
- Step 5 – **augtf** creates an *augmented* matrix which combines the plant model, sensitivity function and weights (i.e. expression (9.6)).
- Step 6 – **hinfopt** finds the stabilising controller, with results returned in the LHS arguments; note that the optimum K_{w_1} was returned as 7.2.
- Step 7 – **branch** extracts elements from the matrix of results (here, only the controller parameters of interest are returned in Gcss).
- Steps 8, 9 – the (state-space) controller parameters are converted to transfer function format for convenience.

Now consider the resulting controller. It is frequently the case with H^∞ controller design that a high-order controller is obtained. In practical implementation on-line, this may be numerically ill-conditioned and will clearly be computationally more demanding than a low-order controller – an issue which is often quite critical in on-line applications. We can of course reduce the order of the controller using, for instance, any one of the methods set out in section 4.4. For example, using the Routh model reduction method, we obtain the following second-order approximation to the controller specified in equation (E9.1.1):

$$G_c(s)|_2 = \frac{7.2(12s + 1)}{461s^2 + 140s + 1} \tag{E9.1.2}$$

We can now consider how robust this solution will be for an uncertain plant gain. Using, once again, a Routh stability criterion with uncertain plant gain K_p, we find that using the second-order controller of equation (E9.1.2) the system will become unstable when the plant gain is 9.6. This compares with 1.66 in the case of PID control; hence the advantage of the robust approach is clear.

As to performance, Figure E9.1.1 compares the simulated step response of the PID controller and the robust specification at the nominal plant gain of unity. Clearly there is a loss of performance with the robust design – evidenced by the slower response and offset. However, it is likely that the offset at least could be improved upon by further refinement of the weighting functions used in the original design.

Of course another much simpler approach to achieving robustness would be to use a *detuned* PID controller, for instance by weighting

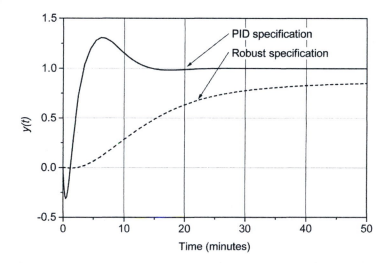

Figure E9.1.1 Performance of PID and robust controllers at the nominal plant gain.

the controller gain tuned at the nominal plant gain as the inverse of the maximum plant gain expected. However the latter would need to be known for this and the problem remains in that uncertainty may stretch beyond the plant gain (i.e. component time constants may also be uncertain).

LQG and H^2 methods

Other methods of robust control design include *linear–quadratic–Gaussian* (LQG) and H^2 methods.

In LQG design, a controller is sought which minimises a quadratic cost function (similar to equation (8.6) but with an input term) in the presence of white Gaussian noise. Various algorithms are available which generally involve the solution of a Diophantine equation in a manner similar to pole-placement design; or the solution of a Ricatti equation (see for example Zaheer-uddin, 1992). Similar methods can also be used to develop adaptive LQG controllers (see Åström & Wittenmark, 1989).

LQG and H^∞ methods have been linked enabling noisy systems to be considered within an H^∞ design framework (Grimble, 1986). In this work a simple LQG–H^∞ linked design procedure is developed through the use of a weighting term to modify system noise. The merit of the method is the simplicity of the algorithm which makes it especially applicable to on-line situations (i.e. robust–adaptive control). In later work, Grimble (1989) introduces a two-degrees-of-freedom controller which enables disturbance rejection and parameter robustness to be optimised as well as maintaining closed-loop performance properties by using a combined LQG and H^∞ cost function.

Design according to an H^2 norm minimisation is an alternative to using the H^∞ norm. Here,

$$\|\mathbf{G}\|_2 \overset{\Delta}{=} \left[\int_{-\infty}^{\infty} \sum_{i=1}^{p} (\sigma(j\omega))^2 \, d\omega \right]^{1/2} \tag{9.9}$$

Using this, stability of the closed loop can be guaranteed but may involve many iterations in the solution algorithm (see for example Ganesh & Pearson, 1989).

Robust design in discrete time

Most of the formative literature on robust design tends to deal with the continuous-time problem. There are basically two possibilities for the more common discrete time case:

1   Develop the continuous-time method for discrete-time system interpretation by mapping the discrete-time (z-domain) model onto the w-plane (see section 5.3). The resulting w-domain controller can then be translated back to the z-domain using the bilinear transformation. (This

procedure in done automatically in the MATLAB robust control toolbox using, for example, the dhinfopt function.)

2   Design the controller in the s-domain $(G_c(s))$ then discretise $(G_c(s) \rightarrow G_c(z))$ using methods in Chapter 5.

Grimble (1987a) has developed a simplified procedure for designing discrete-time robust controllers. He then adapts this for robust self-tuning control applications and the procedures are simple enough to be amenable to manual calculations for low-order plant models (Grimble, 1987b).

Robust control is an area which offers great promise to the HVAC problem; most especially where non-linearities and variations in plant operating conditions render uncertainties in the HVAC plant model which limits the potential of many of the methods we have considered in earlier chapters. In section 8.4 for example, we saw how gain scheduling could be effectively used to deal with variable plant behaviour but only because the nature of the variability was fully understood. Robust design offers the potential to deal with situations in which several aspects of the plant behaviour are uncertain in the presence of disturbances and measurement noise. Despite this, little work has been done to apply the robust methods described here to the HVAC case, though Attia & Rezeka (1994) develop a state observer method to obtain robust multivariable control of temperature and humidity in hot arid climates. Their results, based on simulations, demonstrate improved stability and response times over conventional control, subject to a significant room space participating in the control loops. Chen & Lee (1990) consider robustness in the presence of parameter uncertainties in the adaptive control of HVAC plant.

9.2 Expert systems

The potential for these systems in HVAC applications was first discussed by Dunn (1987) and extensive progress has been reported since. Expert, rule-based and knowledge-based systems are a special case in the field of control in that they tend mainly to find their applications in status monitoring and energy management, rather than in regulatory control. At the heart of these systems is a set of heuristic rules generally based on experience and observation about the process (or a physically similar process) under control. Thus a set of conditions obtained from on-line plant measurements leads to a set of required actions inferred from a database of knowledge which forms the adjustment mechanism for this case. Boolean {IF [condition] AND [condition] OR [condition] THEN [action]} rule inference is used. In the simplest case, the database of possible actions amounts to little more than a look-up table.

A significant amount of progress has been reported in the field of expert systems applied to HVAC problems. Shaw (1987) for example describes an expert system for heating system monitoring and fault diagnosis,

'BREXBAS'; whilst Dexter *et al.* (1993) describe an application of an expert system based on on-line open- and closed-loop response tests for the commissioning stage of HVAC air handling systems. In this, commissioning diagnoses are inferred from one of six possible groups of fault scenarios such as: incorrect installation and set-up; poor controller tuning; incorrect valve size/characteristic; actuator malfunction. In later related work, Dexter & Haves (1994) use *fuzzy logic* to relate a variety of plant performance criteria to performance results obtained using an emulator (emulators are discussed briefly in section 1.3 and fuzzy logic for HVAC control is treated in the next section).

Anderson *et al.* (1989) develop and apply an expert system for detecting and reporting HVAC system malfunctions in a large industrial plant. This work is based on a monthly predictor which uses a form of multiple least-squares regression to predict plant output variables based on the previous year's data. Results from the predictor are compared with on-line measurements from the plant and the resulting deviations form the basis of diagnoses about plant status which are based on a 200-rule expert system. The expert system is based mainly on interviews with plant operating personnel.

Ling & Dexter (1994) report on the design of an expert controller using a rule-based supervisor to reset temperature set point of a predictive control algorithm, concluding that the expert controller can generate energy savings as well as compensate for day-to-day variations in plant operating conditions.

Most applications of these systems lie outside regulatory control of plant. A wide variety of expert and knowledge-based systems have been developed as aids to HVAC design for instance (e.g. Hitchcock, 1991; Yong & Nelson, 1990), and for use in HVAC simulation modelling (Liu & Kelly, 1988). Considerable work has been done on expert system applications in energy management and energy auditing (Haberl *et al.*, 1988; Potter *et al.*, 1991; Kreider *et al.*, 1990; Norford *et al.*, 1990), and on HVAC plant monitoring and diagnostics (Wong & Ferrano, 1990; Klima, 1990; Kaler, 1990; Brothers, 1988).

In many cases, linguistic rule-based inference in itself is too imprecise or vague and *fuzzy logic* techniques can be used to offer a 'spectrum' of decision-making. We will therefore take a separate look at this method in the next section.

9.3 Fuzzy logic control

Throughout this book we have attempted to translate one or more precisely defined input variables into one or more precisely defined control signals using the mechanism of a numerically defined control process or algorithm. The emphasis has been on precise numerical and logical reasoning ultimately based on the physical laws of mechanics, electromagnetism and, especially in the HVAC field, thermodynamics. However people do not conduct their lives according to these scientific precepts in general.

Consider for example the following alternative decision-making processes associated with taking lunch:

- 'I am hungry, so I am going for lunch.'
- 'My body requires 50g of protein and 150g of carbohydrate to sustain my metabolism at the desired 140 W for the next 4h.'

The hungry human follows the first of these two considerations every time! This is one interpretation of *fuzzy logic*: decision-making according to pragmatic human common sense. Interpreted more widely, fuzzy logic has the intrinsic advantage in that it can be applied through the use of knowledge- or experientially based linguistic IF–THEN rules. Fuzzy logic can be applied to virtually any branch of decision-making provided that sufficient human knowledge or experience exists as a basis for that decision-making.

Since most control problems involve a set of inputs, a 'black box', and a set of outputs, it is reasonable to suppose that a *fuzzy logic controller* might occupy the black box and do as good a job or a better job based on human reasoning than, say, a PID controller. In this section, we will restrict ourselves to an overview of the subject applied to control problems and take a look at the potential of fuzzy logic control for HVAC applications. For interested readers, fuzzy logic was developed through the classical work of Lotfi Zadeh during the 1960s and 1970s (see for example Zadeh, 1973).

A general structure of the fuzzy logic controller

A fuzzy logic controller (FLC) consists of three entities (Figure 9.7):

- An (input) *fuzzifier*
- A *fuzzy inference mechanism*
- An (output) *defuzzifier*

The fuzzifier accepts one or more inputs as *crisp points* (i.e. precise numerical values) and maps these onto one or more *fuzzy input sets* (FIS). The fuzzy inference mechanism then maps the resulting FIS onto a single aggregated *fuzzy output set* (FOS) using rule-based reasoning and the defuzzifier translates the aggregated FOS into a crisp output point compatible, for example,

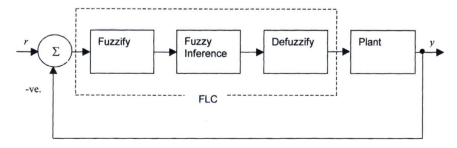

Figure 9.7 General structure of an FLC.

with plant input requirements. Thus an FLC simply maps an input space onto an output space through the mechanism of a set of linguistic rules.

Membership functions and fuzzy sets

The mapping of a crisp point onto an FIS is achieved through *fuzzy membership functions*. Classically, a membership function permits a binary membership of 0 or 1 but a fuzzy membership function can take a value anywhere in the interval $0 \rightarrow 1$. Three types of membership function (MF), μ, are commonly used in fuzzy systems. These are the Gaussian (shown in Figure 9.8(a) about a zero mid-point); the trapezoid (Figure 9.8(b)); and the triangular membership function (Figure 9.8(c)). In some instances, a fuzzy singleton form of membership is useful in that it can considerably simplify subsequent fuzzy inference. The singleton fixes membership at unity at a specified crisp point, and at zero for all other inputs. The Gaussian, trapezoid and triangular MFs have the advantage that they can intrinsically achieve a certain amount of input noise suppression. See Wang (1997) for further details.

A set of fuzzy membership functions defined across a *universe of discourse*, U (i.e. the full range of an input variable), forms a fuzzy set.

As an illustration, suppose we wish to develop an FLC for the control of temperature in a space and a discussion with the occupants of a similar space reveals that they find the room comfortable at 20°C. Conditions are said to be 'chilly' when the temperature drops to 18°C or below, and 'stuffy' when the temperature rises to 22°C and above. For convenience, we express a universe of discourse expressed in terms of the room temperature error with respect to the acceptable value of 20°C. Consider a trapezoidal membership function, which affords a reasonably smooth transition between individual fuzzy subsets. It is also reasonable to assume that the transition from acceptable comfort to 'stuffy' or 'chilly' is not abrupt but merges with the band of acceptable comfort. We might therefore construct the FIS of Figure 9.9 for this situation.

We have created some degree of overlap between the fuzzy subsets, recognising that the transitions between good and stuffy and between good and chilly are not abrupt (i.e. the temperature is moving from good to

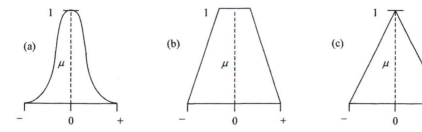

Figure 9.8 (a) Gaussian MF, (b) trapezoid MF and (c) triangular MF.

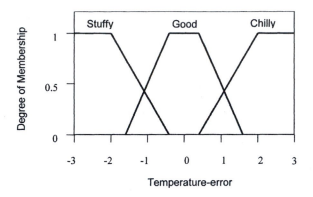

Figure 9.9 FIS for the temperature control illustration.

stuffy, etc.). Moderate overlap for fuzzy input sets is generally desirable to help account for uncertainty in interpretation of the fuzzy subsets but the degree of overlap in practice is usually notional. Note that a *negative* temperature error is generated when the feedback condition is higher than its reference condition (i.e. $\varepsilon = r - y$). Similarly, when conditions are on the chilly side of good, a positive temperature error is evident.

One or more fuzzy input sets generated in the manner described above are now mapped onto a fuzzy output set (FOS) using rule-based reasoning and a fuzzy inference mechanism. There will be as many output sets as there are FLC controlled variables. Here, we confine ourselves to the less complicated SISO/MISO case. However dealing with MIMO fuzzy logic control design is quite straightforward in practice; a single FOS is generated for each control variable from the various interdependent inputs (the rule base accounting for interdependence among the outputs).

Returning to our temperature control problem, we might assume that the plant needs to be able to heat when the room is chilly, cool when the room is stuffy and occupy an idle state when neither of these two states prevails. We therefore construct the following fuzzy output set, again assuming a trapezoidal MF for each fuzzy subset (Figure 9.10). Thus we define an output for the FOS as a control signal ($0 \rightarrow -1$ for cooling and $0 \rightarrow 1$ for heating) which forms the required input to the plant.

Now that we have defined the FIS and FOS, we need to decide how to map the input space onto an output space.

Fuzzy inference and defuzzification

The fuzzy inference mechanism is the heart of an FLC. It consists of a rule base of IF–THEN rules and a means for translating the (one or more) fuzzy input sets onto an aggregated fuzzy output set based on these rules. Each rule in the rule base takes the following general form:

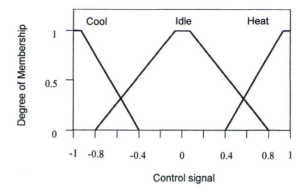

Figure 9.10 FOS for the temperature control illustration.

Rule$_{xy}$: IF   (U_x is A_x) AND... ...AND (U_y is A_y) THEN (V is B)

  (antecedent)   (consequent)   (9.10)

where A_x, A_y are fuzzy subsets of the universes of discourse U_x, U_y (i.e. $A_x \subset U_x$ and $A_y \subset U_y$) and B is a subset of the output space, V (i.e. $B \subset V$). Note that the logical OR operator may also appear (and frequently does) in rule construction.

For our temperature control illustration, we can write down three clear rules governing plant operation:

1   IF (error is stuffy) THEN (plant is cool)
2   IF (error is good) THEN (plant is idle)
3   IF (error is chilly) THEN (plant is heat)

There are several mechanisms for inferring fuzzy outputs from fuzzy inputs based on the defined rules – see for example Wang (1997). Perhaps the most common and enduring one is Mamdani inference (Mamdani & Assilian, 1975).

Firstly, AND and OR combinations in the antecedents of the rules are resolved using normal Boolean logic with fuzzy membership degrees preserved. The *min* operator is used to resolve AND, and the *max* operator used to resolve OR, i.e.

A AND $B \rightarrow \min(A, B)$ (9.11)

A OR $B \rightarrow \max(A, B)$ (9.12)

Secondly, we apply the result from the (resolved) antecedent to the consequent with equality of degree. That is, the consequent enjoys the same degree of fuzzy membership as the antecedent. The following simple illustration shows how a single rule can thus be inferred.

Consider the fuzzy system of Figure 9.11(a) consisting of two fuzzy input sets, U_1 and U_2, and fuzzy output set, V. Suppose the following rule applies:

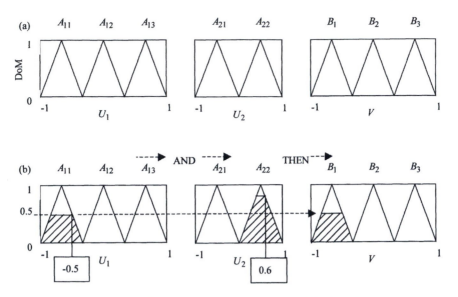

Figure 9.11 Fuzzy inference illustration: (a) the fuzzy system, and (b) the inference.

IF (U_1 is A_{11}) AND (U_2 is A_{22}) THEN (V is B_1)

Suppose also that, on some occasion, crisp input points lie at $U_1(-0.5)$ and $U_2(0.6)$. From the FIS we find that the membership degree of A_{11} is 0.5 and for A_{22} it is 0.8. The resolved antecedent will therefore be

$$\min(A_{11}, A_{22}) = \min(0.5, 0.8) \rightarrow 0.5$$

This degree of membership is 'fired' at the FOS resulting in a trapezoidal patch of the B_1 output fuzzy subset with a height of 0.5 representing our fuzzy output. Figure 9.11(b) illustrates the inference.

Finally, we defuzzify the output by translating the patch of output subset to a crisp point. Note that, in practice, the output will be an aggregate fuzzy set consisting of all patches of fuzzy outputs fired from each rule added together. The crisp point can be deduced using a number of methods – the average of the maximum patch of the aggregated set, the largest maximum, the smallest maximum, the centroid, to name four such methods. Probably the most common is the centroid, in which the centre of the area of the aggregate patch is located by an integration method, though it does involve a calculation effort for each output produced. From the centre point, the crisp output is then read off the horizontal axis – see Figure 9.12, based on the above illustration.

Returning to the temperature control problem, we see from the above that the FLC output given the FIS, FOS and rule base we have established is quite straightforward. However, this procedure must be repeated at each sampling instant which requires on-line computation for a practical application of the method. Here, the MATLAB fuzzy logic toolbox has been used

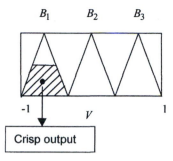

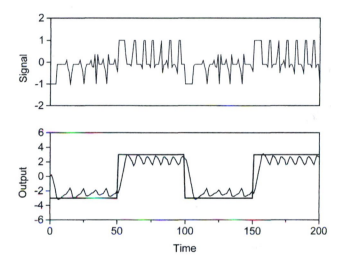

Figure 9.12 Defuzzification to a crisp output.

Figure 9.13 Simulation of the basic temperature control FLC.

to synthesise the problem, forming a *fuzzy logic control block* (FLCB) (Jang & Gulley, 1995). To generate results for our temperature control problem, suppose that the plant has the model used in Examples 8.4 and 9.1 and the plant gain is 10, i.e.

$$G(s) = \frac{10(1 - s)}{\left(10s^2 + 11s + 1\right)}$$

Using the FLCB, the plant model and an (arbitrary) square-wave reference signal as a disturbance, a Simulink model of the resulting system produces the results given in Figure 9.13 (Simulink, 1996). The amplitude of

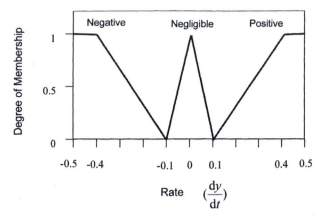

Figure 9.14 Additional rate FIS for the temperature control problem.

the reference signal was set at ± 3K – the same as the range of input error used in the FIS – in order to force the FLC to its limits.

An undesirable degree of oscillation is evident in the regions of near-zero error. This is a not uncommon feature with fuzzy logic controllers due to the tendency of the algorithm to toggle between fuzzy subsets (and, hence, rules) as the response approaches a zero error state. A review of the rules is one possible remedy (note that number and shape of the fuzzy sets tend to be far less influential in FLC performance than the rules themselves). Another possibility is to introduce a second FIS in the form of the time derivative of the plant output. In fact, the use of such a rate input as an accompaniment to an error input is common in FLC design. The FIS shown in Figure 9.14 was therefore added, based on an expected limiting $\mathrm{d}y/\mathrm{d}t$ of $\pm 0.5\,\mathrm{Kmin}^{-1}$.

Additional rules were fixed as follows:

4   IF (error is good) AND (rate is negative) THEN (plant is heat)
5   IF (error is good) AND (rate is positive) THEN (plant is cool)

These rules re-establish control at near-zero error to help force the error to zero. Figure 9.15 shows the Simulink model with the rate input added and Figure 9.16 shows the modified results.

Some improvement is evident with this modification though there is room for further refinement of the rules and the possible addition of further input fuzzy subsets. A method for systematically improving rule construction through the use of a *linguistic plane* is discussed briefly in the next section.

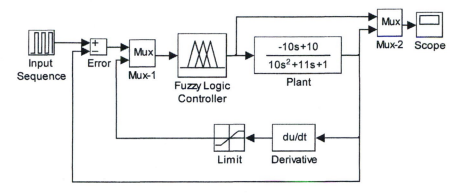

Figure 9.15 Simulink model of the FLC (temperature control illustration).

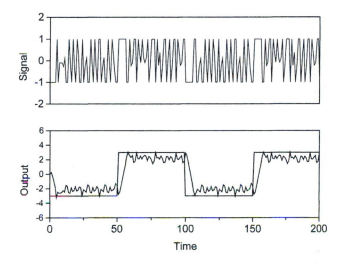

Figure 9.16 FLC system results with additional rate input.

The above illustration has served to demonstrate the key merit of FLC – it is simple to implement especially in situations where operating experience about the plant is available and conventional control is inappropriate or unworkable. Though FLC alone will rarely lead to improvements over, for example, well tuned PID, it is robust, giving it major advantages where there are uncertainties, non-linearities and variations in the operating environment of the plant. Current work in neuro-fuzzy control, allowing the fuzzy inference system to learn and adapt, offers considerable potential for the future. We will look at some of these developments after reviewing some of the applications of FLC to HVAC problems.

Fuzzy logic applications to HVAC problems

Dexter & Trewhella (1990) were among the first to apply fuzzy logic methods to HVAC plant. They develop five fuzzy input sets for the performance assessment of a hot water fan coil unit under PI control. The fuzzy input sets cover maintenance, energy use, control error, valve travel and occupant dissatisfaction. Two fuzzy output sets are generated reflecting occupant response and plant performance. All fuzzy sets were based on trapezoidal membership functions and most of the fuzzy input sets were designed around computer simulations of the plant. The rule base was developed with some advice from practitioners. Crisp output was obtained from the centroid of the aggregated fuzzy output sets. The results revealed that a fuzzy approach can give a plausible assessment of plant performance though a careful interpretation of the results was found to be essential.

For an unusual application in the holistic treatment of FLC, Dounis *et al.* (1995) consider the thermal *and visual* environment resulting in a fuzzy controller which controls heating, cooling and window opening, shading and lighting.

Huang & Nelson (1991) combine a PID controller with an FLC and apply it to a general second-order plant model. Fuzzy input sets are developed for control error, integral error and derivative error (essentially the terms of a PID controller) and an output set for the control signal. Mamdani max–min implications are used with centroid defuzzification of the resulting aggregate fuzzy output set. Results obtained using computer simulation show that the combined controller can virtually eliminate all of the damped oscillation common with tuned conventional PID control as well as achieve faster response.

In later work, Huang & Nelson (1994a) give detailed consideration to rule development of FLCs for HVAC plant. They then go on to test the resulting FLC experimentally (Huang & Nelson, 1994b). In this work, error feedback and error rate are used to form triangular fuzzy input sets. They identify the major influence of the rule base in the frequent problem of oscillation at (or near) zero error experienced by many FLCs (and demonstrated in the illustrative temperature control problem considered earlier). To find a remedy, the rule base is set out on a linguistic plane in which the fuzzy subsets of both universes of discourse (error and rate) are 'plotted' against one another forming a matrix of input membership. Thus it is possible to identify desirable rule trajectories on a linguistic plane and this led to the elimination of near-zero error oscillation in the Huang and Nelson work. In their subsequent experimental work, an FLC is implemented for the control of a steam–air heating coil in an air handling plant. Results show that the FLC achieves superior performance when compared with tuned PID control, most particularly exhibiting far less oscillation than the latter.

So *et al.* (1994) apply an FLC to the control of air handling plant in VAV systems. Using triangular FISs for error and error rate, FLC signals are generated for the air handling plant fresh air damper, fan control, cooling coil, humidifier and reheat coil. Results are compared with tuned and detuned PID control (the latter so that robustness could be considered), using computer simulation. The results show that the FLC compares well with tuned PID control but was more robust and the FLC was superior (in terms of response time and offset) than detuned PID control. In later work, they develop a self-learning FLC using an artificial neural network (ANN) based on the same air handling control problem as their earlier work (So *et al.*, 1997). The ANN is used to monitor the plant and update the parameters of the FLC which permits robustness in spite of changes in operating conditions and non-linearities. We will take a closer look at these methods in the next section.

9.4 Artificial neural networks

The irresistible desire to create systems that emulate the cell-level operation of the brain has inspired work over the past 50 years (but mostly in the last 20) on *artificial neural networks* (ANNs). The brain is made up of millions of neurons, which are able to communicate signals, in the form of small electrical impulses, between one another. Networks of neurons can be 'trained' to perform complex tasks by adjusting the values of intercommunicating signals, and control signals represent just one area of potential application. Others include pattern recognition, image processing, classification and system identification.

In the following, a general overview of the subject is given. Various texts are available with particular reference to control, to which readers are referred for a more substantial treatment of the subject. See for instance Page *et al.* (1993) or Warwick *et al.* (1992).

Neuron models

McCulloch & Pitts (1943) first discussed the structure of the neuron in the context of the nervous system and, from their propositions, the simple model of Figure 9.17 evolved.

The inputs, from real-world sources or from other neurons, $u_1, \ldots, u_n$, are weighted by adjustable scalar weights, $w_1, \ldots, w_n$, and the bias (or offset), which has a constant input of 1, is weighted by adjustable value, b. The sum of the weighted signals is then processed by a filtering function, f, which can take a number of forms. Common examples of output filtering functions are the *hard limit*, *linear* and *sigmoid* functions (Figure 9.18).

With the hard limit filter, the neuron output is limited to either 0 or 1. The linear filter permits any value and the sigmoid filter forces the input (which may take any value from $-\infty$ to $+\infty$) into the interval $0 \rightarrow 1$ or alternatively $-1 \rightarrow 1$. The choice of filtering function is largely determined in relation to the choice of network. Thus the output of a neuron can be expressed as

Inputs: Input weights:

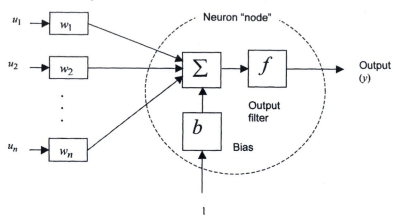

Figure 9.17 Simple neuron model based on McCulloch & Pitts (1943).

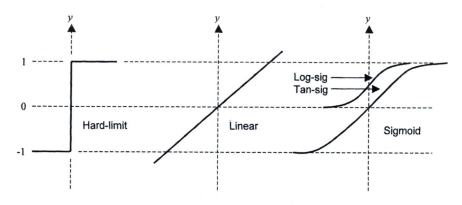

Figure 9.18 Neuron output functions.

$$y = f\left[\sum_{i=1}^{i=n}(w_i u_i) + w_0\right]$$ (9.13)

Feedforward networks

The most common ANN architecture by far is a multiple-layer feedforward network designed for training by *back-propagation*. Typically, this will consist of a weighted input layer, one or two *hidden layers* of neuron nodes and a

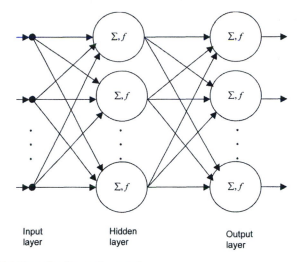

Input layer

Hidden layer

Output layer

Figure 9.19 Multi-layer feedforward network.

weighted output layer of neuron nodes (Figure 9.19). Weighting functions used with such network architectures are most commonly sigmoidal in the hidden layer(s) and linear in the output layer.

Back-propagation

A training set, consisting of an input vector and a corresponding target output vector, forms the basis of network training by back-propagation. For a detailed treatment of network training by back-propagation, see Rumelhart *et al.* (1986).

Briefly, inputs from the input vector enter the network, are weighted, then passed to and processed by the first hidden layer which in turn excites any subsequent hidden layer and so on to output. The resulting outputs and their targets are used to generate errors which are then propagated back through the network from the output side to the input side resulting in error values at each hidden layer neuron. Neuron weights are then adjusted until an overall network error criterion (e.g. a minimum sum of errors squared) is met.

In the most common approach, neuron weights and biases are changed according to the *delta rule*,

$$\Delta(p)w_i \propto -\frac{\partial E(p)}{\partial w_i} \tag{9.14}$$

where $\Delta(p)w_i$ is the change in the ith weight arising because of input sequence, p, and $E(p)$ is the sum of errors squared.

The proportionality of (9.14) is converted to an equation through a constant called the *learning rate*. In effect, this achieves a *gradient descent* in sum of squared error (weights and biases are changed in the opposite direction to the error gradient) until an acceptable criterion for $E(p)$ is reached. The problem with gradient descent methods is that they can converge on a local error minimum, rather than a global error minimum which reflects all weights and biases, when a multi-neuron non-linear hidden layer is used and this may or may not reflect a good network outcome. It is also computationally demanding.

Many techniques are available to overcome these problems. One such method, Levenberg–Marquadt back-propagation, which will be used in examples a little later, uses a form of Newton's method for weight/bias updating which is more accurate than gradient descent in regions of local error minima (Hagan & Menhaj, 1994).

These frequently used networks can be used for non-linear system modelling and identification as well as control applications.

Radial basis networks

Chen *et al.* (1991) discuss an alternative class of ANNs which use *radial basis functions* (RBFs). RBF networks make use of Gaussian-shaped neuron filters in their hidden layer (Figure 9.20). Typically, there are as many neurons in the RBF network hidden layer as there are training input data sets. These networks tend to use more neurons than conventional feedforward networks but are computationally less demanding, using a form of least-squares regression for training.

Recurrent networks

Recurrent networks contain feedback paths between output and input. Feedback makes the recurrent class of network particularly suitable for recognising temporal data. Two examples of recurrent architectures are the Elman (Elman, 1990) and Hopfield networks (Li *et al.*, 1989).

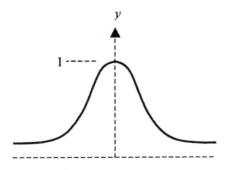

Figure 9.20 Form of an RBF neuron filter.

The *Elman network* resembles a conventional feedforward network with a single hidden layer but there is a feedback connection between the hidden layer and the input layer (Figure 9.21). The feedback connections in the Elman network are stored, and therefore delayed, by one time interval, and they are fixed; so the Elman network is strictly only partially recurrent. Typically, the hidden layer uses sigmoid filtering and the output layer uses linear filtering. Elman networks are trained using back-propagation and there is an intrinsic appeal for their use in control. Elman (1990) amply demonstrates the potential of this network architecture in temporal pattern recognition but, despite the clear potential, there are few instances of the Elman network being used for feedback control in the literature.

In the *Hopfield network*, which is a two-layer network, the inputs are connected to one neuron node only and are only used to initiate training (Figure 9.22). The delayed (and weighted) feedbacks transmit to all neuron nodes whilst the initial inputs pass to only one node each. After initial start-up, the inputs are disconnected and training is achieved by adjusting weights and biases until the change in output meets some prescribed tolerance. The Hopfield network is used only with *saturated* linear filtering

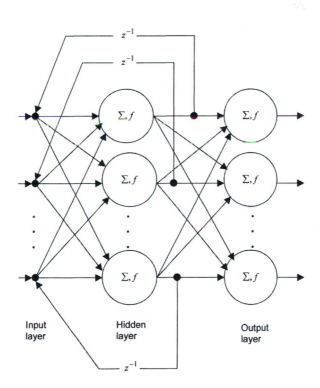

Figure 9.21 The Elman network.

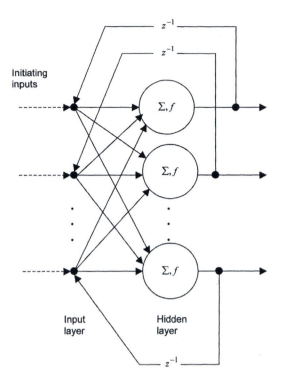

Figure 9.22 The Hopfield network.

(i.e. a linear filter whose output is clipped to ±1). As Li *et al.* (1989) point out, the Hopfield network is not easy to implement in practice but they get round this by implementing the network using analogue circuits.

ANN applications

ANNs are establishing themselves in the HVAC control field in system identification and control – in particular non-linear control. We will consider both applications in two illustrative examples in the following. Example 9.2 deals with a system identification problem, and Example 9.3 deals with a non-linear control application.

Example 9.2

In Example 7.3, an ARX212 model was fitted to the cyclic component of a room temperature transient. The model reproduced some of the cyclic features but was far from perfect (Figure E7.3.5). Now let us consider how well an artificial neural network might be trained to reconstruct the data.

Firstly, we transpose the input data set, **U**, and the output data set, **Y**, to form 174-element row vectors. To establish an ANN, let us assume a single four-neuron hidden layer using tan-sigmoid filters, and based on the current and previous values of the signal, $u(t)$, $u(t-1)$, as inputs. The single output, $y(t)$, prompts a single neuron output layer and a linear filter is assumed for this.

Using the MATLAB neural network toolbox (Demuth & Beale, 1998), we can construct a two-row × 174-column matrix of current and previous signals with the function delaysig,

```
>P=delaysig(U,0,1);
```

(in which P returns the 2 × 174-element matrix). Using a conventional feedforward network, we now fix initial values for the weights and biases of the ANN using the feedforward initialisation function, initff, for four neurons in the hidden layer,

```
>[W1,B1,W2,B2]=initff(P,4,'tansig',Y,'purelin');
```

(in which input weights are returned in W1, the hidden layer biases in B1, hidden layer output weights in W2, the output layer bias in B2 and Y contains the target output data). Training the ANN using Levenberg–Marquadt back-propagation,

```
>[W1,B1,W2,B2]=trainlm(W1,B1,'tansig',W2,B2,'purelin',
               P,Y,[a,b,c,d,e,f,g,h]);
```

where a = training iterations or *epochs* through the network between updating progress (two used), b = maximum training epochs (100 used), c = target sum of squared error (0.1 used), d = minimum gradient (10^{-4} used), e = initial learning rate (10^{-3} used), f = learning rate multiplier for increasing (10 used), g = learning rate multiplier for decreasing (0.1 used) and h = maximum learning rate (10^{10} used). The weights and biases shown in Table E9.2.1 were returned for the ANN. The resulting ANN is simulated using the input data, P, with the feedforward network simulation function, simuff,

```
>Y_ann=simuff(P,W1,B1,'tansig',W2,B2,'purelin');
```

Results are compared with the actual data in Figure E9.2.1. The results are good and certainly a significant improvement on the ARX212 result. However, after 100 epochs, the sum of squared error was 5.1 – nowhere

Table E9.2.1 Weights and biases for the ANN

W1		B1	W2^T	B2
−0.623	0.578	2.903	0.150	23.159
2.260	−1.927	5.904	−0.055	
0.183	−0.182	0.757	−0.379	
3.287	0.115	1.152	−2.001	

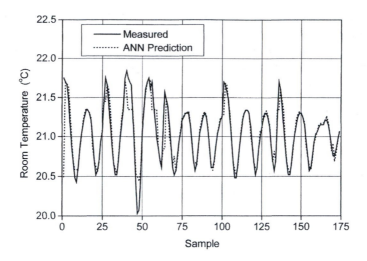

Figure E9.2.1 System identification using a feedforward ANN.

near target. Hence there is room for considerable further improvement by, for instance, experimenting with more neurons in the hidden layer, or a larger number of training epochs, or modified initial weights and biases. Nevertheless, the example illustrates the potential for the use of an ANN in complex function identification.

Now we will take a look at the potential of ANNs in HVAC control.

Example 9.3

In the following, we develop a non-linear model for a common type of HVAC control loop. We consider the problems associated with tuning a PID controller for the loop, then go on to see how an ANN can improve matters.

Consider the heating coil control arrangement as depicted, for instance, by Figure 1.15, but with control from the supply duct instead of the space. For a coil, as any heat exchanger, we can express energy balances for the water, heat exchange material and air. For convenience, we 'lump' together the heat exchange material and water passages (i.e. lumped-parameter modelling), to give

$$C_{wm}\frac{\mathrm{d}\theta_{wo}}{\mathrm{d}t} = m_w c_{pw}(\theta_{wi} - \theta_{wo}) - UA(\theta_{wo} - \theta_{ao}) \qquad (\text{E9.3.1})$$

in which the coil surface temperature is assumed to be θ_{wo}, and $C_{wm} =$ the thermal capacity of the coil water and heat exchange material (kJK^{-1}), $m_w =$ mass flow rate of heating water (kgs^{-1}), $c_{pw} =$ the specific heat capacity of water (kJkg^{-1}K^{-1}), θ_{wi}, $\theta_{wo} =$ inlet and outlet water temperatures (°C),

UA = coil overall thermal transmittance (kWK^{-1}) and θ_{ao} = air outlet temperature (°C).

On the air side, an instantaneous heat balance, q, is appropriate due to the absence of any significant thermal capacity,

$$q = m_a c_{pa}(\theta_{ao} - \theta_{ai})$$ (E9.3.2)

and the heat transfer across the coil heat exchange material

$$q = UA(\theta_{wo} - \theta_{ao})$$ (E9.3.3)

which assumes that the water in the coil is perfectly mixed and equal to the outlet water temperature, and m_a = air mass flow rate (kgs^{-1}), c_{pa} = specific heat capacity of air (kJkg^{-1}K^{-1}) and θ_{ai} = air inlet temperature (°C).

The UA value will mainly vary with the water-side surface heat transfer coefficient, h_w, since air-side flow conditions are constant (or are assumed to be so in this example). Thus,

$$UA = \frac{A}{[1/h_w + 1/h_a]}$$

which neglects the thermal resistance of the coil heat exchange material, and h_a is the (assumed constant) air-side surface heat transfer coefficient which includes a fin efficiency correction. We might reasonably assume that $h_w \propto c_w^{0.8}$, based on the Dittus–Boelter correlation for heat transfer in tubes (e.g. McAdams, 1976) in which fluid properties are assumed constant. Hence, using typical design data in terms of the split between h_w and h_a, a reasonable approximation for UA which accounts for variations in controlled water flow rate emerges as

$$UA \cong \frac{f_1}{[f_2 m_w^{-0.8} + 1]}$$ (E9.3.4)

where f_1 and f_2 are constants.

Combining equations (E9.3.2) and (E9.3.3),

$$\theta_{ao} = \frac{(m_a c_{pa}\theta_{ai} + UA\theta_{wo})}{(m_a c_{pa} + UA)}$$ (E9.3.5)

Substituting for θ_{ao} in equation (E9.3.1),

$$C_{wm}\frac{d\theta_{wo}}{dt} = m_w c_{pw}(\theta_{wi} - \theta_{wo}) - UA\theta_{wo} + \frac{UA m_a c_{pa}\theta_{ai}}{(m_a c_{pa} + UA)} + \frac{(UA)^2 \theta_{wo}}{(m_a c_{pa} + UA)}$$ (E9.3.6)

In a deliberate attempt to maximise the non-linear problem, a linear valve characteristic is chosen instead of the more common equal-percentage valve. Using a first-order representation for the valve and actuator,

$$\tau_w \frac{dm_w}{dt} = m_{wd}\psi - m_w$$

where τ_v is the valve actuator time constant, m_{wd} is the design mass water flow rate (kgs^{-1}) and ψ is the installed characteristic of the valve. Based on equations (2.9) and (2.13) in section 2.2, ψ can be expressed as an incoming control signal for the linear valve. Using typical practical values for valve authority and valve let-by (0.5 and 0.005 respectively), we can write

$$\tau_v \frac{dm_w}{dt} = \frac{m_{wd}(0.005 + 0.995u)}{\left\{0.5\left[1 + (0.005 + 0.995u)^2\right]\right\}^{0.5}} - m_w \qquad \text{(E9.3.7)}$$

where u is the incoming signal. Finally, a first-order representation is assumed for the duct-mounted temperature sensor in order to complete the model,

$$\tau_d \frac{d\phi}{dt} = \theta_{ao} - \phi$$

(in which τ_d is the sensor time constant and ϕ the sensor output signal). Substituting for θ_{ao} using equation (E9.3.5),

$$\tau_d \frac{d\phi}{dt} = \frac{m_a c_{pa} \theta_{ai}}{(m_a c_{pa} + UA)} + \frac{UA\theta_{wo}}{(m_a c_{pa} + UA)} - \phi \qquad \text{(E9.3.8)}$$

We can now summarise the model equations using the usual short-hand notation,

$$\dot{\theta}_{wo} = \frac{\theta_{wo}}{C_{wm}} \times \left[\frac{(UA)^2}{(m_a c_{pa} + UA)} - m_w c_{pw} - UA\right] + \frac{m_w c_{pw}\theta_{wi}}{C_{wm}} + \frac{UAm_a c_{pa}\theta_{ai}}{C_{wm}(m_a c_{pa} + UA)}$$

$$\text{(E9.3.9)}$$

$$\dot{m}_w = \frac{m_{wd}}{\tau_v} \times \frac{(0.005 + 0.995u)}{\left\{0.5\left[1 + (0.005 + 0.995u)^2\right]\right\}^{0.5}} - \frac{m_w}{\tau_v} \qquad \text{(E9.3.10)}$$

$$\dot{\phi} = \frac{m_a c_{pa}\theta_{ai}}{\tau_d(m_a c_{pa} + UA)} + \frac{UA\theta_{wo}}{\tau_d(m_a c_{pa} + UA)} - \frac{\phi}{\tau_d} \qquad \text{(E9.3.11)}$$

$$UA = \frac{f_1}{\left[f_2 m_w^{-0.8} + 1\right]} \qquad \text{(E9.3.12)}$$

output $\Rightarrow$ ϕ
constants $\Rightarrow$ $m_a, c_{pa}, c_{pw}, C_{wm}, \tau_d, \tau_v, f_1, f_2$
inputs $\Rightarrow$ u (when open loop), θ_{ai}
variables $\Rightarrow$ $\theta_{wo}, m_w, \phi, UA$

Firstly, we construct the non-linear model in Simulink (Simulink, 1996). Figure E9.3.1 shows the resulting open-loop case. The following boundary data and constants have been applied for this illustration: q (at design) = 100 kW; θ_{wi}, θ_{wo} (at design) = 80°C and 70°C; raising 10.84 m^{3}s^{-1} of air from 8.5°C to 16°C; m_{wd} = 2.38 kgs^{-1}; f_1 = 2.22, f_2 = 0.4; c_{pw} = 4.2 kJkg^{-1}K^{-1}, c_{pa} = 1.025 kJkg^{-1}K^{-1}; τ_v = 30s, τ_d = 40s. And, based on

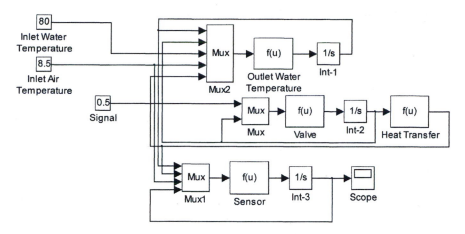

Figure E9.3.1 Simulink model of the open-loop non-linear heating coil.

copper tubes, aluminium fins of conventional spacing, $C_{wm} = 250.3\text{kJK}^{-1}$ (including the water content).

We can initially appreciate the extent of non-linearity of the plant. By simulating open-loop responses of various valve signals, u, and noting steady-state results at these signals, we can examine the extent of the problem since, under linear control, the resulting static responses in outlet air temperature should describe a linear relationship between outlet temperature and signal. For comparison, the static response based on an equal-percentage valve was also generated and both sets of results are given in Figure E9.3.2.

For the coil problem, the disturbance variable is the inlet air temperature and changes in set point are unlikely to be frequently called for in normal operation. Hence if the influence of the inlet air temperature can be removed from the closed loop, the feedback path and controller become redundant, save for the periodic modification of set points when needed and to account for any unmeasured disturbances in the practical situation. Before we consider how an ANN can help us with this problem, let us first consider how we might tune a fixed-parameter PID controller to deal with the non-linear loop we have created.

A conventional PID controller can be designed by tuning the controller parameters using, for example, Ziegler–Nichols rules. The difficulty with this non-linear case lies in deciding in which region of operation to tune with reference to. A glance at Figure E9.3.2 confirms that the rate of change of plant gain, K_p, with respect to signal, u (that is, dK_p/du), is greatest when the signal is at its lowest (i.e. where the valve is operating close to its seat). Conversely, dK_p/du is low when the signal is high and the plant is close to capacity. This means that if we tune with respect to plant conditions at near-

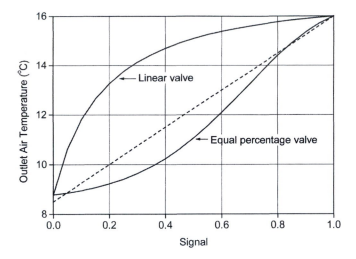

Figure E9.3.2 Static linearity of the coil model.

capacity, we are likely to experience instability when the plant is required to operate at light load. On the other hand, if we tune for stable operation at light load, we may experience inferior loop performance when the plant operates at near-capacity (i.e. unresponsiveness and, possibly, offset). Nevertheless, as a priority we must ensure stability across range so we must tune to plant conditions at low u.

This we can do by 'measuring' the plant open-loop response to a low step change in u. Figure E9.3.3 gives the open-loop response in outlet temperature arising from a step change in u from $0 \rightarrow 0.2$. Using the methods set out in section 7.1, the following linear model parameters are identified based on this response:

$$K_p = 23.75 \qquad \tau = 96\,\text{s} \qquad t_d = 48\,\text{s}$$

Using any of the methods set out in section 7.3, we can now obtain controller parameter estimates. For example, using Ziegler–Nichols rules (Table 7.3) the following are obtained:

$$K_c = 0.1\,\text{K}^{-1} \qquad i_t = 96\text{s} \qquad d_t = 24\text{s}$$

The influence of plant non-linearity under this cautiously tuned PID strategy can now be appreciated. Figure E9.3.4 gives simulated closed-loop responses when the entering air temperature is stepped (at 50 s) from an initial 12.25°C (mid-range) to 16°C (i.e. zero load), and then stepped from 12.25°C to 8.5°C (full capacity). As expected, control action is responsive and good when operating at light load but far less responsive when operating towards capacity.

We can train an ANN to compensate for the disturbance by treating the ANN as a feedforward controller in the disturbance path. In section 4.3, we

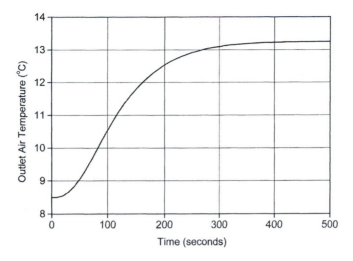

Figure E9.3.3 Open-loop step response of the non-linear heating coil at low *u*.

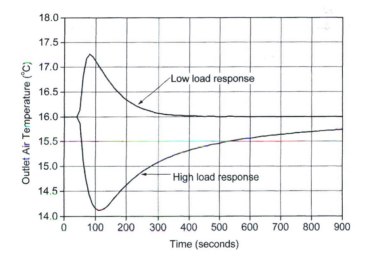

Figure E9.3.4 Response of the non-linear coil under fixed parameter PID control.

saw how to design a feedforward controller based on the inverse plant model, but this of course requires an accurate model of the plant. An ANN can be trained using measurements of the plant response. Here, we will use our non-linear model to provide the 'measurements', for illustration.

Figure E9.3.5 shows where we need the ANN to be located. This is based on the design of a hybrid feedback proportional/ANN controller proposed

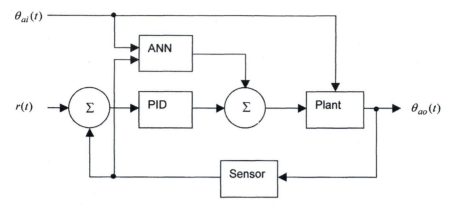

Figure E9.3.5 ANN feedforward compensator.

by Hepworth *et al.* (1994). The simplest basis for the ANN will be to train it to learn the static inverse of the plant. That is, we require it to predict the required steady-state control signal from the disturbance variable value and the plant output. Since this essentially ignores the plant dynamics, it should give an improvement in control rather than perfect results.

In the following, the MATLAB neural network toolbox is used to train the required ANN (Demuth & Beale, 1998).

1 *Generating the training data.* The training data are obtained by simulating the plant steady-state response to a range of control signals and inlet air temperature values to obtain a set of data in the form given in Figure E9.3.2. In general, best results in ANN training arise when as much data as possible are presented. Thus steady-state plant responses are 'measured' using the open-loop non-linear coil model at control signal increments of 0.05 (in the interval $0 \rightarrow 1$), and inlet air temperature increments of 0.5 K (in the interval 8.5°C (design) $\rightarrow$ 16°C (zero load)). Hence 21 control signal values were applied and 16 values of inlet air temperature, resulting in a 336-element matrix of plant output 'measurements'.

2 *Formatting the training data.* The ANN training data consist of a 2×336-element matrix, **P**, of inputs, comprising all possible combinations of inlet air temperature and resulting plant outlet temperature as a function of control signal. The ANN target data consist of a 336-element row vector, **U**, of corresponding control signal values.

3 *Training the network.* The initial weights and biases of the network are first set using the function initff. Like the previous example, we will use a standard feedforward network, trained using Levenberg–Marquadt back-propagation. A hidden layer consisting of six tansigmoid neurons is used with an output layer of one linear neuron node (i.e. one output). Thus,

Table E9.3.1 Weights and biases for the ANN

W1	B1	W2^{T}	B2
1.607	−0.472	−1.744	2.367
0.696	−0.778	7.507	
0.812	−0.210	−6.312	
−0.934	0.714	4.931	
1.547	−0.150	−5.775	
−0.425	0.469	−2.549	

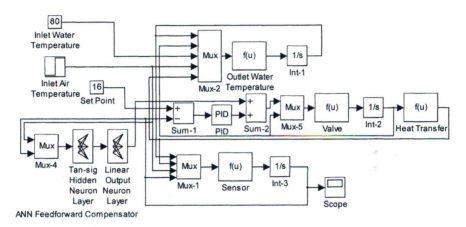

Figure E9.3.6 Non-linear coil control with ANN disturbance compensation.

```
>[W1,B1,W2,B2]=initff(P,6,'tansig',U,'purelin');
```

The network is now trained using the trainlm function,

```
>[W1,B1,W2,B2]=trainlm(W1,B1,'tansig',W2,B2,'purelin',
              P,U,[a,b,c,d,e,f,g,h]);
```

where a = epochs between updating (five used), b = maximum epochs (200 used), c = target sum of squared error (0.02 used), d = minimum gradient (10^{-4} used), e, h = initial and maximum learning rate (10^{-3}, 10^{10} used) and f, g = learning rate multipliers (10, 0.1 used).

4   *Results.* The network reached a sum of squared error of 0.0198 after 49 epochs, returning the weights and biases shown in Table E9.3.1.

The performance of the compensator was assessed using a Simulink simulation (Figure E9.3.6). Results are presented in Figure E9.3.7, showing the plant step response in the region known to give PID control the most difficulty (i.e. operating at or near capacity). The ANN gives a substantial improvement in response. Note that we could not expect a perfect response

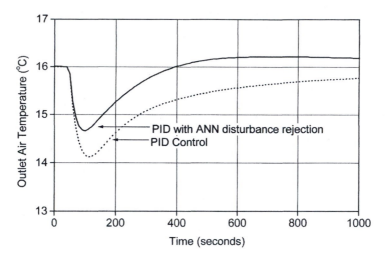

Figure E9.3.7 Non-linear coil control – response using ANN-based disturbance rejection.

with feedforward compensation based on the *static* response of the plant but better results may be possible by extending the training data, trying different neuron filters or trying an increased number of hidden layer neurons.

Nevertheless, this example has demonstrated the potential that ANNs offer HVAC control, in terms of a relatively simple problem. Dynamic control using an ANN in the feedback path is a real possibility though these tend to work best when they are adaptively trained on-line. We will take a look at some of the established applications for ANNs in HVAC control in the next section.

Progress in HVAC control applications of ANNs

Early work applying ANNs to HVAC problems mostly concerned identification and modelling problems. Miller & Seem (1991) use a three-layer feedforward ANN, trained using back-propagation, to predict the start-up time for heating plant during set-back – in effect 'optimum start control'. They compare the results of the ANN prediction with a conventional recursive least-squares (RLS) method, concluding that there was no significant advantage for the ANN in performance terms. However, the ANN required less data for training than the RLS method required for model fitting and the ANN proved to be more robust. Anstett & Kreider (1993) apply an ANN to the prediction of energy use in a large institutional building and, like Miller & Seem (1991), they compare this approach with a conventional approach to energy management.

Curtiss *et al.* (1994) use a back-propagated feedforward ANN with two hidden layers each of 10 neuron nodes for energy management in central plant, concluding that ANNs can be successfully used to model energy use as well as to carry out energy management tasks, such as set point resetting. Huang & Nelson (1994c) train an ANN to determine delay times in HVAC plant (an accurate knowledge of plant dead time can be used to improve control). They also use a four-layer (two-hidden-layer) architecture, trained using delta rule back-propagation. Besides successfully predicting delay time, the ANN was found to be capable of tolerating different levels of input measurement noise.

HVAC control applications of ANNs have received some attention, mostly recently. Ahmed *et al.* (1996) apply a simplified ANN architecture, the *general regression neural network* (GRNN), to identify and control a plant involving a heating coil and a valve. The GRNN is simpler to implement than the feedforward back-propagated network since the need to train weights and biases is replaced with the minimisation of a single parameter. In this application, the GRNN is used to identify static plant characteristics for use in feedforward compensation. Ding & Wong (1990) use an ANN for the control of valves in a complex heating and cooling network based on simulated operating data.

Several workers have trained ANNs to act as compensators to improve conventional feedback control (Curtiss *et al.*, 1996; Hepworth *et al.*, 1994). In the first of these, an artificial neural PID 'ANPID' takes PID controller parameters and control error as its inputs and produces an output which is added to the feedback PID controller. Hepworth and co-workers use a radial-basis-function-based ANN which is trained to learn the static inverse characteristics of a non-linear heating coil in order to provide disturbance compensation (Example 9.3 is based partly on this). They conclude that this relatively simple network design results in more consistent control over range than with conventional feedback control. They also use a simple P controller in the feedback path to compensate for plant dynamics and unmeasured disturbances.

Curtiss *et al.* (1993) and So *et al.* (1995) develop ANN-based predictive controllers as alternatives to conventional PID control. Curtiss and co-workers develop a 'look-ahead' network adaptive ANN in which plant error is back-propagated through the ANN in a Hopfield-like fashion and the network used to predict future plant behaviour (i.e. future error) as a basis for control action. The resulting controller was found to be at least as good as well tuned PID control but this crucially depended on the choice of network learning rate. In later work, Curtiss (1996) goes on to demonstrate the implementation of a network-assisted PID controller on a laboratory-scale heating coil control loop. So *et al.* (1995) develop a combined identifier/controller for a MISO application in which control of an entire air handling plant is considered using an ANN. They compare this with well tuned PID and a fuzzy logic controller concluding that the response rate of the ANN-based method was inferior to the fuzzy logic controller and steady-state accuracy was slightly inferior to well tuned PID. However, the ANN training data were relatively easily obtained, the controller required no tuning, nor did it require any expert knowledge (as did the fuzzy logic controller).

Clearly, the potential for ANNs in both HVAC prediction and control is substantial and we might expect considerable attention in this field for some time to come.

References

Ahmed, O., Mitchell, J.W., Klein, S.A. (1996) Application of general regression neural network (GRNN) in HVAC process identification and control. *ASHRAE Transactions*, **102** (1), 1147–1156.

Anderson, D., Graves, L., Reinert, W., Kreider, J.F., Dow, J., Wubbena, H. (1989) A quasi-real-time expert system for commercial building HVAC diagnostics. *ASHRAE Transactions*, **95** (2), 954–960.

Anstett, M., Kreider, J.F. (1993) Application of neural networking models to predict energy use. *ASHRAE Transactions*, **99** (1), 505–517.

Åström, K.J., Wittenmark, B. (1989) *Adaptive Control*. Addison-Wesley, Reading, MA.

Attia, A.E., Rezeka, S.F. (1994) Quantitative robust control of temperature and humidity in hot and dry climates. *ASME Transactions: Journal of Dynamic Systems, Measurement and Control*, **116** (2), 286–292.

Brothers, P.W. (1988) Knowledge engineering for HVAC expert systems. *ASHRAE Transactions*, **94** (1), 1063–1073.

Chen, S., Cowan, C.F.N., Grant, P.M. (1991) Orthogonal least squares learning algorithm for radial basis function networks. *IEEE Transactions on Neural Networks*, **2** (2), 302–309.

Chen, Y.H., Lee, K.M. (1990) Adaptive robust control scheme applied to a single-zone HVAC system. *ASHRAE Transactions*, **96** (2), 896–903.

Chiang, R.Y., Safonov, M.G. (1992) *Robust Control Toolbox for Use with MATLAB*. The Mathworks Inc., Natick, MA.

Curtiss, P.S. (1996) Experimental results from a network-assisted PID controller. *ASHRAE Transactions*, **102** (1), 1157–1168.

Curtiss, P.S., Kreider, J.F., Brandemuehl, M.J. (1993) Adaptive control of HVAC processes using predictive neural networks. *ASHRAE Transactions*, **99** (1), 496–504.

Curtiss, P.S., Brandemuehl, M.J., Kreider, J.F. (1994) Energy management in central HVAC plants using neural networks. *ASHRAE Transactions*, **100** (1), 476–493.

Curtiss, P.S., Shavit, G., Kreider, J.F. (1996) Neural networks applied to buildings – a tutorial and case studies in prediction and adaptive control. *ASHRAE Transactions*, **102** (1), 1141–1146.

Demuth, H., Beale, M. (1998) *Neural Network Toolbox for Use with MATLAB*. The Mathworks Inc., Natick, MA.

Dexter, A.L., Haves, P. (1994) Building control systems: evaluation of performance using an emulator. *Building Services Engineering Research and Technology*, **15** (3), 131–140.

Dexter, A.L., Trewhella, D.W. (1990) Building control systems: fuzzy rule-based approach to performance assessment. *Building Services Engineering Research and Technology*, **11** (4), 115–124.

Dexter, A.L., Haves, P., Jørgensen, D.R. (1993) Automatic commissioning of HVAC systems. *Proceedings of the CLIMA 2000 Conference*, London.

Ding, Y., Wong, K.V. (1990) Control of a simulated dual temperature hydronic system using a neural network approach. *ASHRAE Transactions*, **96** (2), 727–732.

Dounis, A.I., Santamouris, M.J., Lefas, C.C., Argiriou, A. (1995) Design of a fuzzy set environment comfort system. *Energy and Buildings*, **22**, 81–87.

Dunn, A. (1987) Expert systems: their application to the building services industry. *Building Services Engineering Research and Technology*, **8** (4), 73–77.

Elman, J.E. (1990) Finding structure in time. *Cognitive Science*, **14**, 179–211.

Ganesh, C., Pearson, J.B. (1989) H^2 optimisation with stable controllers. *Automatica*, **25** (4), 629–634.

Grimble, M.J. (1986) Optimal H^∞ robustness and the relationship to LQG design problems. *International Journal of Control*, **43** (2), 351–372.

Grimble, M.J. (1987a) H^∞ robust controller for self-tuning control applications. Part 1: Controller design. *International Journal of Control*, **46** (4), 1429–1444.

Grimble, M.J. (1987b) H^∞ robust controller for self-tuning control applications. Part 2: Self-tuning and robustness. *International Journal of Control*, **46** (5), 1819–1840.

Grimble, M.J. (1989) Minimisation of a combined H^∞ and LQG cost function for a two-degrees-of-freedom control design. *Automatica*, **25** (4), 635–638.

Haberl, J.S., Smith, L.K., Cooney, K.P., Stern, F.D. (1988) An expert system for building energy consumption analysis: applications at a university campus. *ASHRAE Transactions*, **94** (1), 1037–1063.

Hagan, M.T., Menhaj, M.B. (1994) Training feedforward networks with the Marquadt algorithm. *IEEE Transactions on Neural Networks*, **5** (6), 989–993.

Hepworth, S.J., Dexter, A.L., Willis, S.T.P. (1994) Neural network control of a nonlinear heater battery. *Building Services Engineering Research and Technology*, **15** (3), 119–129.

Hitchcock, R.J. (1991) Knowledge-based system design guide tools. *ASHRAE Transactions*, **97** (2), 676–684.

Horowitz, I.M., Sidi, M. (1972) Synthesis of feedback systems with large plant ignorance for prescribed plant tolerances. *International Journal of Control*, **16** (2), 287–309.

Huang, S., Nelson, R.M. (1991) A PID-law-combining fuzzy controller for HVAC applications. *ASHRAE Transactions*, **97** (2), 768–774.

Huang, S., Nelson, R.M. (1994a) Rule development and adjustment strategies of a fuzzy logic controller for an HVAC system: part one – analysis. *ASHRAE Transactions*, **100** (1), 841–850.

Huang, S., Nelson, R.M. (1994b) Rule development and adjustment strategies of a fuzzy logic controller for an HVAC system: part two – experiment. *ASHRAE Transactions*, **100** (1), 851–856.

Huang, S.-H., Nelson, R.M. (1994c) Delay time determination using an artificial neural network. *ASHRAE Transactions*, **100** (1), 831–840.

Jang, J.S.R., Gulley, N. (1995) *MATLAB Fuzzy Logic Toolbox User's Guide*. The Mathworks Inc., Natick, MA.

Kaler, G.M. (1990) Embedded expert system development for monitoring packaged HVAC equipment. *ASHRAE Transactions*, **96** (2), 733–742.

Klima, J. (1990) An expert system to aid troubleshooting of operational problems in solar domestic hot water systems. *ASHRAE Transactions*, **96** (1), 1530–1538.

Kreider, J.F., Cooney, K., Graves, L., Meadows, K., Stern, F., Weilert, L. (1990) An expert system for commercial building HVAC and energy audits – a progress report. *ASHRAE Transactions*, **96** (1), 1549–1553.

Kwakernaak, H. (1985) Minimax frequency domain performance and robustness optimisation for linear feedback systems. *IEEE Transactions on Automatic Control,* **AC-30** (10), 994–1004.

Li, J.H., Michael, A.N., Porod, W. (1989) Analysis and synthesis of a class of neural networks: linear systems operating on a closed hypercube. *IEEE Transactions on Circuits and Systems,* **36** (11), 1405–1422.

Ling, K.-V., Dexter, A.L. (1994) Expert control of air conditioning plant. *Automatica,* **50** (5), 761–773.

Liu, S.T., Kelly, G.E. (1988) Knowledge-based front-end input generating program for building system simulation. *ASHRAE Transactions,* **94** (1), 1074–1084.

Lundström, P., Skogestad, S., Wang, Z.-Q. (1991) Weight selection for *H*-infinity and Mu-control methods – insights and examples from process control. *Robust Control System Design Using H-infinity and Related Methods* (ed. Hammond, P.H.), 139–157. Institute of Measurement and Control, London.

Mamdani, E.H., Assilian, S. (1975) An experiment in linguistic synthesis with a fuzzy logic controller. *International Journal of Man–Machine Studies,* **7** (1), 1–13.

McAdams, W.H. (1976) *Heat Transmission.* McGraw-Hill, New York.

McCulloch, W.S., Pitts, W.H. (1943) A logical calculus of the ideas immanent in nervous activity. *Bulletin of Mathematical Biophysics,* **15**, 115–133.

Miller, R.C., Seem, J.E. (1991) Comparison of artificial neural networks with traditional methods of predicting return time from night or weekend setback. *ASHRAE Transactions,* **97** (2), 500–508.

Norford, L.K., Allgeier, A., Spadaro, J.V. (1990) Improved energy information for a building operator: exploring the possibilities of a quasi-real-time knowledge-based system. *ASHRAE Transactions,* **96** (1), 1515–1523.

Page, G.F., Gomm, J.B., Williams, D. (1993) *Application of Neural Networks to Modelling and Control.* Chapman & Hall, London.

Poslethwaite, I. (1991) Robust control of multivariable systems using *H*-infinity optimisation. *Robust Control System Design Using H-infinity and Related Methods* (ed. Hammond, P.H.), 2–32. Institute of Measurement and Control, London.

Potter, R.A., Kreider, J.F., Brandemuehl, M.J., Windingland, L.M. (1991) Development of a knowledge-based system for cooling load demand management at large installations. *ASHRAE Transactions,* **97** (2), 669–675.

Rumelhart, D.E., Hinton, G.E., Williams, R.J. (1986) Learning internal representations by error propagation. *Parallel Data Processing* (ed. Rumelhart, D.E., McClelland, J.), vol. 1. MIT Press, Cambridge, MA.

Shaw, M.R. (1987) Applying expert systems to environmental management and control systems. *Proceedings of the Unicom Seminar on Intelligent Buildings,* London.

Simulink (1996) *SIMULINK 2 Dynamic System Simulation for MATLAB.* The Mathworks Inc., Natick, MA.

So, A.T.P., Chow, T.T., Chan, W.L., Tse, W.L. (1994) Fuzzy air handling system controller. *Building Services Engineering Research and Technology,* **15** (2), 95–105.

So, A.T.P., Chow, T.T., Chan, W.L., Tse, W.L. (1995) A neural-network-based identifier/controller for modern HVAC control. *ASHRAE Transactions,* **101** (2), 14–31.

So, A.T.P., Chan, W.L., Tse, W.L. (1997) Self-tuning fuzzy air handling system controller. *Building Services Engineering Research and Technology,* **18** (2), 99–108.

Wang, L.-X. (1997) *A Course in Fuzzy Systems and Control.* Prentice-Hall, Englewood Cliffs, NJ.

Warwick, K., Irwin, G.W., Hunt, K.J. (eds) (1992) *Neural Networks for Control and Systems*. Peter Peregrinus, Stevenage (on behalf of the Institution of Electrical Engineers, London).

Wong, K.-F.V., Ferrano, F.J. (1990) Availability-based computer management of a cold thermal storage system. *ASHRAE Transactions*, **96** (1), 1524–1529.

Yong, L.-F., Nelson, R.M. (1990) An expert system to select heat exchangers for waste heat recovery applications. *ASHRAE Transactions*, **96** (1), 1539–1548.

Zadeh, L.A. (1973) Outline of a new approach to the analysis of complex system and decision making processes. *IEEE Transactions on Systems, Man and Cybernetics*, **SMC-3** (1), 28–44.

Zaheer-uddin, M. (1992) Optimal control of a single zone environmental space. *Building and Environment*, **27** (1), 93–103.

Zames, G. (1981) Feedback and optimal sensitivity: model reference transformations, multiplicative seminorms and approximate inverses. *IEEE Transactions on Automatic Control*, **AC-26** (2), 301–320.

Index